MEMS Reference Shelf

MEMS Reference Shelf

For other titles in this series, go to:
www.springer.com/series/7724

Ai Qun Liu

RF MEMS Switches and Integrated Switching Circuits

Design, Fabrication, and Test

Chapter Co-Authors

Aibin Yu
Institute of Microelectronics, Singapore

Ming Tang
Institute of Microelectronics, Singapore

Muhammad Faeyz Karim
Institute for Infocomm Research, Singapore

Selin Teo Hwee Gee
Institute of Microelectronics, Singapore

‡ The co-authors are PhD students of Ai Qun Liu

Springer

Dr. Ai Qun Liu
School of Electrical & Electronic Engineering
Nanyang Technological University
Nanyang Avenue
Singapore 639798
eaqliu@ntu.edu.sg

ISSN 1936-4407
ISBN 978-0-387-46261-5 e-ISBN 978-0-387-46262-2
DOI 10.1007/978-0-387-46262-2
Springer New York Dordrecht Heidelberg London

Library of Congress Control Number: 2010930386

Printed on acid-free paper

Springer is part of Springer Science+Business Media (www.springer.com)

This book is dedicated to
my beloved mother and late father

Preface

Wireless communication has led to an explosive technological advancement and breakthrough in radio frequency microelectromechanical systems (RF MEMS) technology. The technology is seen to have the potential of replacing the mechanical, wireless, and semiconductor technology in future communication systems as well as communication satellites. In many cases, the RF MEMS technology does not only offer devices with substantial reduction in size, weight, cost, and power consumption but also promises more superior performance compared to those offered by the conventional technology. By combining high RF performance, low cost, and low power consumption, RF MEMS devices allow one to explore new architectures and configurations which were not possible with the traditional technology and expand its potential applications from defense-related products to personal communication devices.

This book, *RF MEMS Switches and Integrated Switching Circuits – Design, Fabrication and Test*, presents our latest accomplishments in RF MEMS switches and integrated switching circuits. The state of the art of RF MEMS switches and integrated switching circuits are described. Through this book, we wish to acquaint readers with the basics of RF MEMS and knowledge of how to design practical RF switches and switching circuits.

This book owes much to many.

I am grateful to all my PhD students Aibin, Min, and Faeyz for their excellent research work and dedicated contributions. I am particularly indebted to Selin for her help in coordinating the book project. Thanks also go to my excellent colleagues Alphones, Geok Ing, Zhong Xiang, Yi Long, Aditya, Chao Lu, Radhakrishnan, Hong Wang, and Joachim (KTH, Sweden) for their expert advice and friendly help. I am also not forgetting my debts to Joseph Ting and Gim Pew from DSO National Laboratories for directing me to the research field of RF MEMS while it was still in its infancy. I would also like to acknowledge my debts to Steven Elliot, the Senior Engineering Editor of Springer, who has been so persistent and supportive in giving me his support to accomplish the writing of this book. Working with him is a pleasure, even when the pressure is on. My final particular thanks are to my research group members, both past and present, for contributing their effort and full support to me.

Although I have made my best effort in writing this book, I have undoubtedly made errors or omission and sincerely apologize for them. I welcome comments and suggestions from readers about such errors and key concepts that may have been omitted. Updates, corrections, and other information will be available at my web site, http://nocweba.ntu.edu.sg/laq_mems.

Singapore
March 18, 2010

Ai Qun Liu

Contents

Chapter 1
Introduction

This book compiles the latest advancements in the field of radio frequency (RF) microelectromechanical systems (MEMS) technology and devices, spanning a comprehensive range of RF MEMS topics. Highlights of contributions include areas such as (1) lateral series switches, (2) capacitive shunt switches, (3) coplanar waveguide transmission line switches, (4) multi-throw switches, (5) tunable electromagnetic bandstop and bandpass filters, (6) capacitive switching filters, and (7) different types of switches and fabrication processes.

In the following sections, the background and literature necessary as foundations for each of the subsequent chapters detailing the various technologies developed are organized as follows: Section 1.1 gives the overview and literature background of RF MEMS switches, especially for their relevance to lateral series switches as will be presented in detail in Chapter 2. Section 1.2 provides the ground works for research into areas of improvement for capacitive shunt switches – such as in reliability, power handling and degradation – which will be presented in Chapter 3. Section 1.3 gives the background on coplanar waveguide transmission lines before Chapter 4 presents new designs, optimizations, and their results. Section 1.4 introduces the single-pole-multi-throw switching RF circuits, so that works on SP3T and SP4T switching circuits may be developed fully in Chapter 5. Sections 1.5 and 1.6 give the background for electromagnetic band gap (EBG) circuits, which is the foundation of the technologies developed in Chapters 6 and 7, respectively, for both tunable bandstop and bandpass filter circuits. Section 1.7 provides the introduction to capacitive switching bandpass filters which research developments made will be covered in Chapter 8. Finally, Sections 1.8 and 1.9 present the initiation for fabrication technologies necessary to the realization of RF MEMS devices, in particular by substrate transfer processes and surface planarization techniques. The intricacies required of such technologies are then accordingly detailed in Chapters 9 and Chapters 10.

Selin Teo Hwee Gee and Ai Qun Liu

A.Q. Liu, *RF MEMS Switches and Integrated Switching Circuits*,
MEMS Reference Shelf 5, DOI 10.1007/978-0-387-46262-2_1,

1.1 RF MEMS Switches

A radio frequency (RF) microelectromechanical switch is a switching device that is fabricated using the micromachining technology, where the switching between the on- and off-states is achieved via the mechanical displacement of a freely movable structure. The displacement is induced by a microactuator. Various actuation mechanisms then exist, including electrostatic [1–3], electrothermal [4], magnetostatic [5], and electromagnetic [6–10] means. Most often, the RF MEMS switches rely on electrostatic actuation, which is based on the attractively electrostatic force existing between charges of opposite polarity. An electrostatic drive offers extremely low power consumption since power is required only during switching. The other advantages of using electrostatic actuation are its simple fabrication technology, compared to electromagnetic excitation: the high degree of compatibility with a standard IC process line and its ease of integration with planar and microstrip transmission lines. For many applications, the main drawback is to overcome the high driving voltage. When the available supply voltage is limited, for instance, within the range of 3–5 V in mobile phones, on-chip high-voltage generators [11] may be incorporated either monolithically or in a hybrid fashion [12].

The RF MEMS switches can be classified as vertical switches and lateral switches based on the motion direction of the switching structure. The vertical switches perform out of wafer plane displacement and surface contact. The lateral switches perform in wafer plane displacement and sidewall contact. The vertical RF MEMS switches are usually fabricated using surface micromachining process and metal is used as its structural material. On the other hand, the lateral RF MEMS switches are usually fabricated using bulk micromachining process, using single-crystal silicon or polysilicon as its structural material. Most RF MEMS switches developed in the last decades are vertical switches due to their excellent RF performance. However, the lateral switches have shown some promising performance in dynamics, together with fabrication processes which are simpler compared to its vertical switching counterparts. As such, much recent research interest has been geared toward the lateral switches.

The lateral switches have the benefit of co-fabrication. The movable switching structure, the contacts, the transmission line, and the support structures can be fabricated in a single lithographic step. Besides, the actuator design is more flexible. It is easy to get a mechanical force in opposing directions even when electrostatic designs are used. The lateral RF MEMS switches can use different elements as the structural material, for instance, nickel, polysilicon, and single-crystal silicon. To date, three main types of lateral switches have been reported. All of them are the resistive series switches.

The first type of lateral switch is the comb-drive or cantilever beam-actuated lateral switch which uses electroless plated nickel as its structure material [13–14]. The schematic illustration of a comb-drive-actuated lateral switch [13] is shown in Fig. 1.1; in the plating process, the nickel height is restricted to 10–50 μm and the smallest possible width is limited to 5 μm. The actuation voltage is within the range of 35–150 V. The contact resistance is 5–20 Ω.

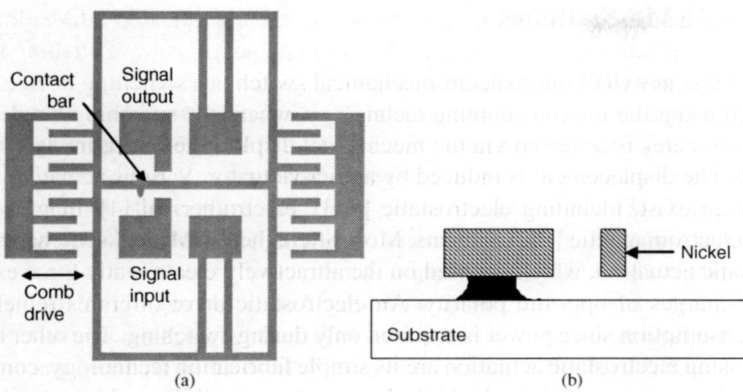

Fig. 1.1 The schematic illustration of a comb-drive-actuated lateral switch fabricated using nickel surface micromachining process: **(a)** top view and **(b)** cross-sectional view [13]

The second type of the lateral switch is the electrothermally or electrostatically actuated switch using polysilicon as the structural material [15–24]. The electrothermally actuated switch has demonstrated RF performance up to 50 GHz [22–24], as shown in Fig. 1.2. The switch utilizes a parallel six-beam thermal actuator which requires actuation voltage of 2.5–3.5 V. The mechanical structures are fabricated using 2-μm polysilicon surface micromachining process and four masks. Silicon nitride is used as a structural connection to isolate the RF signal and the DC bias. Gold of thickness 0.3–0.5 μm is deposited on the polysilicon to form a transmission line and contact bumps. The switch has an isolation of –20 dB at 40 GHz. The insertion loss is –0.1 dB at 50 GHz after de-embedding the substrate effect and –1.0 dB before de-embedding the substrate effect. The switching "on" time is 300 μs. Although this type of lateral switch exhibits high RF performance and

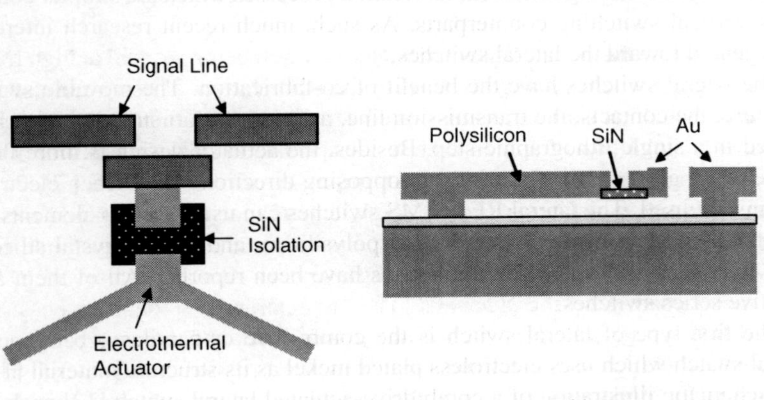

Fig. 1.2 The schematic illustration of a thermally actuated lateral switch: **(a)** 3D view and **(b)** cross-sectional view [22]

requires low actuation voltage, its drawbacks are the complicated fabrication process, the high power consumption due to thermal actuation, slow switching speed, and also a fragile switching structure.

The third type of the lateral switch is an electrostatically actuated switch using single-crystal silicon as its structural material [25–27] and fabricated using a single-mask bulk micromachining process. The electrostatic actuator can be either a comb-drive or a cantilever beam. The switching structures are typically fabricated using deep reactive ion etching (DRIE) process on silicon-on-insulator (SOI) wafer. The metal contact is realized by depositing a thin layer of metal directly on the entire surface of the switch structure, as shown in Fig. 1.3. Compared to other RF MEMS switches, the fabrication process of this type of lateral switch is the simplest. Single-crystal silicon has negligible biaxial stress and vertical stress gradient, and superior thermal characteristics. Adapting a silicon process in fabricating the lateral switch can reduce the deformation of the switch structure caused by either thermal effects or stress gradients. As a result, the third type of lateral switch can provide better mechanical characteristics. However, most of the time, such lateral switches can only work at DC.

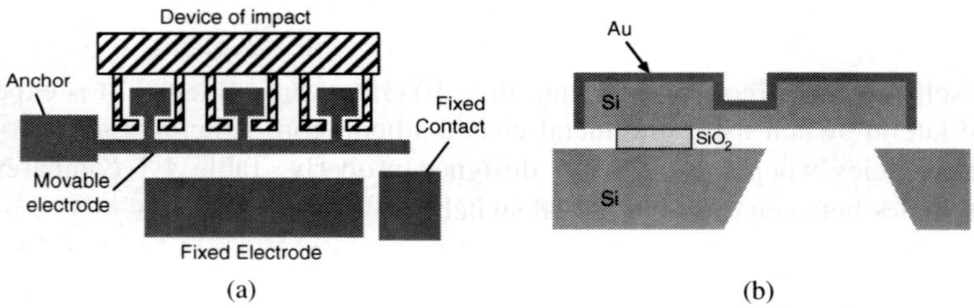

Fig. 1.3 The schematic illustration of an electrostatically actuated lateral switch: **(a)** top view and **(b)** cross-sectional view [25]

To investigate the possibility of such lateral switch being worked at high frequencies, the literature survey was extended to the RF behavior of non-solid-metal-core structures which consist of metal-coated structures with non-metal material as the core. A metal-coated silicon-core high-aspect-ratio transmission line was first developed in 1995 [28], followed by an improved version in 1998 [29, 30]. The transmission line was fabricated by a single-crystal reactive etching and metallization (SCREAM) process. It consists of pairs of parallel-plate waveguide formed from two deep (150-μm) suspended single-crystal silicon beams with 0.5-μm-thick metal coating on the sidewalls. The attenuation was found to be less than –0.18 dB/cm at 10–50 GHz. A step impedance filter and a continuous microactuated phase shifter were realized using this transmission line [28–30]. High Q (quality factor) inductors with polysilicon as the core material and copper as the surface metal were demonstrated in [31]. The spiral structure of an inductor was formed using polysilicon surface micromachining process and suspended over a 30-μm-deep cavity in the silicon substrate beneath. Copper was electrolessly plated onto the polysilicon spiral to achieve low resistance. High quality factors (Qs) over

Table 1.1 Comparison of characteristics of lateral RF MEMS switches

References	[13]	[22]	[25]	[27]	[28]
Actuation mechanism	Electrostatic	Thermal	Electrostatic	Magnetic+ electrostatic	Electrostatic
Structure material	Nickel	Poly-Si	Silicon	Silicon	Silicon
Power consumption	μW	18–99 mW	μW	11.3 nJ/cycle	μW
Frequency (GHz)	DC	DC to 40	DC	DC	DC to 25
Insertion loss (dB)	–	0.1	–	–	0.1–1.0
Return loss (dB)	–	–	–	–	22
Isolation (dB)	–	20	–	–	22
Switching voltage (V)	35–150	2.5–3.5	50–260	10	23.3
Contact resistance (Ω)	5–20	0.1–0.3	1000	2–8	1–2
Switching speed (μs)	–	300	30	100 (close) 20 (open)	36
Size (mm^2)	–	0.8 × 0.34	>1.2 × 0.6	>2 × 2	0.8 × 0.4

30 and self-resonant frequencies higher than 10 GHz were achieved. It is expected that the lateral switch using the metal-coated silicon-core structures may work at high frequencies when the device is designed properly. Table 1.1 compares the characteristics between different lateral switches.

1.2 Capacitive Shunt Switches

Capacitive switches are key elements in various applications such as tunable capacitors and tunable bandpass filter. Most of the capacitive switches are implemented in shunt configuration. In this construction, when the metal bridge is at the up-state position, the RF signal can go through from an input port to an output port, which is called switch on-state; when the metal bridge is driven down, then most of the RF signal is shorted to ground, which is called switch off-state, as shown in Fig. 1.4.

An important figure of merit of the RF MEMS capacitive switch is the ratio of on-state capacitance to off-state capacitance, that is, $R = C_d / C_u$, where C_d is the on-state capacitance with a typical value of several pico-farads and C_u is the off-state capacitance with a typical value of tens of femto-farads.

The larger the capacitance ratio, the smaller the off-state insertion loss and the higher the on-state isolation, and therefore broader frequency band can be obtained. In most cases, the RF capacitive switches are more suitable for high-frequency (>10 GHz) applications because of its capacitive coupling nature. Especially, at W-band, only the capacitive switches can be implemented because the DC-contact switches have large contact resistance, which gives rise to high loss at W-band. Therefore, to develop capacitive switches suitable for low-frequency (<10 GHz)

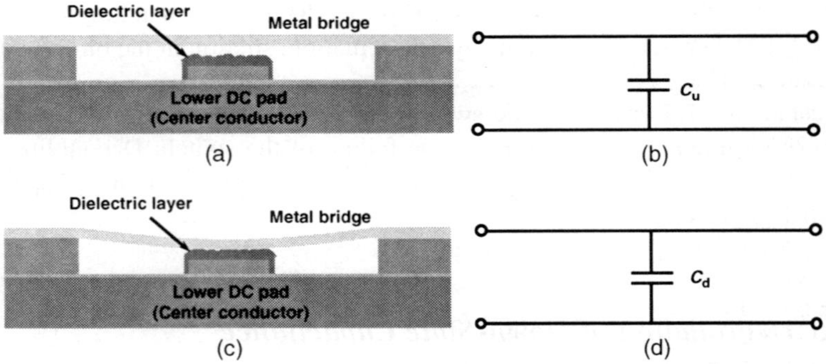

Fig. 1.4 Schematic of shunt capacitive switch with its equivalent circuits: **(a)** the *on*-state of the switch; **(b)** the *on*-state of the capacitance; **(c)** the *off*-state of the switch; and **(d)** the *off*-state capacitance

applications, one method is to increase the capacitance ratio by using dielectric material with high dielectric constant.

1.2.1 Reliability of Capacitive Switches

The reliability of MEMS switches is of major concern for long-term applications. The reliability of the capacitive switch is mainly determined by stiction between the dielectric layer and the metal layer, which is due to charge injection and charge trapping in the dielectric layer [32]. There are two cases of failure in capacitive switches [33]. The first one is that the metal bridge actuates and then returns to the off-state position, irrespective of the polarity of the applied voltage; the second failure mechanism is due to charge injection in the dielectric layer and results in the metal bridge remaining in the down-state position when the actuation voltage is removed (sticking) or the metal bridge does not actuate under an applied bias voltage.

The dielectric charging problem can be minimized by (i) using silicon dioxide as the dielectric layer. However, using silicon dioxide instead of silicon nitride as the dielectric layer results in smaller on-state capacitance; (ii) another solution to minimize the dielectric charging problem is to use a bipolar voltage when biasing the capacitive switch; and (iii) finally, using side pull-down electrodes can also minimize the dielectric charging problem.

1.2.2 Power Handling of Capacitive Switches

Power handling capabilities of RF MEMS switches are another major concern for the application of RF MEMS switches. Compared with that of p-i-n diode and FET,

RF MEMS switches can handle less RF power. The power handling capability of an RF MEMS switch is determine by self-actuation, stiction in the on-state, and hot switching [34].

Besides the DC driven of the switch, the RF signal can also lead to the pull-down of the metal bridge and hence the failure of the switch. Decoupling the RF and DC-actuation pads can substantially increase the power handling capability of the MEMS switch.

1.2.3 Degradation of Down-State Capacitance

As mentioned, large down/up capacitance ratio is important for a capacitive switch to achieve low insertion loss and high isolation. Unfortunately, because of the on-state capacitance degradation problem, which means the actual on-state capacitance is always smaller than the designed value, it is difficult to obtain the capacitance ratio which is larger than 150 for conventional capacitive switches. The on-state capacitance degradation is mainly caused by nonplanarization of metal bridge, surface roughness of capacitance area, etching hole in the metal bridge, etc. For the capacitive switch, it is believed that a flat metal bridge and small roughness of dielectric layer are important to reduce the on-state capacitance degradation [35].

1.3 Coplanar Waveguide Transmission Lines

At RF and microwave frequencies, transmission lines are used to carry electrical signal from one point to another. Transmission lines have to be impedance matched, that is, all transmission lines in a circuit are designed to be 50 Ω of characteristic impedance to prevent unwanted reflections in high-frequency circuits. To do this, the signal line and the ground lines are arranged in a certain configuration where dielectric is in between. Coplanar waveguide (CPW) is a planar transmission line which is compatible with microwave and millimeter wave-integrated circuit technology. CPW is relatively insensitive to variations in substrate thickness. It has low radiation loss and allows circuit elements to be easily connected in shunt as well as in series. These characteristics have made CPW an exceptional candidate for high-frequency low-cost high-performance circuits.

In coplanar lines, the field is tightly concentrated in the apertures between conductors, leading to current crowding for the conventional thin film techniques. When the conductor thickness is larger, the current can be distributed and the loss can be reduced. This makes it possible to work in high power applications due to the increased conduction interface. A new type of CPW transmission line with Si-core and metal overcoat is introduced in this book, in which the signal and ground lines' thickness is 50–75 μm. The Si-core CPW transmission line demonstrates all advantages that include balanced propagation and coplanar configuration. The Si-core

CPW can be implemented using a silicon back-bone structure patterned in desired shapes followed by a metal overcoat. The design and characterization of the Si-core CPW transmission line will be discussed.

1.4 Single-Pole-Multi-Throw Switching Circuit (SPMT)

Integrating RF switches in a transmission line network forms an RF switching circuit, such as an SPMT switching circuit, a tunable filter, and a phase shifter. The SPMT switching circuits are the key components in transmitting/receiving (T/R) circuits, switchable filters, reconfigurable antennas, switching matrices, etc. In the SPMT switching circuit, a total of m switches are incorporated in an $(m + 1)$ port junction device. When an SPMT switching circuit works, one switch is closed and the remaining $(m - 1)$ switches are open.

The major part of SPMT switching circuits is the single-pole-double-throw (SP2T) switching circuit. So far, six types of RF MEMS SP2T switching circuits have been developed. The first type is the most common one with two capacitive shunt MEMS switches placed a quarter wavelength from the center of the T-junction [36, 37], as shown in Fig. 1.5a. When one switch is actuated, the virtual RF short is transformed to an open at the T-junction, thus blocking almost all the signals from passing to that port. An insertion loss of 0.81 dB and an isolation of 20.3 dB at X-band and an insertion loss of 0.43 dB and an isolation of 28.7 dB at K-band have been demonstrated. A pull-in voltage of 9 V is achieved through changing the shape of the beam. The second design is a monolithic SP2T MEMS switching circuit [38, 39], which places two MEMS resistive series switches at two output armatures to operate in a 2.3-GHz diversity antenna, as shown in Fig. 1.5b. This switching circuit has an insertion loss of 0.2 dB and an isolation of 50 dB from DC to 4 GHz. The pull-in voltage is 30 V and the switching time is 20 μs. The mechanical life span is 10^9 switching cycles. The third design is a Ku-band SP2T switching circuit based on the toggle switch [40], as shown in Fig. 1.5c. Two toggle switches are perpendicular to each other where both have fixed connections with a flexible metal band to two output ports. When one toggle switch is on, the switch can be connected to the input port and the signal is routed to the output port. This SP2T switching circuit exhibits the insertion loss of 0.96 dB and the isolation of 30 dB from 7 to 20 GHz. The switching voltages are 30 V to close and 35 V to open. The fourth design is a double cantilever beam MEMS switching circuit developed for wireless applications, as shown in Fig. 1.5d. Two beams are controlled by a single actuation electrode. An RF performance with 50-dB isolation below 5 GHz and <0.18-dB insertion loss up to 30 GHz was observed [41]. The fifth design is a 35- to 60-GHz SP2T MEMS switching circuit [42] in which two resistive series switches are placed at the end of each output line and a short-ended 50 Ω line is connected at the cross-junction, as shown in Fig. 1.5e. When one switch is actuated, the open-ended line formed by the open state of the switch and the short-ended line

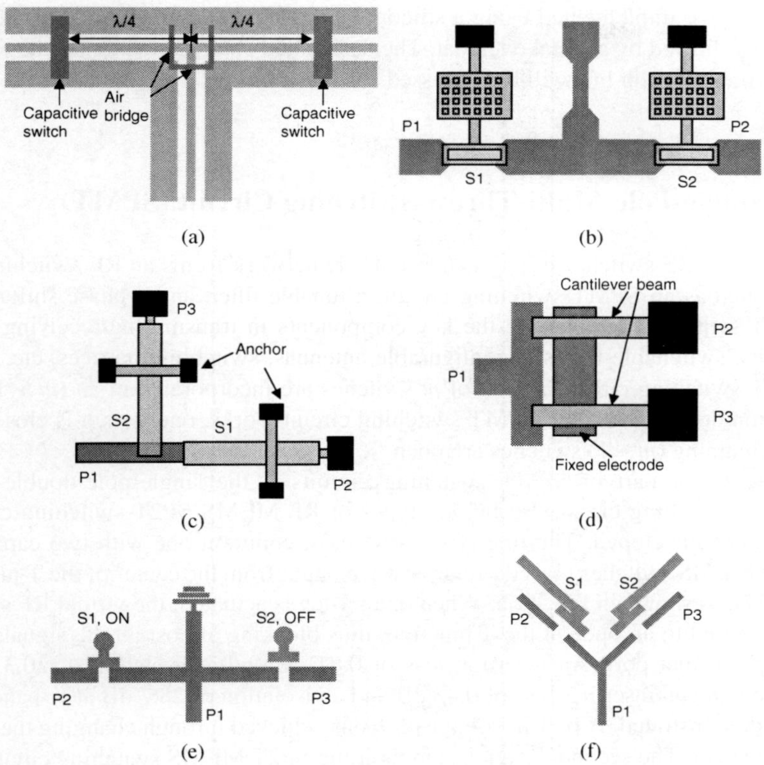

Fig. 1.5 Schematic of different types of SP2T switching circuits

comprises double resonance pair which transforms to be open at the cross-junction. This circuit presents the insertion loss below 1 dB and the isolation higher than 19 dB from 35 to 60 GHz. The actuation voltage is 35 V. All SP2T switching circuits use vertical MEMS switches. Only one lateral SP2T MEMS switching circuit was reported for satellite-based communications [43], as shown in Fig. 1.5 f. It uses thermal-actuated lateral MEMS switch fabricated by silicon deep reactive ion etching (DRIE) process. The microstrip transmission is fabricated on a glass wafer separately. The two wafers are bonded together and the device is packaged using an alumina cavity package, with DC and RF contacts on opposite sides. The isolation of this SP2T switching circuit is better than 50 dB at 1–6 GHz. The insertion loss is 1 dB at 2 GHz and 1.7 dB at 6 GHz. Table 1.2 compares the characteristics of different types of RF MEMS SPDT switching circuits.

There is very limited work on other single-pole-multiple-throw switching circuits. A single-pole-three-throw (SP3T) switching circuit has been implemented in hybrid form where three cantilever beams are micromachined separately and then

Table 1.2 Comparison of characteristics between different MEMS SP2T switching circuits

References	[1]	[1]	[2]	[3]	[5]	[7]	[8]	[9]	This work
Actuation mechanism	ES	ES	ES	ES	ES	ES	ES	Therm	ES
Frequency (GHz)	8–12	18–26	26–40	DC to 15	12–18	DC to 30	35–60	1–6	0.05–20
Insertion loss (dB)	0.81	0.43	0.3	0.1–0.2	0.96	0.18	0.5–1	0.2–1.7	0.2–0.9
Return loss (dB)	–	–	40	–	17	30	10–24	–	16.3–33
Isolation (dB)	20.3	28.7	30	40	30	50	19–26	50	65–26
Switching voltage (V)	9	9	–	30	30/on 35/off	15	35	3	23
Switching speed (μs)	52/on 213/off	–	–	20	–	60	–	–	35
Size (mm^2)	17.4	4.4	–	–	–	–	2	64	1.48

ES – electrostatic; therm – thermal

integrated on an alumina substrate using flip-chip technology [44]. It exhibits the insertion loss of 0.5 dB at 16 GHz and the isolation of 20 dB at 18 GH. An SP3T switching circuit is proposed in [45], where the circuit consists of three resistive series switches based on fixed–fixed beam architecture in a coplanar environment. A single-pole-four-throw (SP4T) switching circuit using four series switches is developed in [46], which shows good match up to 20 GHz with an isolation of 50 dB at 10 GHz. A single-pole-eight-throw (SP8T) switching circuit using the rotary switch is reported [47], which has an insertion loss of 2.16 dB and an isolation of 31 dB at 20 GHz.

1.5 Electromagnetic Band Gap (EBG) Structures for RF Circuits

Electromagnetic band gap (EBG) structures are periodic structures that have attracted much attention in the microwave and millimeter-wave community due to their filtering properties or inhibition of signal propagation in certain directions. Electromagnetic crystal structures are invented for microwave and optical wavelengths [48–51]. Planar EBG structures offer tremendous application possibilities for active and passive devices because these structures exhibit wide bandpass and band rejection properties at microwave and millimeter wave frequencies [52–56]. For compact design, the passband of the EBG structures is used as a slow wave medium. On the other hand, the stopband is used to suppress the surface waves, leakage, and spurious transmission. EBG structures can have various design structures such as bumpy or corrugated surfaces, metal pads or high impedances surface, and planar EBG structures. Generally, the ground plane is perturbed by EBG structures with different shape and different lattice structures. The shape may be uniform or non-uniform circular, square, rectangular, triangular patterned and the structures are named on the basis of the grid arrangement such as square, rectangular, triangular, and honey-comb. Different shapes and sizes of EBG structures provide different S-parameter performances. Due to their unique properties, the EBG structures can be used to design microwave components and devices for different applications such as filters [57–60], mixers, antennas [61, 62], oscillator, power amplifiers [63, 64], and phase array.

Tunable band gap and bandpass performances are desirable in many applications. Semiconductor varactor diodes have been considered in tunable EBGs. Although semiconductor varactors offer integration possibilities, EBG structures based on such varactors suffer from high losses and small tunability at high microwave frequencies, where integration is potentially possible due to the small wavelengths and sizes. By incorporating RF MEMS switches and the EBG filters, a new paradigm has been opened for the development of state-of-the-art tunable filters. These tunable filters are miniaturized with high Q, wider rejection level, low insertion loss and enable the generation of tunable bandstop filters.

1.6 Tunable EBG Bandpass Filters and Reconfigurable Circuits

The goal of many commercial and military systems is the ability to easily reconfigure or switch the frequency of operation, the output power, and the input impedance of the system [65–67]. Changing the frequency of operation means the resonant frequency must be reconfigured, together with the bandpass filter central frequency, the local oscillator frequency, and the matching networks for the low noise amplifiers and power amplifiers. In satellite systems, the switching networks are used for coaxial switches, while in base station systems, they are implemented using PIN diodes except after the power amplifiers. PIN diode switches are less expensive compared to coaxial switches, but require input and output amplifiers to compensate for the loss introduced by the switching networks. Coaxial switches result in outstanding isolation, insertion loss and can handle high power, but they are heavy, bulky, and expensive. RF MEMS switches can easily meet the isolation requirement of $N \times N$ switching matrices making the system smaller and lighter which is essential for satellite communications [68].

In many portable applications, the antenna input impedance is strongly dependent on the position of the portable device, and low-loss reconfigurable matching network at the input of the antenna would result in substantial performance improvement. Reconfigurable MEMS switching circuits can be used to generate a large range of impedance loci which are necessary for transistor and diode characterization (gain, noise, conversion loss, etc.). Moreover, the RF MEMS devices generate low intermodulation products which are essential when the reconfigurable circuit comes before the low noise amplifier/mixer chain.

The filters are custom-machined, carefully assembled, tuned, and calibrated. Typically tracking filters are mechanically tuned by adjusting the cavity dimensions of the resonator or magnetically altering the resonant frequency of a ferromagnetic YIG element, which have multioctave bandwidths and high quality factor (Q) resonators [69, 70]. They also consume considerable amount of DC power (0.75–3 W) and their linearity is not high. Neither of these approaches can result in miniaturization or produced in large volumes for wireless communication systems. Solid-state varactors can provide a wide tuning range, but there are losses and linearity problems at microwave frequencies [71, 72]. The semiconductor-based varactors have a quality factor (Q) as low as 2 or 3 which can be associated with fixed capacitors in series in order to increase the Q of the resulting resonator, but in this case the tuning range is decreasing. In general, losses can reduced by using a weak coupling between the resonator and the varactor; higher Q means lower tuning. In contrast, employing MEMS technology enables the construction of miniaturized tunable filters that exhibit low-loss, low power consumption, and excellent linearity. The successful application of MEMS fabrication processes can reduce the overall size, weight, and cost of RF-integrated systems and making them attractive for many commercial and military applications [73, 74].

The RF MEMS switching circuits may also allow the development of low power systems based on tunable antennas, low-noise tunable oscillators, tunable filters, and tunable matching networks. The ultimate goal is to integrate the whole front-end system on a chip. The main bottleneck is represented not by the inductors or switches but by the low-loss (high Q) filters, diplexers, and crystal reference. The filters are currently being addressed using film bulk acoustic resonator (FBAR) technology and MEMS resonators. However, using the RF MEMS fabrication allows the elimination of off-chip inductor in the oscillator circuit making the integration of the tunable filter on a single chip possible. EBG structures and MEMS switches will be used to design tunable bandpass and reconfigurable filters.

1.7 Capacitive Switching Bandpass Filters

When frequency spectrum gets crowded with a large number of communication devices operating in a particular band, filter is becoming more and more critical. Tunable bandpass filters greatly simplify the design of transceiver and play a major role, especially in the area of wideband and multimode transceivers. In signal processing and communication systems, it is often desirable to reject unwanted frequencies from a signal. However, it is undesirable to use multiple filters since large surface area are occupied. Therefore, it is desirous to use a tunable filter to track blocks for multi-band telecommunication systems, radiometers, and wideband radar systems.

Many civil and military telecommunication and radar systems, such as electronic warfare, software programmable RF front-end and digital receivers, require RF tunable bandpass filters to be as flexible as possible in terms of center frequencies and bandwidth. The tunable bandpass filter which is tunable over a wide frequency range also offers high rejection [75, 76]. Generally, the tunable bandpass filters can be classified into three basic categories: mechanically tunable filters, magnetically tunable filters, and electronically tunable filters [77–89]. However, none of these satisfies the requirements of miniaturization and mass production.

In recent years, tunable bandpass filters based on MEMS technology which brings much improvement in the trade-offs between the tuning range and the losses in filter designs are widely studied. Capacitive switch and DC-contact switch can be used to construct these MEMS tunable bandpass filters.

1.8 Substrate Transfer Process

High aspect ratio (HAR) is desired to provide sufficient lateral capacitance, to increase the sensitivity of the sensor and to suppress the out-of-plane motion for many MEMS devices and components, such as the comb-drive actuators, the inertial sensors, and the variable capacitors. Different fabrication processes such as DRIE for silicon-on-insulator (SOI)-based process and silicon-on-glass (SOG)-based process are developed for high aspect ratio suspended silicon structures.

The high aspect ratio structures are etched in a silicon wafer using reactive ion etching (RIE) technique, followed by the deposition of a thin layer of silicon dioxide

(SiO_2). After removal of SiO_2 at the trench bottom, MEMS structures are released using a silicon isotropic dry etching. Finally, metal is deposited [90, 91]. Only a normal silicon wafer and single mask are needed to form suspended structures with thickness of 20 μm and aspect ratio greater than 10. However, this process has two shortcomings. First, the etching depth varies with the trench width of the structures due to the aspect-ratio-depending-etching (ARDE) effect of the DRIE. Second, the release process sacrifices some silicon of the structures and results in irregular shape at the bottom of the silicon structures. The two problems lead to the deviation of the mechanical performance of the device. SOI-based process can avoid those problems well. In the SOI wafer, the device silicon layer with uniform thickness is separated from the handle silicon layer by a buried oxide layer. After deposition and patterning of the hard mask, the device layer is etched using DRIE. Then, the MEMS structures are released using wet etching of the buried oxide or DRIE over-etching [92, 93]. A glass substrate is more favorable than a silicon substrate for high-frequency applications because of its low-loss characteristic [94–96]. The straightforward method to construct devices or circuits is to first build RF circuits on a silicon wafer or a SOI wafer and then transfer them to a glass substrate [94]. Bulk silicon dissolved wafer process is developed in such a way to fabricate 1- to 25-μm-thick MEMS structures on a glass substrate, in which device structures are etched on a silicon wafer and heavily boron doped, then the silicon wafer is anodically bonded to a glass followed by the silicon dissolving to leave heavily boron-doped devices attached to the glass substrate [97]. However, it is difficult to fabricate thick MEMS structures using the silicon-dissolved wafer process due to the limitation of the diffusion process. The typical diffusion time is 15–20 h for a diffusion depth of 15–20 μm. In order to supply thick device silicon layer (>30 μm) on the glass, the conventional SOG process is widely used [98–100]. In the conventional SOG process, some shallow trenches are etched in a glass wafer or the backside of a silicon wafer first. After the silicon wafer is anodically bonded to a glass wafer, the silicon wafer is thinned from the backside using either mechanically grinding or chemically etching with KOH. Then, the MEMS structures are aligned to the shallow trenches through a double-side alignment, followed by the formation of the high-aspect-ratio structures by etching through the silicon layer using DRIE process. However, the conventional SOG-based process has three shortcomings. First, the Si-glass stack always bows or warps after anodic bonding due to residual stress of the bonding. Therefore, double-side alignment and yield are restrained. Second, since the glass is a poor thermal conductor, the substrate temperature ramps rapidly during the DRIE process. Therefore, it worsens the DRIE etching quality of the high-aspect-ratio structures. Third, since glass is an insulating material, the notching effect is hash. Ionization of low frequencies (380 kHz) in the high-density inductively coupled plasma (ICP) etch tool can reduce the notching effect by removing the accumulated charges on the insulating surfaces. However, when the etching depth is large (>100 μm) and over-etching lasts long time, notching effect damages structures. To avoid the notching effect, a metal layer can be deposited and patterned on glass or on silicon before anodic bonding [100].

In Chapter 9, a new substrate transfer process is developed to fabricate high-aspect-ratio silicon-suspended structures with large thickness (>30 μm) on a glass substrate. This fabrication process is to first pattern the high-aspect-ratio silicon structure in desired shapes using the DRIE process. Then, the silicon structure is transferred to the glass substrate using a Si/thin film/glass anodic bonding process followed by two silicon thinning processes. Finally, the suspended structure is released by a self-aligned etching of the glass. This fabrication process voids all the problems in the conventional SOG-based process. The design of the fabrication process is described in Section 9.1. The three key unit processes – silicon/thin film/glass anodic bonding, silicon thinning using KOH etch, and metal deposition using a shadow mask – are explored in Sections 9.2, 9.3, and 9.4, respectively. The summary is drawn in Section 9.5.

1.9 Surface Planarization Process

MEMS fabrication processes can be divided into surface micromachining process and bulk micromachining process. In this work, the surface micromachining process is used for the fabrication of the capacitive switches and the switch circuits. The planarization of the metal bridge plays a significant role in reducing the contact area between the bridge and the dielectric layer. Specially, the surface planarization processes are developed for a flat metal bridge and reduces the roughness of the dielectric layer [101–104]. In this book, a new planarization process is proposed by filling the CPW slot with photoresist that can effectively improve the flatness of the bridge. Besides the surface planarization, surface roughness is another important consideration to improve the intimate contact between the metal bridges and the dielectric layer when the metal bridges are driven down. Therefore, all fabrication issues related to the RF effects of the metal bridge flatness and the surface roughness of the dielectric layer will be discussed in Chapter 10.

References

1. K. E. Petersen, "Micromechanical membrane switches on silicon," IEEE Trans. Electron Devices Vol. 23(4), 1976, pp. 376–386.
2. K. E. Petersen, "Dynamic micromechanics on silicon: techniques and devices," IEEE Trans. Electron Devices Vol. ED-25, 1978, pp.1241–1250.
3. M.-A. Grétillat, F. Grétillat and N. F. de Rooij, "Micromechanical relay with electrostatic actuation and metallic contacts," J. Micromech. Microeng. Vol. 9, 1999, pp. 324–331.
4. H. Sehr, A. G. R. Evans, A. Brunnschweiler, G. J. Ensell, C. G. J. Schabmueller, "Design and fabrication of a thermally actuated microrelay based on vertical bimorphs," Eurosensors XVI, the 16th European Conference on Solid-State Transducers, Prague, Czech Republic, 2002, pp. 752–755.
5. J. A. Wright and Y.-C. Tai, "Magnetostatic MEMS relays for the miniaturization of brushless DC motor controllers," Tech. Digest, 12th IEEE Int. Conf. on Microelectromechanical Systems, Orlando Florida, USA, 1999, pp. 594–599.

6. H. Hosaka, K. Kuwano and K. Yanagisawa, "Electromagnetic microrelays: concepts and fundamental characteristic," Sensor. Actuator. A Vol. 40, 1994, pp. 41–47.
7. W. P. Taylor, M. G. Allen and C. R. Dauwalter, "Batch fabricated electromagnetic microrelays," Proc. 45th Relay Conf., Lake Buena Vista Florida, USA, 1997, pp. 8.1–8.6.
8. W. P. Taylor and M. G. Allen, "Integrated magnetic microrelays: normally open, normally closed, and multi-pole devices," Tech. Digest 1997 Int. Conf. on Solid-State Sensors and Actuators, Chicago, USA, 1997, pp. 1149–1152.
9. H. A. C. Tilmans et al., "A fully-packaged electromagnetic microrelay," Tech. Digest. 12th IEEE Int. Conf. on Microelectromechanical Systems, Orlando Florida, USA, 1999, pp. 25–30.
10. S. N. Qiu, C. X. Qiu, I. Shih, M. Yu, G. Brassard and G. Seguin, "Development of a microwave microelectromechanical system switch," J. Vac. Sci. Technol. A Vol. 20(3), 2002, pp. 966–970.
11. J. F. Dickson, "On-chip high-voltage generation in NMOS integrated circuits using an improved voltage multiplier technique," IEEE J. Solid State Circuits Vol. SC-11(3), 1976, pp. 374–378.
12. C. Vaughan, "RF MEMS components for radio architecture," 2001 Sensors Expo, Chicago, IL, USA, 5–7 June, 2001.
13. S. Roy and M. Mehregany, "Fabrication of electrostatic nickel microrelays by nickel surface micromachining," Proc. IEEE Int. Conf. Micro-electro-mechanical Systems, Amsterdam, Netherlands, 1995, pp. 353–357.
14. S. Roth, C. Marxer, G. Feusier and N. F. de Rooij, "One mask nickel micro-fabricated reed relay," Proc. 13th IEEE Int. Conf. Micro-electro-mechanical Systems, Miyazaki, Japan, 2000, pp. 176–180.
15. R. Wood, R. Mahadevan, V. Dhuler, B. Dudley, A. Cowen, E. Hill and K. Markus, "MEMS microrelay," Mechatronics Vol. 8, 1998, pp. 535–547.
16. J. Simon, S. Saffer and C.-J. Kim, "A micromechanical relay with a thermally-driven mercury micro-drop," Proc. IEEE, 9th Ann. Int. Workshop on Microelectromechanical Systems, San Diego CA, USA, 1996, pp. 515–520.
17. S. Saffer, J. Simon and C.-J. Kim, "Mercury-contact switching with gap-closing microcantilever," Proc. SPIE, Vol. 2882, 1996, pp. 204–208.
18. J. Simon, S. Saffer, F. Sherman and C.-J. Kim, "Lateral polysilicon microrelay with a mercury microdrop contact," IEEE Trans. Ind. Electron. Vol. 45(6), 1998, pp. 854–860.
19. J. Kim, W. Shen, L. Latorre and C.-J. Kim, "A micromechanical switch with electrostatically driven liquid-metal droplet," Sensor. Actuator. Vol. 97–98, 2002, pp. 672–679.
20. N. D. Masters and L. L. Howell, "A self-retracting fully compliant bistable micromechanism," J. MEMS Vol. 12(3), 2003, pp. 273–280.
21. C.-L. Dai, K. Yen and P.-Z. Chang, "Applied electrostatic parallelogram actuators for microwave switches using the standard CMOS process, J. Micromech. Microeng. Vol. 11, 2001, pp. 697–702.
22. Y. Wang, Z. Li, D. T. McCormick and N. C. Tien, "A low-voltage lateral MEMS switch with high RF performance," J. MEMS Vol. 13(6), 2004, pp. 902–911.
23. Y. Wang, Z. Li, D. T. McCormick and N. C. Tien, "A micromachined RF microrelay with electrothermal actuation," Sensor. Actuator. A Vol. 103, 2003, pp. 231–236.
24. Y. Wang, Z. Li, D. T. McCormick and N. C. Tien, "Low-voltage lateral-contact microrelays for RF applications," 15th IEEE Int. Conf. on Micro-electro-mechanical Systems, 2002, pp. 645–648.
25. I. Schiele and B. Hillerich, "Comparison of lateral and vertical switches for application as microrelays," J. Micromech. Microeng. Vol. 10, 1999, pp. 146–150.
26. Z. Li, D. Zhang, T. Li, W. Wang and G. Wu, "Bulk micromachined relay with lateral contact," J. Micromech. Microeng. Vol. 10, 2000, pp. 329–333.
27. R. L. Borwick III, P. A. Stupar and J. De Natale, "A hybrid approach to low-voltage MEMS switches," Proc. 12th Int. Conf. Solid-State Sensors, Actuators and Microsystems (Transducer'03), Boston, MA, 2003, pp. 859–862.

28. A. A. Ayon, N. J. Kolias and N. C. MacDonald, "Tunable, micromachined parallel-plate transmission lines," Proc. IEEE, 1995, pp. 201–208.
29. T. D. Kudrle, H. P. Neves and N. C. MacDonald, "Microfabricated single crystal silicon transmission lines," RAWCON′98 Proc., Colorado Springs CO, USA, 1998, pp. 269–272.
30. A. Q. Liu, M. Tang, A. Agarwal and A. Alphones, Low-loss lateral micromachined switches for high frequency applications, J. Micromech. Microeng. Vol. 15, 2005, pp. 157–167.
31. T. D. Kudrle, H. P. Neves and N. C. MacDonald, "A micromachined millimeter wave phase shifter," Transducers'99, Sendai, Japan, 1999, pp. 1276–1279.
32. J. R. Reid, "Dielectric charging effects on capacitive MEMS actuators," IEEE MTT-S International Microwave Symposium, RF MEMS Workshop, Phoenix, AZ, May 2001.
33. G. M. Rebeiz, RF MEMS Theory, Design and Technology, Wiley-Interscience, New York, NY, 2003.
34. D. Peroulis, S. P. Pacheco and L. P. B. Katehi, "RF MEMS switches with enhanced power-handling capabilities," IEEE Trans. Microwave Theory Tech. Vol. 52(1), 2004, pp. 59–68.
35. A. B. Yu, A. Q. Liu, Q. X. Zhang, A. Alphones, L. Zhu and S. A. Peter, "Improvement of isolation MEMS capacitive switch via membrane planarization," Sensor. Actuator. A: Phys. Vol. 119, 2005, pp. 206–213.
36. S. P. Pacheco, D. Peroullis and L. P. B. Katehi, "MEMS single-pole double-throw (SPDT) X and K-band switching circuits," IEEE MTT-S Int. Microwave Symp. Dig., Phoenix AZ, USA, 2001, pp. 321–324.
37. M. C. Scardelletti, G. E. Ponchak and N. C. Varaljay, "MEMS, Ka-band single-pole double-throw (SPDT) switch for switched line phase shifters," IEEE Antenna. Propagat. Soc. Int. Symp., Kanagawa, Japan, June 2002, Vol. 2, pp. 2–5.
38. D. Sievenpiper, H. J. Song, H. P. Hsu, G. Tangonan, R. Y. Loo and J. Schaffner, "MEMS-based switched diversity antenna at 2.3 GHz for automotive applications," The 5th Int. Symp. Wireless Personal Multimedia Communications, Honolulu Hawaii, USA, October 2002, Vol. 2, pp. 762–765.
39. D. Hyman, A. Schmitz, B. Warneke, T. Y. Hsu, J. Lam, J. Brown, J. Schaffner, A. Walston, R. Y. Loo, G. L. Tangonan, M. Mehregany and J. Lee, "GaAs-compatible surface-micromachined RF MEMS switches," Electron. Lett. Vol. 35(3), 1999, pp. 224–226.
40. B. Schauwecher, K. M. Strohm, T. Mack, W. Simon and J.-F. Luy, "Single-pole double-throw switch based on toggle switch," Electron. Lett. Vol. 39(8), 2003, pp. 668–670.
41. K. Lee, R. Liu and B. C. Kim, "RF MEMS switch for wireless LAN applications," Proc. Int. Conf. MEMS, NANO Smart Systems (ICMENS′03), Banff Alberta, Canada, 2003.
42. J.-H. Park, S. Lee and J.-M. Kim, "A 35-60 GHz single-pole double-throw (SPDT) switching circuit using direct contact MEMS switches and double resonance technique," 12th Int. Conf. Transducers, Solid-State Sensors, Actuators and Microsystems, Boston, USA, Vol. 2, June 2003, pp. 1796–1799.
43. R. W. Moseley, E. M. Yeatman, A. S. Holmes, R. R. A. Syms, A. P. Finlay and P. Boniface, "Laterally actuated, low voltage, 3-port RF MEMS switch," Proc. MEMS, Turkey, Istanbul, 22–26 January 2006, pp. 878–880.
44. M. Daneshmand and R. R. Mansour, "Fabrication and modeling of an SP3T RF MEMS switch," IEEE AP-Symposium and North American Radio Science Meeting, Columbus Ohio, USA, 2003, pp. 391–394.
45. E. K. I. Hamad, G. E. Nadim and A. S. Omar, "A proposed SP3T wideband RF MEMS switch," Proc. IEEE Vol. 3, 2004, pp. 2839–2842.
46. G. L. Tan, R. E. Mihailovich, J. B. Hacker, J. F. De Natale and G. M. Rebeiz, "Low-loss 2- and 4-bit TTD MEMS phase shifters based on SP4T switches," IEEE Trans. Microwave Theory Tech. Vol. 51(1), 2003, pp. 297–304.
47. S. Pranonsatit, A. S. Holmes, I. D. Robertson and S. Lucyszyn, "Single-pole eight-throw RF MEMS rotary switch," IEEE J. MEMS Vol. 15(6), 2006, pp. 1735–1744.
48. L. Brillouin, Wave Propagation in Periodic Structures, 2nd edn., Dover Publications, Inc., New York, NY, 1953.

49. D. Sievenpiper, L. Zhang, R. F. J. Broas, N. G. Alexopolous and E. Yablonovitch, "High impedance electromagnetic surface with a forbidden frequency band," IEEE Trans. Microwave Theory Tech. Vol. 47(11), 1999, pp. 2059–2074.
50. A. Bayers, I. Rumsey, Z. Popvic and M.-P. May, "Surface wave guiding using periodic structures," IEEE–APS Dig., Salt Lake City Utah, USA, Vol. 1, 2000, pp. 342–345.
51. D. Sivenpiper and E. Yablonovitch, "Eliminating surface currents with metallodielectric photonics crystal," IEEE-MTT-S Int. Microwave Symp. Dig., Baltimore MD, USA, Vol. 1, 1998, pp. 663–666.
52. N. C. Karmakar, "Improved performance of photonic bandgap microstripline structures using Chebyshev distribution," Microwave Opt. Technol. Lett. Vol. 33(1), 2002, pp. 1–5.
53. V. Radisic, Y. Qian, R. Coccioli and T. Itoh, "A novel 2-D photonic bandgap structures for microstrip lines," IEEE Microwave Guide Wave Lett. Vol. 8, 1998, pp. 66–71.
54. B. Elamaran, I. M. Chio, L. Y. Chen and J. C. Chiao, "A beam-steerer using reconfigurable PBG ground plane," IEEE MTT-S Int. Microwave Symp. Dig., Boston MA, USA, 2000, pp. 835 –838.
55. I. Chang and B. Lee, "Design of defected ground structures for harmonic control of active microstrip antenna," IEEE AP-S Int. Symp. Dig., Texas, USA, 2002, pp. 852–855.
56. N. C. Karmakar and M. N. Mollah, "Investigations into nonuniform photonic-bandgap microstripline low-pass filters," IEEE Trans. Microwave Theory Tech. Vol. 51, 2003, pp. 564–572.
57. M. L. Her, Y. Z. Wang, C. M. Chang and K. Y. Lin, "Coplanar waveguide (CPW) defected ground plane structure (DGS) for bandpass filter applications," Microwave Opt. Tech. Lett. Vol. 42, 2004, pp. 331–334.
58. C. S. Kim, J. S. Park, D. Ahn and J. Lim, "A novel 1-D periodic defected ground structure for planar circuits," IEEE Microwave Wireless Comp. Lett. Vol. 10(4), 2000, pp. 131–133.
59. D. Ahn, J. S. Park, C. S. Kim, J. Kim, Y. Qian and I. Itoh, "A design of low pass filter using novel microstrip defected ground structure," IEEE Trans. Microwave Theory Tech. Vol. 49(1), 2001, pp. 86–93.
60. L. Garde, M. J. Yabar and C. D. Rio, "Simple modeling of DGS to design 1D PBG low pass filter," Microwave Opt. Tech. Lett. Vol. 37(3), 2003, pp. 228–232.
61. H. Y. D. Yang, N. G. Alexopoulos and E. Yablonovitch, "Photonic bandgap materials for high gain printed circuit antennas," IEEE Trans. Antenna. Propagat. Vol. 45(1), 1997, pp. 185–187.
62. S. Y. Lin and K. L. Wong, "A conical pattern annular ring microstrip antenna with a photonic bandgap ground plane," Microwave and Optical Tech. Lett. Vol. 30(3), 2001, pp. 159–161.
63. V. Radisic, Y. Qian and I. Itoh, "Novel architectures for high efficiency amplifiers for wireless applications," IEEE Trans. Microwave Theory. Tech. Vol. 46(11), 1998, pp. 1901–1909.
64. C. Y. Hang, W. R. Deal, Y. Qian and I. Itoh, "High efficiency transmitter front end integrated with planar antennas and PBG," Asia-Pacific Microwave Conf., Australia, 2000.
65. R. E. Collin, Foundations for Microwave Engineering, McGraw-Hill, New York, NY, 1992.
66. D. M. Pozar, Microwave Engineering, Wiley, New York, NY, 1998.
67. G. L. Matthaei, L. Yong and E. M. T. Jones, Microwave Filters, Impedance-Matching Networks, and Coupling Structures, McGraw-Hill, New York , NY, 1964.
68. G. M. Rebeiz, RF MEMS — Theory, Design, and Technology, 1st edn., Wiley, New York, NY, 2002.
69. Y. Ishikawa, T. Nishikawa, T. Okada and S. Shinmura, "Mechanically tunable MSW bandpass filter with combined magnetic units," IEEE-MTT-S Int. Microwave. Symp. Dig., 1990, pp. 143–146.
70. W. J. Keane, "YIG filters aid wide open receivers," Microwave J. Vol. 17(8), 1980, pp. 50–54.
71. S. R. Chandler, I. C. Hunter and J. C. Gardiner, "Active varactor tunable bandpass filters," IEEE Microwave Guided Wave Lett. Vol. 3(3), 1993, pp. 70–71.

72. A. R. Brown and G. M. Rebeiz, "A varactor-tuned RF filter," IEEE Trans. Microwave Theory Tech. Vol. 48(7), 2000, pp. 1157–1160.
73. H. J. D. Santos, Introduction to Microelectromechanical (MEMS) Microwave Systems, Artech House, Norwood, MA, 1999.
74. H. J. D. Santos, "MEMS-A wireless vision," Proc. Int. MEMS Workshop, Singapore, 2001.
75. I. C. Hunter, L. Biloner, B. Jarry and P. Guillan, "Microwave filters – application and technology," IEEE Trans. Microwave Theory Tech. Vol. 50, 2002, pp. 794–805.
76. J. Uher and J. R. Hofer, "Tunable microwave and millimeter-wave bandpass filters," IEEE Trans. Microwave Theory Tech. Vol. 39, 1991, pp. 643–653.
77. J. B. Tsui, Microwave Receivers with Electronic Warfare Applications, Wiley, New York, NY, 1992.
78. H. Tanbakachi et al., "A broadband tracking YIG-tuned mixer for a state of the art spectrum analyzer," European Microwave Conference Dig., Rome, Italy, Sept. 1987, pp. 482–490.
79. R. W. de Greese, "Low-loss gyromagnetic coupling through single crystal garnets," J. Appl. Phys. Vol. 30, 1958, pp. 1555–1559.
80. P. S. Carter, "Equivalent circuit of orthogonal-loop-coupled magnetic resonance filters and bandwidth narrowing due to coupling resonance," IEEE Trans. Microwave Theory Tech. Vol. 18, 1970, pp. 100–105.
81. R. F. Fierstal, "Some design considerations and realizations of iris-coupled YIG-tuned filters in the 12–40 GHz region," IEEE Trans. Microwave Theory Tech. Vol. 18, 1970, pp. 205–212.
82. H. Tanbakuchi D. Nicholson, B. Kunz and W. Ishak, "Magnetically tunable oscillators and filters," IEEE Trans. Magn. Vol. 25, 1989, pp. 3248–3253.
83. S. R. Chandler, L. C. Hunter and J. C. Gordiner, "Active varactor tunable bandpass filters," IEEE Microwave Guided Wave Lett. Vol. 3, 1993, pp. 70–71.
84. A. R. Brown and G. M. Rebeiz, "A varactor-tuned RF filter," IEEE Trans. Microwave Theory Tech. Vol. 48, 2000, pp. 1157–1160.
85. I. C. Munter and J. D. Rhodes, "Electronically tunable microwave bandpass filters," IEEE Trans. Microwave Theory Tech. Vol. 30, 1982, pp. 1354–1360.
86. K. Shen, F. Hui, W. Wong, Z. Chen, J. Lau, P. C. H. Chan and P. K. Ko, "A three-terminal SOI gated varactor for RF applications," IEEE Trans. Electron Devices Vol. 48, 2001, pp. 289–293.
87. B. H. Mockly and Y. Zhang, "Strontium titanate thin films for tunable YBa2Ca3O7 microwave filters," IEEE Trans. Appl. Supercond. Vol. 11, 2001, pp. 450–453.
88. A. Kozyrev, A. Ivanov, V. Keis, M. Khazov, V. Osadchy, T. Samoilova, O. Soldatenkov, A. Pavlov, G. Koepf, C. Mueller, D. Galt, T. Rivkin, "Ferroelectric films: nonlinear properties and applications in microwave devices," IEEE MTT-S Int. Microwave Symp. Dig., Baltimore MD, USA, Vol. 2, 1998, pp. 985–988.
89. A. Tombak, J. Maria, F. T. Ayguavives, Z. Z. Jin, G. T. Stauf, A. I. Kingo and A. Mortazawi, "Voltage-controlled RF filter employing thin-film Barium-Strontium-titanate tunable capacitors," IEEE Trans. Microwave Theory Tech. Vol. 51, 2005 pp. 462–467.
90. Z. L. Zhang and N. C. MacDonald, "A RIE process for submicron, silicon electromechanical structures," J. Micromech. Microeng. Vol. 2, 1992, pp. 31–38.
91. K. A. Shaw, Z. L. Zhang and N. C. MacDonald, "SCREAM I: a single mask, single-crystal silicon, reactive ion etching process for microelectromechanical structures," Sensor. Actuator. A Vol. 40(1), 1994, pp. 63–70.
92. S. A. McAuley, H. Ashraf, L. Atabo, A. Chambers, S. Hall, J. Hopkins and G. Nicholls, "Silicon micromachining using a high-density plasma source," J. Phys. D: Appl. Phys. Vol. 34, 2001, pp. 2769–2774.
93. A. A. Ayon, K. Ishihara, R. A. Braff, H. H. Sawin and M. A. Schmidt, "Microfabrication and testing of suspended structures compatible with silicon-on-insulator technology," J. Vac. Sci. Technol. B Vol. 17, 1999, pp. 1589–1593.
94. R. Dekker, P. G. M. Baltus and H. G. R. Maas, "Substrate transfer for RF technology," IEEE Trans. Electron Devices Vol. 50(3), 2003, pp. 747–757.

95. S. Seok, W. Choi and K. Chun, "A novel linearly tunable MEMS variable capacitor," J. Micromech. Microeng. Vol. 12, 2002, pp. 82–86.
96. J. Kim, J. Park, C. Baek and Y. Kim, "The SiOG-based single-crystalline silicon (SCS) RF MEMS switch with uniform characteristics," J. Microelectromech. Syst. Vol. 13(6), 2004, pp. 1036–1042.
97. Y. B. Gianchandani, K. Najafi, "A bulk silicon dissolved wafer process for microelectromechanical devices," J. Microelectromech. Syst. Vol. 1(2), 1992, pp. 77–85.
98. B. Y. Majis, B. Bais and A. S. Tamsir, "Area-changed capacitive accelerometer using 3-mask fabrication process," Proc. SPIE Vol. 5276, 2004, pp. 482–488.
99. J. Chae, H. Kulah, and K. Najafi, "A CMOS-compatible high aspect ratio silicon-on-glass in-plane micro-accelerometer," J. Micromech. Microeng. Vol. 15, 2005, pp. 336–345.
100. M. C. Lee, S. J. Kang, K. D. Jung, S. Choa and Y. C. Cho, "A high yield rate MEMS gyroscope with a packaged SiOG process," J. Micromech. Microeng. Vol. 15, 2005, pp. 2003–2010.
101. J. Y. Park, G. H. Kim, K. W. Chung and J. U. Bu, "Monolithically integrated micromachined RF MEMS capacitive switches," Sensor. Actuator. A: Phys. Vol. 89, 2001, pp. 88–94.
102. E. J. J. Kruglick and K. S. J. Pister, "Lateral MEMS microcontact considerations," IEEE J. Microelectromech. Syst. Vol. 8, 1999, pp. 264–271.
103. S. L. Firebaugh, H. K. Charles, R. L. Edwards, A. C. Keeney and S. F. Wilderson, "Fabrication and characterization of a capacitive micromachined shunt switch," J. Vac. Sci. Technol. A Vol. 22(4), 2004, pp. 1383–1387.
104. Z. J. Yao, S. Chen, S. Eshelman, D. Denniston and C. Goldsmith, "Micromachined low-loss microwave switches," J. Microelectromech. Syst. Vol. 8(2), 1999, pp. 129–134.

Chapter 2
Lateral Series Switches

The objective of this chapter is to study the lateral RF MEMS series switch [1–14]. The switch consists of a silicon-core (Si-core) transmission line and a cantilever beam. The switch connects or disconnects with the RF circuit by the in-plane motion of the cantilever beam. The Si-core coplanar waveguide (CPW) transmission line is used to configure the switch. The Si-core CPW has the same advantage as the normal CPW, that is, it enables series and shunt elements connection without metal coating. The main advantages of the lateral switch are reliable mechanical performance and simple fabrication process.

This chapter is organized as follows. First, the RF design, circuit modeling and simulation of the lateral RF MEMS switch are presented. The RF performance of the switch can be improved by optimizing the electrical parameters of the switch. Second, the mechanical design and optimization of the cantilever beam of the lateral switch are discussed. The dynamic responses such as the switching time and the release time of the switch are studied. Finally, the experimental results of the lateral switch such as its insertion loss, return loss, isolation, threshold voltage, switching speed are discussed. Comprehensive modeling and design of the lateral switch are verified by extensive experiments for both electrical and mechanical characteristics.

2.1 Electrical Design and Simulation

In this section, the RF circuit design and the lumped-element modeling of the lateral switch are discussed. The main purpose is to study the effect of the design parameters and to realize lateral switches with low insertion loss, high return loss, and high isolation. Since the switch is a metal-contact series switch, the on-state and the down-state of the switch and the off-state and the up-state of the switch are used interchangeably.

Min Tang and Ai Qun Liu

A.Q. Liu, *RF MEMS Switches and Integrated Switching Circuits*,
MEMS Reference Shelf 5, DOI 10.1007/978-0-387-46262-2_2,

2.1.1 RF Circuit Design of the Lateral Switch

A lateral switch consists of a Si-core CPW and an electrostatic actuator, as shown in Fig. 2.1a. A cantilever beam is fixed at one port. The free end of the cantilever beam comes into contact with the contact bump at the other port upon turning on the switch. The cantilever beam serves as the signal line alone. The ground lines beside the cantilever beam are extended toward the cantilever beam to avoid drastic increase in the characteristic impedance. The width of the gaps between the cantilever beam and the ground lines is 20–30 μm. The characteristic impedance of the cantilever beam section, Z_l, is about 78 Ω simulated by a 3D full-wave finite element method (FEM) analysis tool – Ansoft's high-frequency structure simulator (HFSS) V8.0 [15]. At the free end of the cantilever beam, one ground line protrudes toward the cantilever beam further to serve as a fixed electrode. Therefore, no additional fixed electrode is required. When sufficient DC bias voltage is applied between the cantilever beam and the ground line, the cantilever beam is pulled toward the fixed electrode by electrostatic force until its free end hits the contact bump, resulting in the on-state of the switch. When DC bias voltage is removed, the mechanical stress of the beam overcomes the stiction forces and pulls the cantilever beam away, resulting in the off-state of the switch. Due to the asymmetrical layout of the two ports, the S-parameters obtained from the two ports are not reciprocal. The return loss of port 2 is better than that of port 1 at the off-state since the open stub at port 2 is shorter than port 1. Hence, generally port 2 acts as the input port and port 1 acts as the output port to block more RF signal at the off-state of the switch. Figure 2.1 (b) shows the cross sectional view of the lateral switch. The switch is on a SOI wafer.

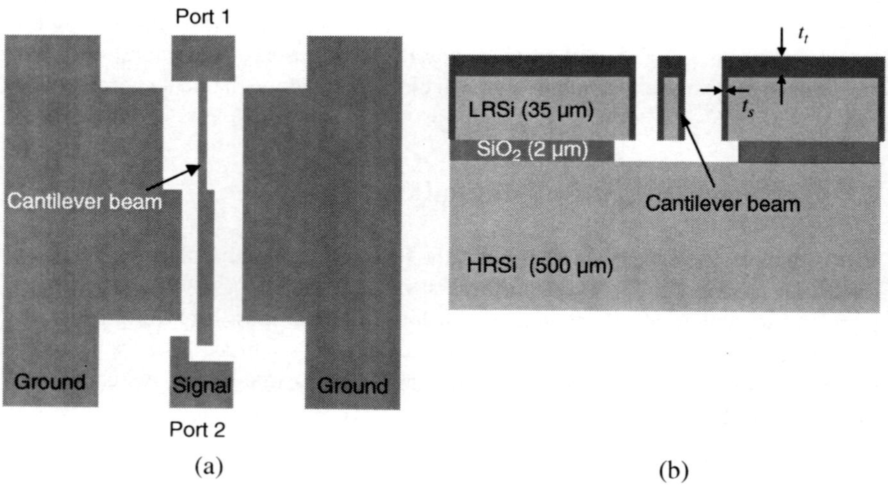

Fig. 2.1 Schematics of a lateral switch: **(a)** top view and **(b)** cross-sectional view. Copyright/used with permission of/courtesy of Institute of Physics and IOP Publishing Limited

The substrate is 500-μm-thick high-resistivity silicon (HRSi). The device layer is 35-μm-thick low-resistivity silicon (LRSi). The switch structures are made of LRSi. A thin layer of metal is coated on the top and sidewalls of the switch structures to propagate RF signal.

2.1.2 RF Circuit Model of the Lateral Switch

Figure 2.2 shows the equivalent circuit of the lateral switch. The model consists of a characteristic impedance, Z_0, for the input and output sections of the Si-core CPW; a line resistor, R_l, of the cantilever beam; a line inductor, L, of the cantilever beam; a switch series capacitor, C_s (off-state) or a contact resistor, R_c (on-state); and a shunt coupling capacitor, C_g. Except Z_0, other parameters are allowed to vary to fit the measurement results or the simulation results. The equivalent circuit is modeled using the design tool of Agilent EESof's Advanced Design System (ADS).

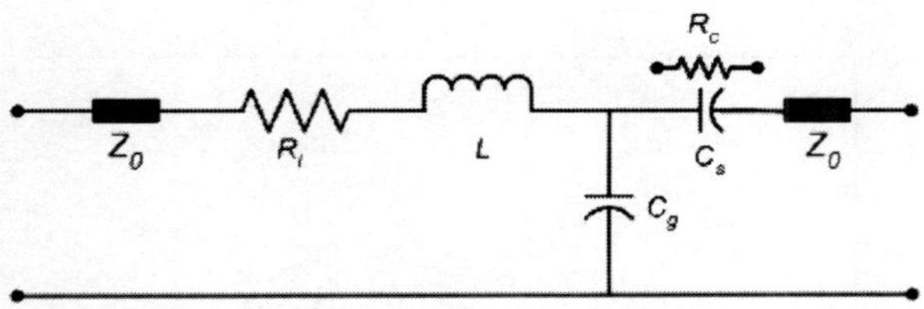

Fig. 2.2 The equivalent circuit of the lateral switch. Copyright/used with permission of/courtesy of Institute of Physics and IOP Publishing Limited

According to the T-equivalent circuit model, S_{21} of the circuit can be given by [16]

$$S_{21} = \frac{2}{2 + (Z_0 + Z_1 + Z_2)/Z_3 + \left(Z_1 + Z_2 + \frac{Z_1 Z_2}{Z_3}\right)\Big/ Z_0} \tag{2.1a}$$

where

$$Z_1 = R_l + j\omega L \tag{2.1b}$$

$$Z_2 = \begin{cases} \frac{1}{j\omega C_s} & \text{at the open state} \\ R_c & \text{at the close state} \end{cases} \tag{2.1c}$$

$$Z_3 = \frac{1}{j\omega C_g} \tag{2.1d}$$

ω is the angular frequency ($\omega = 2\pi f, f$ is the signal frequency).

At the off-state of the switch, the switch capacitance, C_s, is an important factor that affects the isolation of the switch. When

$$S_{21} << -10\,\mathrm{dB} \text{ and } \omega C_s Z_0 \left[2 - \omega^2 C_g L + \frac{C_g}{C_s} + \frac{R_l}{Z_0}\left(1 + \frac{C_g}{C_s}\right)\right] << 1$$

the isolation of switch can be approximately expressed as

$$S_{21} \approx j2\omega C_s Z_0 \tag{2.2}$$

Therefore, the series capacitance, C_s, of the off-state switch can be extracted from the simulated or measured isolation of the switch using Eq. (2.2). Figure 2.3 shows the simulated isolation of an off-state switch with various series capacitances of C_s. The isolation of the off-state switch increases with the decrease in C_s. The equivalent series capacitance C_s of our practical lateral switches is 3–10 fF. Up to 25 GHz, the isolation is higher than 25 dB for $C_s = 3$ fF and 15 dB for $C_s = 10$ fF.

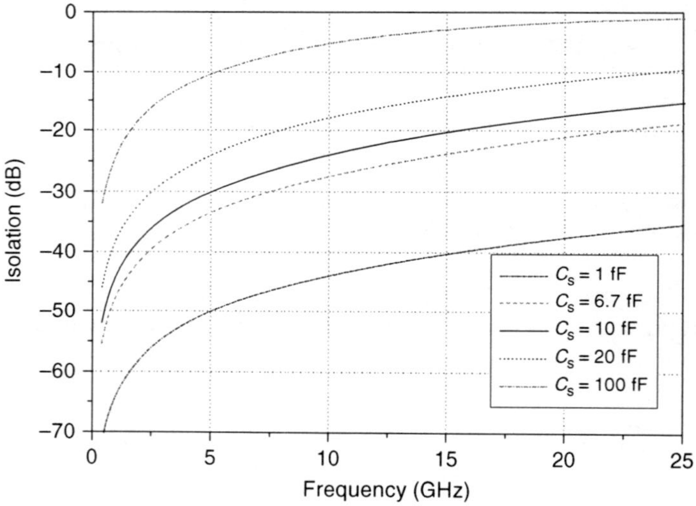

Fig. 2.3 Simulation results of *S*-parameters with various capacitances C_s at the off-state of the lateral switch ($R_l = 1\ \Omega$, $L = 148$ pH, $C_g = 30$ fF)

At the on-state of the switch, the insertion loss and the return loss can be expressed as

$$S_{21} = \frac{2}{2 + K_1 + j \cdot K_2} \tag{2.3a}$$

$$S_{11} = \frac{K_1 + j \cdot K_3}{2 + K_1 + j \cdot K_2} \tag{2.3b}$$

where

$$K_1 = \frac{R_l + R_c}{Z_0} - \omega^2 C_g L \left(1 + \frac{R_c}{Z_0}\right) \tag{2.3c}$$

$$K_2 = \omega \left[C_g \left(R_l + R_c + Z_0 + \frac{R_l R_c}{Z_0} \right) + \frac{L}{Z_0} \right] \tag{2.3d}$$

$$K_3 = \omega \left[C_g \left(R_l - R_c - Z_0 + \frac{R_l R_c}{Z_0} \right) + \frac{L}{Z_0} \right] \tag{2.3e}$$

At low frequencies, when $2+K_1 >> 0$ and $K_2 \ll 2+(R_l + R_c)/Z_0$, the insertion loss and the return loss of the switch can be simplified as

$$S_{21} \approx 2/[2 + (R_l + R_c)/Z_0] \tag{2.4a}$$

$$S_{11} \approx (R_l + R_c)/[2Z_0 + (R_l + R_c)] \tag{2.4b}$$

Figure 2.4 shows the simulated results of the insertion loss and the return loss of the switch at the on-state with various (R_l + R_c). The RF performances of the switch at the on-state deteriorate with the increase in (R_l + R_c). When R_l + R_c < 2 Ω, the insertion loss is less than 0.2 dB up to 10 GHz. When (R_l + R_c) increases to 10 Ω, the insertion loss is larger than 0.8 dB at 10 GHz. The return loss decrease with (R_l + R_c) Therefore, the resistance sum (R_l + R_c) should be as small as possible to achieve low insertion loss and high return loss.

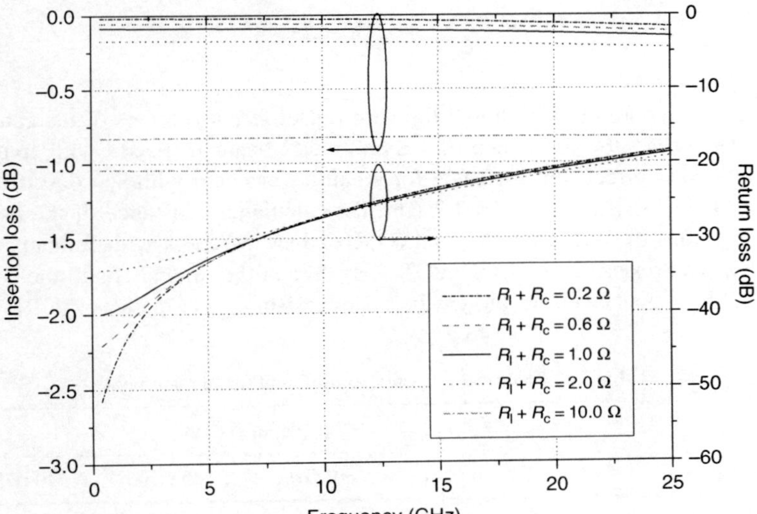

Fig. 2.4 Simulation results of *S*-parameters with various resistance sum at the on-state of the lateral switch ($L = 148$ pH, $C_g = 30$ fF)

The resistance sum ($R_l + R_c$) of the switch at the on-state can also be extracted from the simulated or measured insertion loss of the switch using Eqs. (2.4a) and (2.4b). It is found that the measured resistance sum ($R_l + R_c$) decreases with the increase in the metal thickness of the coating layer since the cantilever beam resistance, R_l, is determined by the thickness of the coating metal layer and the contact resistance, R_c, is largely affected by it too. The effect of the metal thickness on the beam resistance, R_l, is discussed. Due to the skin effect of the metal layer at high frequencies, the fields decay by an amount of $1/e$ in a depth of one skin depth. The skin depth, δ_s, of metal is given by [16]

$$\delta_s = \sqrt{\frac{2}{\omega \mu_0 \sigma}} \tag{2.5}$$

where μ_0 is the permeability of the vacuum ($\mu_0 = 4\pi \times 10^{-7}$H/m) and σ is the conductivity of the metal layer. The conductivity and the skin depth of the metal layer used in our experiments are listed in Table 2.1. The skin depth of aluminum (Al) is 0.83 μm at 10 GHz and 0.53 μm at 25 GHz. The surface resistivity, R_s, of the metal layer is given by

$$R_s = \begin{cases} 1/(\sigma \delta_s) & (t \geq \delta_s) \\ 1/(\sigma t) & (t < \delta_s) \end{cases} \tag{2.6}$$

where t is the metal thickness. Then, the resistance of a cantilever beam, R_l, can be expressed as

$$R_l = R_s \sum_i \frac{l_i}{w_i} \tag{2.7}$$

where l_i and w_i are the length and the width of different sections of the cantilever beam. Therefore, the resistance of the cantilever beam increases with frequency due to the skin effect. For instance, for a cantilever beam with l = 460 μm, w = 3.3 μm, t_t = 1.2 μm, and t_s = 1.2 μm, the calculated resistance of the beam R_l keeps constant of 3 Ω from DC to 4.6 GHz, then increases with frequency after 4.6 GHz. The resistance is 7.1 Ω at 25 GHz. When the thickness and the conductivity of the metal layer are larger, the beam resistance is smaller. The RF power

Table 2.1 Electrical properties of different metal materials

		Skin depth, δ (μm)		
Material	Conductivity, σ(S/m)	10 GHz	25 GHz	40 GHz
Al	3.80×10^7	0.82	0.52	0.4
Au	4.55×10^7	0.75	0.47	0.37
Cu	5.88×10^7	0.66	0.42	0.33

dissipation decreases and the insertion loss of the switch is lower. Therefore, the metal layer with high conductivity has to be coated as thickly as possible to achieve good RF performance.

The series inductance L can be calculated by [16]

$$L = \frac{Z_l \beta l}{\omega} = \frac{Z_l l \sqrt{\varepsilon_{\text{eff}}}}{c} \tag{2.8}$$

where Z_l is the impedance of the cantilever beam, l is the whole length of the cantilever beam, β is the phase constant, ε_{eff} is the relative effective permittivity, and c is the speed of the light in vacuum ($c = 3.0 \times 10^8$ m/s). In this design, $l = 400$–500 µm, $\varepsilon_{\text{eff}} \approx 1.66$, $Z_l = 50$–$78\ \Omega$, and the equivalent series inductance is 86–167 pH. In Fig. 2.5, the simulated results show that the RF performances of the switch at the on-state become better when the inductance increases from 10 to 100 pH. However, the insertion loss and the return loss begin to deteriorate when the inductance L increases further, especially at a high-frequency range.

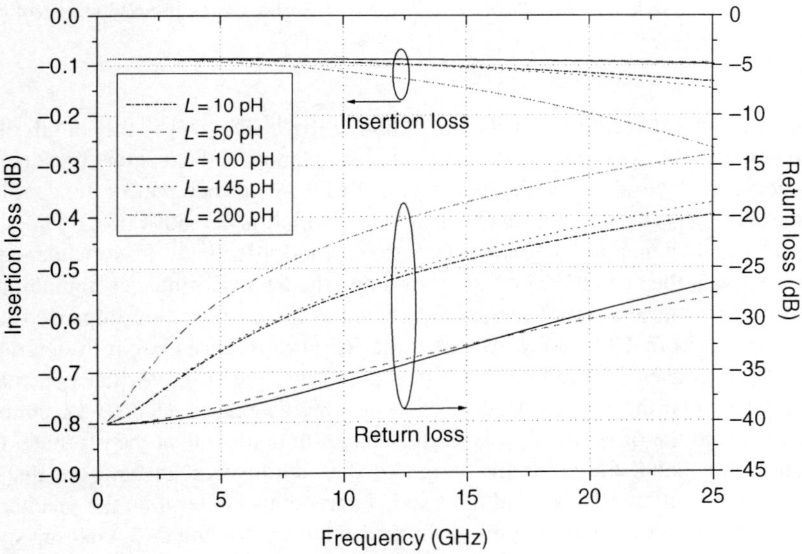

Fig. 2.5 Simulation results of S-parameters with various inductances L at the on-state of the lateral switch ($R_l = 1$ W, $R_c = 1.8$ W, $C_g = 30$ fF). Copyright/used with permission of/courtesy of Institute of Physics and IOP Publishing Limited

C_g is the shunt coupling capacitance between the cantilever beam and the fixed electrode, which can be estimated as

$$C_g = \frac{\varepsilon_0 l_2 h}{g_0 - y} + C_f \tag{2.9}$$

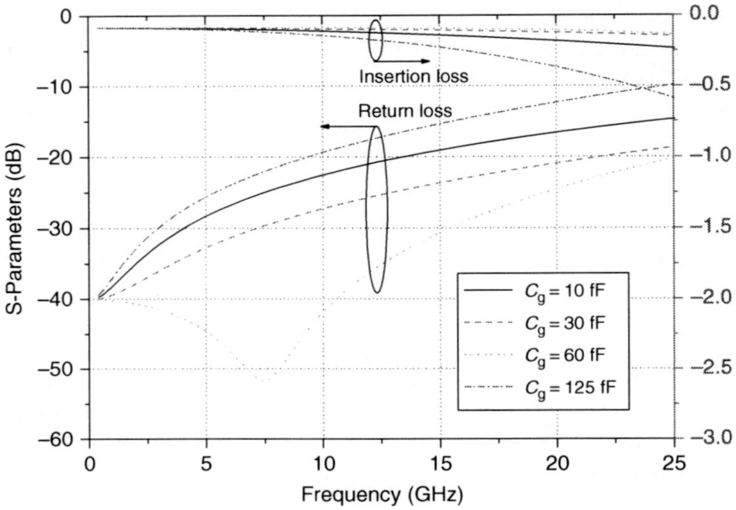

Fig. 2.6 Simulation results of S-parameters with various capacitances C_g when the switch is at the on-state of the switch ($R_l = 1$ W, $R_c = 1.8$ W, $L = 148$ pH). Copyright/used with permission of/courtesy of Institute of Physics and IOP Publishing Limited

where ε_0 is the permittivity of the air (8.854×10^{-12} F/m), l_2 is the length of the fixed electrode, h is the height of the cantilever beam, g_0 is the original gap distance between two electrodes, y is the displacement of the electrode part of the cantilever beam, and C_f is the fringing field capacitance which is 20–60% of C_g [17]. This coupling capacitance is fairly large ($C_g \gg C_s$) and affects the loss mechanism at the on-state of the switch. Figure 2.6 shows that the RF performances, including the insertion loss and the return loss of the switch at the on-state, are improved when C_g increases from 10 to 60 fF, whereas the RF performances begin to deteriorate when C_g increases further to 125 fF. It is noted that when the switch is actuated, the gap between the two electrodes, $g(= g_0 - y)$, decreases. Hence, the coupling capacitance at the on-state, $C_{g\,c}$, is slightly larger than the one at the off-state, $C_{g\,o}$. On the other hand, the structure design of the cantilever beam and the distance between the cantilever beam and the fixed electrode also determine the mechanical performance of the switch in terms of the threshold voltage and the switching speed. Therefore, it is important to select the proper values of l_2 and g_0 for the switch design to compromise between electrical and mechanical performances.

2.1.3 Double-Beam Lateral Switch

A double-beam lateral switch is designed and shown in Fig. 2.7 to achieve low insertion loss and high power handling. Two cantilever beams are used in the signal route together to propagate RF signal. Both fixed connections of two cantilever

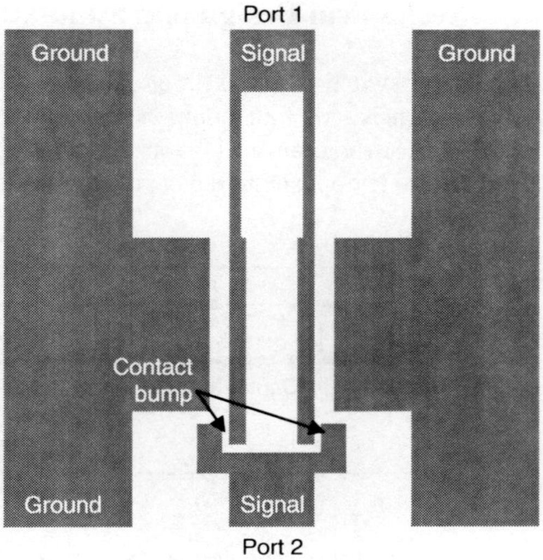

Fig. 2.7 The schematic of the top view of a double-beam lateral switch. Copyright/used with permission of/courtesy of Institute of Physics and IOP Publishing Limited

beams are on port 1 and two contact bumps are on port 2. At the free end of two cantilever beams, both ground lines extend toward the nearby cantilever beams to serve as their fixed electrodes, respectively. Therefore, when sufficient DC bias voltage is applied between the signal line and two ground lines at port 1, both cantilever beams are pulled by the electrostatic force and move toward two ground lines, respectively, until they hit two contact bumps at port 2. Generally, port 2 acts as the input port and port 1 acts as the output port for better signal isolation from the source at the off-state of the switch. Similar to the single-beam switch, the two-port *S*-parameters of the double-beam switch are not reciprocal due to the asymmetrical layout.

Figure 2.8 is the equivalent circuit of the double-beam lateral switch. For the double-beam switch, the resistor $R_l = R_{l0}/2$, the inductor $L = L_0/2$, the series capacitor of the switch at the off-state $C_s = 2C_{s0}$, the contact resistor of the switch at the on-state $R_c = R_{c0}/2$, and the shunt coupling capacitor $C_g = 2C_{g0}$, assuming that two cantilever beams are identical. Hence, the electrical model of the single-beam switch can also be used for the double-beam switch.

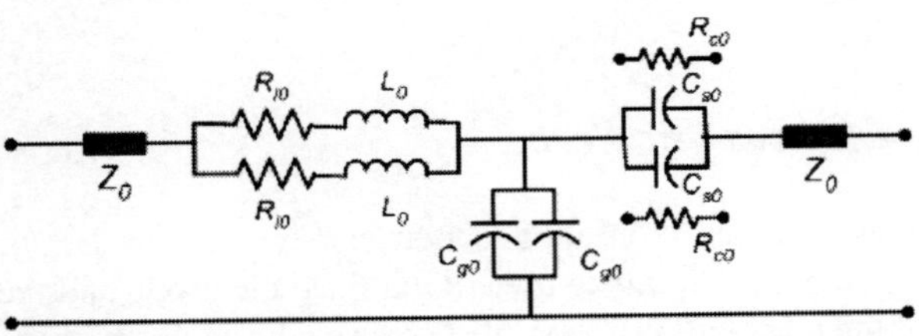

Fig. 2.8 The equivalent circuit of the double-beam lateral switch. Copyright/used with permission of/courtesy of Institute of Physics and IOP Publishing Limited

2.2 Mechanical Design and Simulation

The lateral switch is based on an electrostatic actuator as shown in Fig. 2.9. The actuator consists of four components: a suspended cantilever beam serving as a movable electrode, an anchor on the substrate to support the cantilever beam, a fixed electrode opposite to the cantilever beam, and a contact bump.

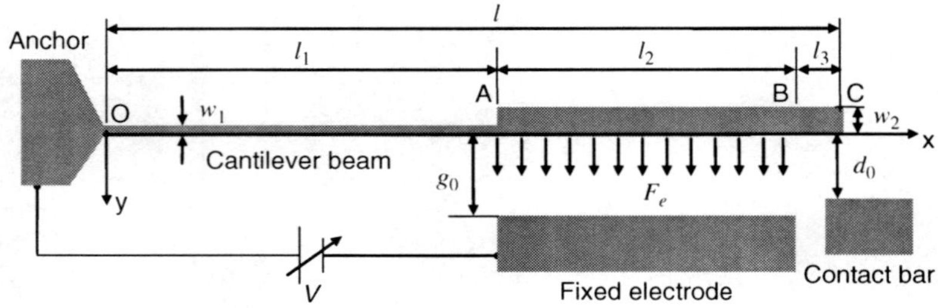

Fig. 2.9 The schematic of the top view of an electrostatic actuator. Copyright/used with permission of/courtesy of Institute of Physics and IOP Publishing Limited

The cantilever beam OC is a beam mass structure as shown in Fig 2.9. For the part of the beam OA, the width is w_1 and the length is l_1. For the part of the mass AC, the width is w_2 and the length is (l_2+l_3) in which l_2 is the length of the electrode section AB and l_3 is the length of the remaining part BC. The mass width, w_2, is designed to be relatively wider than the beam width, w_1, so that low threshold voltage can be maintained and greater deformation of the electrode section may be avoided. Hence, no separate stop bumpers or landing pads are required to avoid short circuit between the two electrodes.

The original distance between the two electrodes is g_0 and the distance between the contact bump and the mass part is d_0. d_0 is relatively smaller than g_0 so that the cantilever beam can touch the contact bump when the switch is actuated. In Fig 2.9, the X- and Z-axes are oriented parallel to the length and the depth of the cantilever beam respectively, while the Y-axis is directed toward the fixed electrode.

For the design of the lateral switch in static behavior, low threshold voltage is always desired. The threshold voltage is determined by two forces – the electrostatic force, F_e, and the restoring force, F_r – which will be discussed.

2.2.1 *Static Electrostatic Force*

When a DC bias voltage, V, is applied between two electrodes, the electrostatic force, F_e, causes the mass to move toward the fixed electrode and the beam is bent. The bending of the mass part is negligible because it has much higher flexure rigidity than the beam. The displacement of the mass increases with voltage until pull-in occurs. If the displacement of the beam OA and mass AC from its original position are designated as $y_1|_x$ and $y_2|_x$, respectively, the electrostatic force, F_e, on the mass can be derived as

$$F_e = \int_{l_1}^{l_1+l_2} \frac{\varepsilon_0 h V^2 dx}{2(g_0 - y_2)^2} = \frac{\varepsilon_0 h l_2 V^2}{2(g_0 - \alpha - \theta l_2)(g_0 - \alpha)} \tag{2.10a}$$

where

$$y_2|_x = \alpha + \theta(x - l_1), \quad l_1 \le x \le l_1 + l_2 \tag{2.10b}$$

$$\alpha = y_2|_{x=l_1} = y_1|_{x=l_1} \tag{2.10c}$$

$$\theta = y_2'|_x = y_1'|_{x=l_1} \tag{2.10d}$$

where ($'$) denotes the derivative with respect to position x. When the electrostatic force is not uniform, the bending moment, M_0, caused by the electrostatic force can be calculated as

$$M_0 = \int_{l_1}^{l_1+l_2} \frac{\varepsilon_0 h V^2 (x - l_1)\, dx}{2(g_0 - \alpha + \theta l_1 - \theta x)^2} = \frac{\varepsilon_0 h V^2}{2\theta^2}\left(\ln \frac{g_0 - \alpha - \theta l_2}{g_0 - \alpha} + \frac{\theta l_2}{g_0 - \alpha - \theta l_2}\right) \tag{2.11a}$$

For small θ, M_0 can be approximated as

$$M_0 = \frac{\varepsilon_0 h V^2 l_2^2}{2(g_0 - \alpha)}\left(\frac{1}{g_0 - \alpha - \theta l_2} - \frac{1}{2(g_0 - \alpha)}\right) \tag{2.11b}$$

The equation for the displacement of the beam $y_1|_x$ is determined by [18]

$$E_1 I_1 y_1''|_x = M_0 + F_e(l_1 - x), \quad 0 \le x \le l_1 \tag{2.12}$$

where E_1 is Young's modulus of the beam and I_1 is the moment of inertia of the cross-sectional area of the beam. At $x = 0$, the boundary conditions are

$$y_1|_{x=0} = y_1'|_{x=0} = 0 \tag{2.13}$$

By solving Eqs. (2.10c), (2.10d), (2.12), and (2.13), α and θ can be obtained as

$$\alpha = \frac{3M_0 l_1^2 + 2F_e l_1^3}{6E_1 I_1} \tag{2.14a}$$

$$\theta = \frac{2M_0 l_1 + F_e l_1^2}{2E_1 I_1} \tag{2.14b}$$

Therefore, at a specific applied DC bias voltage V, F_e, and M_0 are determined by α and θ from Eqs. (2.10a), and (2.11a) and (2.11b), whereas α and θ are determined by M_0 and F_e from Eqs. (2.14a) and (2.14b). Thus, F_e, M_0, α and θ can be found through numerical iterations. For a voltage equal to or larger than a specific value, the iteration results become divergent. This specific voltage is called threshold voltage. For the threshold voltage, once α and θ are known, the displacement of the mass is found by Eq. (2.10b). For an actuator with $l_1 = 275$ μm, $l_2 = 165$ μm, $l_3 = 10$ μm, $w_1 = 2.4$ μm, $w_2 = 5$ μm, $w_{Al} = 0.45$ μm, $g_0 = 4.8$ μm, $d_0 = 2.8$ μm, the threshold voltage is calculated to be 23.7 V. The maximum stable displacement of the mass center is 1.37 μm, which is a little smaller than $1/3g_0$.

2.2.2 Static Restoring Force

When the mass part is displaced, an elastic restoring force by the beam tends to pull the mass back toward its original position. At the end of the electrode part AB of the mass, the restoring force, F_r, can be written as

$$F_r = -ky_2|_{x=l_1+l_2+l_3} = -k\left[\alpha + \theta(l_2 + l_3)\right] \tag{2.15}$$

where k is the equivalent stiffness of the cantilever beam. Suppose that the mass is subject to a concentrated force at the midpoint of the electrode section, the equivalent stiffness, k can be expressed as

$$k = \frac{12E_1I_1E_2I_2}{E_2I_2[4l_1^3 + 9l_1^2l_2 + 6(l_1^2l_3 + l_1l_2^2 + l_1l_2l_3)] + E_1I_1(5l_2^3 + 6l_2^2l_3)/4} \tag{2.16}$$

where E_2 is Young's modulus of the mass and I_2 is the moments of inertia of the cross-sectional area of the mass. Before the metal coating, the beam is merely made up of single-crystal silicon. E_1, E_2, I_1, and I_2 are given by

$$E_1 = E_2 = E_{Si} \tag{2.17a}$$

$$I_1 = \frac{1}{12}w_1^3h \tag{2.17b}$$

$$I_2 = \frac{1}{12}w_2^3h \tag{2.17c}$$

where E_{Si} is Young's modulus of the single-crystal silicon. After the metal coating, the beam is made of single-crystal silicon covered by metal on the top and sidewalls. Therefore, E_1, E_2, I_1, and I_2 can be expressed as

$$E_1 = \frac{E_{Si}w_1 + 2E_mw_m}{w_1 + 2w_m} \tag{2.18a}$$

$$E_2 = \frac{E_{Si}w_2 + 2E_m w_m}{w_2 + 2w_m} \tag{2.18b}$$

$$I_1 = \frac{1}{12}(w_1 + 2w_m)^3 h \tag{2.18c}$$

$$I_2 = \frac{1}{12}(w_2 + 2w_m)^3 h \tag{2.18d}$$

where E_m is Young's modulus of the metal, w_m is the thickness of the metal coated at sidewalls of the silicon beam. The material properties of the lateral switches are given in Table 2.2. Typically, $w_1 = 2.5$–$3\ \mu m$, $w_2 = 5$–$15\ \mu m$, $w_m \leq 1\ \mu m$. Since $E_{Si}w_1 > 5E_m w_m$, the equivalent stiffness is dominated by silicon structures. After the metal coating, the initial gap distance between two electrodes, g_0, and the initial gap distance between the beam end and the contact bump, d_0, are expressed as

$$g_0 = g_{Si} - 2w_m \tag{2.19a}$$

$$d_0 = d_{Si} - 2w_m \tag{2.19b}$$

where g_{Si} and d_{Si} are the initial gap distance between the two silicon electrodes and the initial gap distance between the silicon beam and the silicon contact bump before metal coating, respectively.

Table 2.2 Material properties of the lateral switches

Material	Density (kg/m^3)	Young's modulus (GPa)	Poisson's ratio
Si	2330	162	0.27
Al	2700	70	0.35
Au	19, 280	80	0.44
Cu	8960	128	0.36

2.2.3 Static Threshold Voltage

The balanced position of the cantilever beam can be found as the force is balanced

$$F = F_e + F_r = 0 \tag{2.20}$$

The curves of normalized F_e and $|F_r|$ as functions of the normalized displacement y/g_0 are shown in Fig. 2.10. For a specific mechanical structure, k is a constant. Therefore, the curve for the restoring force, F_r, is a straight line starting from the origin of the coordinates. The curve for the electrostatic force, F_e, is a hyperbola. When the applied bias voltage, V, is less than a specific voltage V_{th}, the two curves have two intersections. Equation (2.20) has two solutions. When the bias voltage V increases, F_e increases and the two intersections move closer. When $V = V_{th}$, the

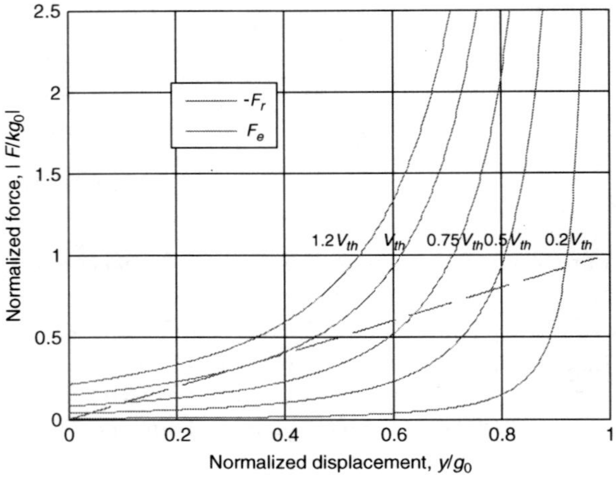

Fig. 2.10 Normalized electrostatic force and restoring force on the movable cantilever beam with various applied bias voltages

intersections coincide. When $V > V_{th}$, the two curves do not intersect since $F_e = -F_r$ at all displacements, that is, the mass part always collapses to touch the contact bump. V_{th} is the threshold voltage, which can be calculated using the iteration method, as discussed in Section 2.2.1.

Figure 2.11 shows the shape of a cantilever beam (l_1 = 275 μm, l_2 = 165 μm, l_3 = 10 μm, w_1= 2.4 μm, w_2 = 5 μm, w_{Al} = 0.45 μm, g_0 = 4.8 μm, d_0= 2.8 μm) with various bias voltages, which is simulated by commercial 3D simulation software – LS DYNA. The deformation of the beam OA of the cantilever beam

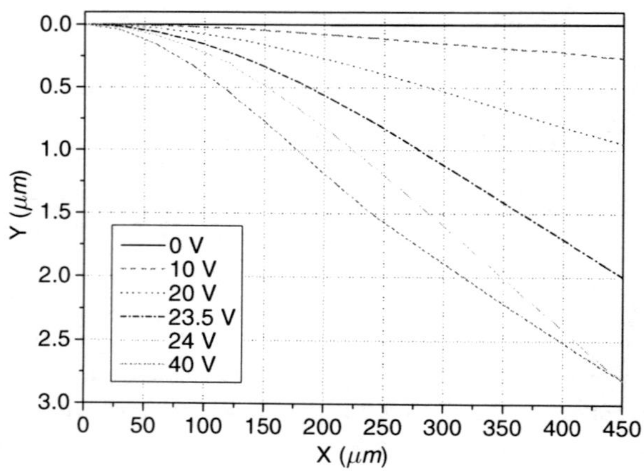

Fig. 2.11 The shape of cantilever beam with various applied bias voltages. Copyright/used with permission of/courtesy of Institute of Physics and IOP Publishing Limited

increases with the bias voltage, whereas the mass part AC of the cantilever beam moves almost without deformation. The simulated threshold voltage is 24 V, which is close to the calculated result, 23.7 V, using the iteration method.

The threshold voltage, V_{th}, is determined by the cantilever beam structure and the original gap between two electrodes, g_0. Figure 2.12 shows that the threshold voltage, V_{th}, decreases when the original gap between the two electrodes g_0 decreases or the length sum$(l_1 + l_2)$ increases. When the cantilever beam length ratio, $l_2/(l_1 + l_2)$, is within the range of 30 – 75 %, V_{th} only changes within 10% of the minimum value of V_{th}, which is referred to as $[V_{th}]_{min}$. The corresponding length ratio $[l_2/(l_1 + l_2)]_{min}$ to $[V_{th}]_{min}$ is 50% when $w_1 = 2.4\ \mu m$, $w_2 = 5\ \mu m$, and $w_{Al} = 0$. It also shows that $[l_2/(l_1 + l_2)]_{min}$ is almost independent of the initial gap, g_0, and the length sum $(l_1 + l_2)$.

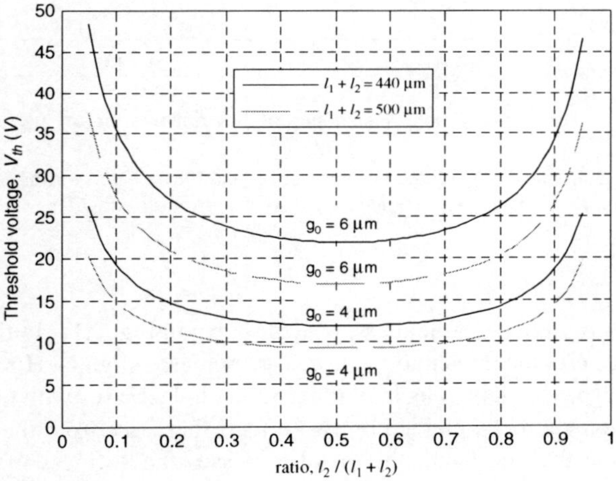

Fig. 2.12 Calculated threshold voltage V_{th} with various lengths $(l_1 + l_2)$, $l_2/(l_1 + l_2)$ ratio, and initial gap distance g_0 ($l_3 = 10\ \mu m$, $w_1 = 2.4\ \mu m$, $w_2 = 5\ \mu m$, $w_m = 0$, $g_{Si} = 6\ \mu m$). Copyright/used with permission of/courtesy of Institute of Physics and IOP Publishing Limited

Figure 2.13 shows that V_{th} is more dependent on the beam width, w_1, than the mass width, w_2. The effect of the mass width, w_2, is negligible. V_{th} increases with beam width, w_1. The mass structure can be designed as a hole mass structure as shown in Fig. 2.21. The hole mass has three advantages compared to the solid mass. First, the solid-mass with the width of more than 5 μm is not easy for release due to the constriction in the fabrication process. This problem can be solved by hole mass. Second, the etched holes reduce the effective mass of the mass structure and increase the flexibility of the whole cantilever beam structure. Third, the lateral switch with the hole mass structure can provide better RF performance than that with the solid mass structure. Section 2.4 provides a more detailed discussion on the cantilever beam effect on the RF performance of the lateral switch.

The effect of the metal thickness, w_m, of sidewalls on the threshold voltage is more complicated. For further elaboration, aluminum is used as a coating metal. The

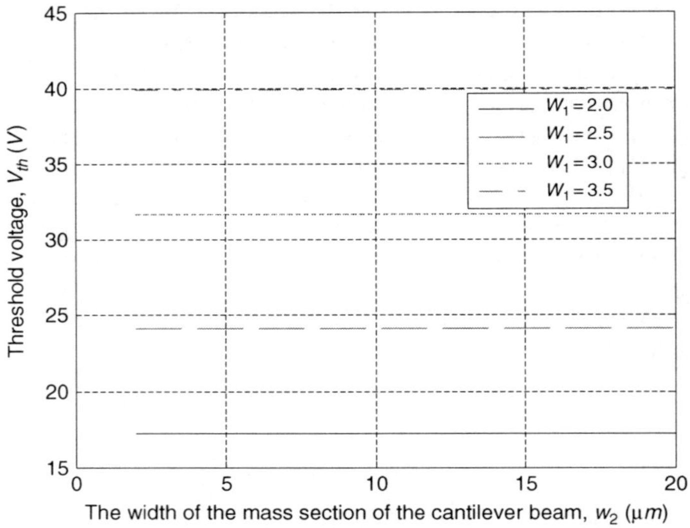

Fig. 2.13 Calculated threshold voltage V_{th} with various cantilever beam widths (w_1, w_2) ($l_1 = 275$ μm, $l_2 = 165$ μm, $l_3 = 10$ μm, $g_{\text{Si}} = 6$ μm, $w_{\text{m}} = 0$). Copyright/used with permission of/courtesy of Elsevier B.V

effect of the aluminum (Al) thickness, w_{Al}, is shown in Fig. 2.14. Initially when w_{Al} increases from zero, the threshold voltage, V_{th}, increases slightly. However, after the threshold voltage, V_{th}, gets to its maximum value, it decreases with increase in w_{Al}. This observation can be explained by two effects that arise from the metal coating at sidewalls. On the one hand, the metal increases the stiffness of the cantilever beam which tends to increase the restoring force, F_{r}, and the threshold voltage, V_{th}. On the other hand, the metal coating reduces the original gap between the two electrodes from g_{Si} to (g_{Si}–2w_{m}), which tends to increase the electrostatic force, F_{e}, and reduce the threshold voltage, V_{th}. When w_{Al} is small, the increase in F_{r} dominates the increase in F_{e}. As a result, V_{th} increases. However, once w_{Al} exceeds a specific value, the increase in F_{e} dominates the increase in F_{r}. Therefore, V_{th} falls. In general, the change of the threshold voltage, V_{th}, due to the metal coating is less than 5 V since w_{Al} is less than 1 μm. The effect of the beam parameters of w_1, w_2, and w_{Al} on the threshold voltage, V_{th}, is summarized in Table 2.3 when $l_1 = 275$ μm, l_2= 165 μm, l_3= 10 μm, and $g_{\text{Si}} = 6$ μm.

2.2.4 Dynamic Frequency Response

The frequency response of the cantilever beam is useful to determine the switching time and the mechanical bandwidth of the lateral switch. The frequency response can be determined by d'Alembert's principle as [19]

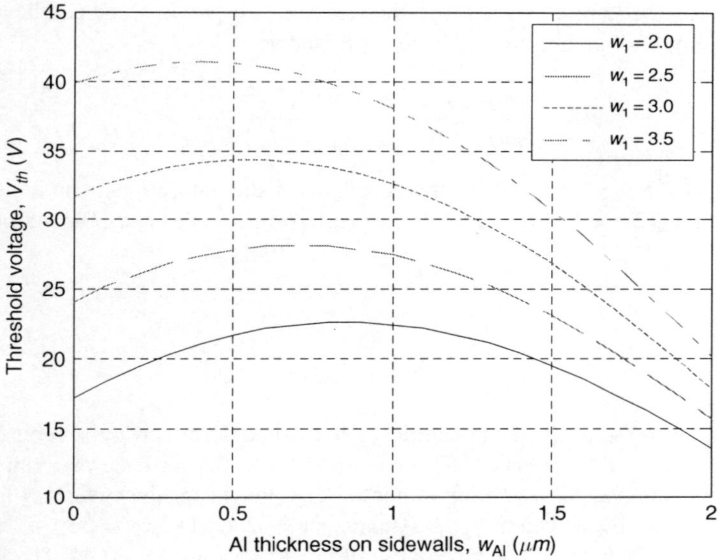

Fig. 2.14 Calculated V_{th} versus thicknesses of Al coated at sidewalls w_{Al} ($l_1 = 275$ μm, $l_2 = 165$ μm, $l_3 = 10$ μm, $g_{Si} = 6$ μm, $w_2 = 5$ μm)

Table 2.3 The effect of the cantilever beam parameters on the threshold voltage, V_{th}

	Mass width, w_2 (μm) ($w_{Al} = 0$)			Al thickness at sidewalls, w_{Al} (μm) ($w_2 = 5$ μm)		
Beam width, w_1 (μm)	2.5	5	15	0.2	0.6	1.0
2.0	17.25	17.25	17.25	19.45	22.15	22.45
2.4	22.65	22.65	22.65	24.7	26.8	26.35
3.0	31.65	31.65	31.65	33.25	34.35	32.6
3.5	39.9	39.9	39.9	41.05	41.05	38.1

$$my'' + by' + ky = f_{ext} \tag{2.21}$$

where (′) denotes the derivative with respect to time t, y is the lateral displacement of the cantilever beam relative to the origin of the fixed electrode, m is the effective mass, k and b are the effective stiffness and the damping coefficient of the simplified system, and $f_{ext}(t)$ is the electrostatic force. The electrostatic force, $f_{ext}(t)$, between the two electrodes generated by a bias voltage, V, can be simplified as

$$f_{ext}(t) = \frac{\varepsilon_0 h l_2 V^2}{2(g_0 - y)^2} \tag{2.22}$$

Based on the Laplace transforms, the frequency response of the cantilever beam with small vibration amplitude can be expressed as

$$\frac{Y(j\omega)}{F(j\omega)} = \frac{1/k}{1 - (\omega/\omega_0)^2 + j\omega/(Q\omega_0)} \tag{2.23}$$

where ω is the working angular frequency, ω_0 is the natural resonant angular frequency, and Q is the quality factor of the cantilever beam. ω_0 and Q are expressed as

$$\omega_0 = \sqrt{k/m} \tag{2.24}$$

$$Q = k/(\omega_0 b) \tag{2.25}$$

The quality factor Q of the cantilever beam is determined by several different variables, such as the pressure, the temperature, and the intrinsic material dissipation. The quality factor is also an important component for the switching time. For example, a cantilever beam ($l_3 = 10$ μm, $w_1 = 2.4$ μm, $w_2 = 5$ μm, $w_m = 0$, $g_{Si} = 6$ μm) has the frequency response of $f = 15$ kHz and $k = 0.94$. The variation in response amplitude of the cantilever beam is simulated with different quality factors. Figure 2.15 shows that the response amplitude at 15 kHz is increased when the quality factor ranges from 0.2 to 2.0. When $Q \leq 0.5$, it has a slow switching time; when $Q \geq 2$, it has a long settling time. In practice, it is beneficial for the switching time that the quality factor of the cantilever beam is designed by $0.5 \leq Q \leq 2$.

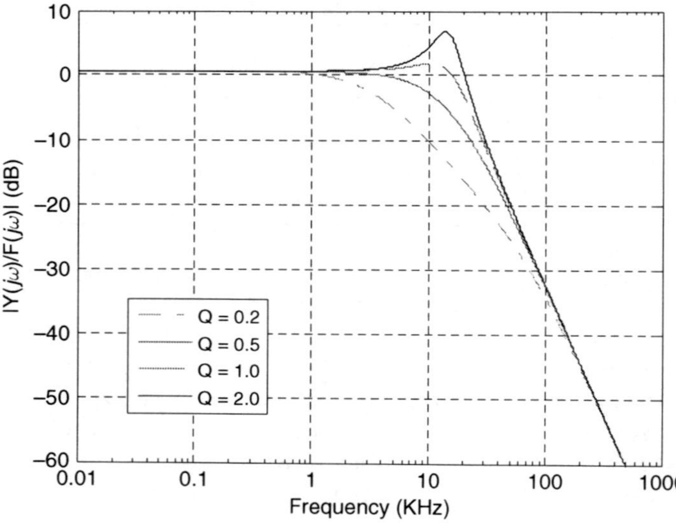

Fig. 2.15 Frequency response of a cantilever beam with resonant frequency $f = 15$ kHz and $k = 0.94$

2.2.5 Dynamic Effective Mass

It is noted that the effective mass of the cantilever beam is not equal to the actual mass of the cantilever beam since only the end portion of the cantilever beam is moving. The effective mass, m, can be estimated by Rayleigh's energy method [19]. Assume the cantilever beam is subject to a concentrated load, P, at the center of the electrode section of the cantilever beam. Referring to Fig. 2.9, we can consider the displacement y_x and kinetic energy E_k of the cantilever beam at three portions, respectively.

The first part is the beam ($0 < x \leq l_1$). Its kinetic energy $E_{\mathrm{k}1}$ is given by

$$\begin{aligned} E_{\mathrm{k}1} &= \frac{1}{2}(\rho_{\mathrm{Si}} w_1 + 2\rho_\mathrm{m} w_\mathrm{m})h \int_0^{l_1} y_1'^2 dx \\ &= \frac{1}{2} m_1 \left(\frac{P}{2E_1 I_1}\right)^2 \frac{-4l_1^4 + 25 l_1^3 l_2 + 10 l_1^2 l_2^2}{30} \end{aligned} \tag{2.26a}$$

where

$$y_1 = \frac{Px^2}{6E_1 I_1}(3l_1 - x) + \frac{P l_2 x^2}{2E_1 I_1} = \frac{Px^2}{6E_1 I_1}(3l_1 + 3l_2 - x) \tag{2.26b}$$

$$m_1 = (\rho_{\mathrm{Si}} w_1 + 2\rho_\mathrm{m} w_\mathrm{m}) l_1 h \tag{2.26c}$$

The second part is from the beginning of the electrode to the center of the electrode of the cantilever beam ($l_1 < x \leq l_1 + l_2/2$). The kinetic energy $E_{\mathrm{k}2}$ is given by

$$\begin{aligned} E_{\mathrm{k}2} &= \frac{1}{2}(\rho_{\mathrm{Si}} w_2 + 2\rho_\mathrm{m} w_\mathrm{m})h \int_{l_1}^{l_1 + l_2/2} y_2'^2 dx \\ &= \frac{1}{2} m_2 \left[\left(\frac{P}{E_1 I_1}\right)^2 \frac{l_1^2 (l_1 + l_2)^2}{8} + \frac{P^2}{E_1 I_1 E_2 I_2} \frac{l_1 l_2^2 (l_1 + l_2)}{24} + \left(\frac{P}{E_2 I_2}\right)^2 \frac{49 l_2{}^4}{5760}\right] \end{aligned} \tag{2.27a}$$

where

$$y_2 = \frac{P l_1^3}{3E_1 I_1} + \frac{P l_2 l_1^2}{2E_1 I_1} + \frac{P(l_1^2 + 2 l_1 l_2)(x - l_1)}{2E_1 I_1} + \frac{P(x - l_1)^2 [3l_2 - (x - l_1)]}{6E_2 I_2} \tag{2.27b}$$

$$m_2 = (\rho_{\mathrm{Si}} w_2 + 2\rho_\mathrm{m} w_\mathrm{m}) l_2 h \tag{2.27c}$$

The third part is from the center of the electrode to the end of the cantilever beam ($l_1 + l_2/2 < x \leq l_1 + l_2 + l_3$). The kinetic energy $E_{\mathrm{k}3}$ is given by

$$E_{k3} = \frac{1}{2}(\rho_{Si}w_2 + 2\rho_m w_m)h \int_{l_1+l_2/2}^{l_1+l_2+l_3} y_3'^2 dx$$
$$= \frac{1}{2}m_2\left[\frac{P\left(l_1^2 + l_1 l_2\right)}{2E_1 I_1} + \frac{Pl_2^2}{8E_2 I_2}\right]^2\left(\frac{1}{2} + \frac{l_3}{l_2}\right) \quad (2.28a)$$

where

$$y_3 = \frac{P(4l_1^3 + 3l_2 l_1^2)}{12E_1 I_1} + \frac{P(l_1^2 + l_1 l_2)(x - l_1)}{2E_1 I_1} + \frac{Pl_2^3/4}{6E_2 I_2} + \frac{P\frac{l_2^2}{4}\left(x - l_1 - \frac{l_2}{2}\right)}{2E_2 I_2} \quad (2.28b)$$

Therefore, the total kinetic energy E_k are given by

$$E_k = E_{k1} + E_{k2} + E_{k3} = \frac{1}{2}my'^2_{max} \quad (2.29a)$$

where the velocity y'_{max} at the end of the cantilever beam is

$$y'_{max} = y'_3\,|_{x=l_1+l_2+l_3} = \frac{P(l_1^2 + l_1 l_2)}{2E_1 I_1} + \frac{Pl_2^2}{8E_2 I_2} \quad (2.29b)$$

The effective mass, m, can be obtained by solving Eq. (2.29a). Figure 2.16 shows the portion mass m_1, m_2 and the effective mass, m, of a cantilever beam changes with the ratio of $(l_2/(l_1+l_2))$ when $l_1 + l_2 = 440\,\mu$m and $l_3 = 10\,\mu$m. It shows

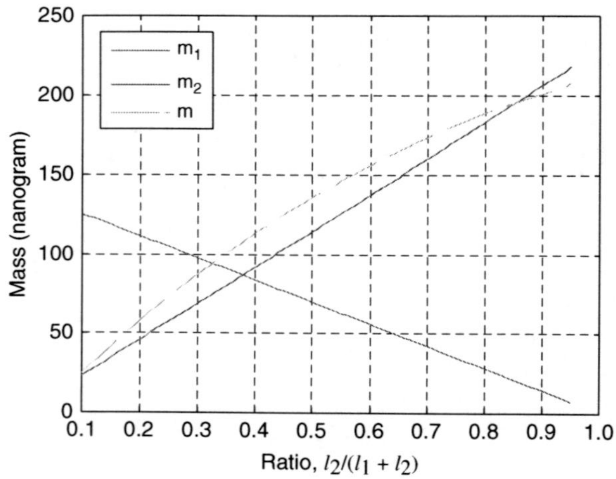

Fig. 2.16 Effective mass and part mass of the cantilever beam versus the ratio of $l_2/(l_1+l_2)$ (w_1 = 2.4 μm, w_2 = 5 μm, w_{Al} = 0.6 μm, l_1+l_2 = 440 μm, l_3 = 10 μm, and h = 35 μm)

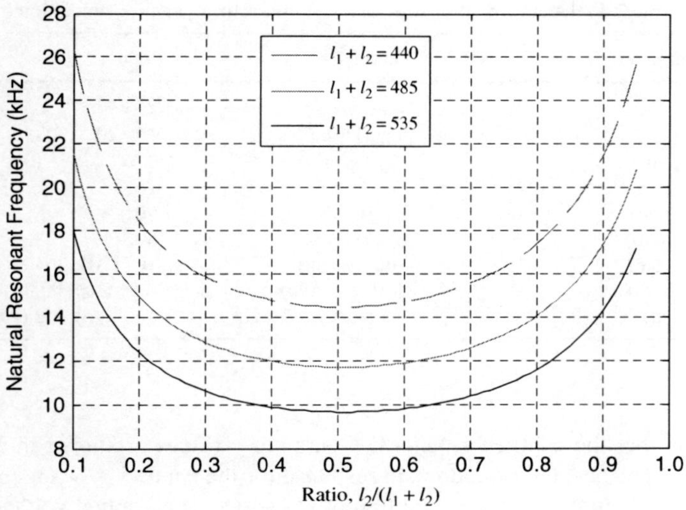

Fig. 2.17 Natural resonant frequency versus the ratio of l_2 /(l_1+l_2) (w_1 = 2.4 μm, w_2 = 5 μm, w_{Al} = 0.6 μm, h = 35 μm)

that the effective mass, m, is mainly determined by the mass of the electrode part, m_2. The effective mass, m, is 5–85% of the actual total mass of the cantilever beam $[m_1 + m_2(1 + l_3/l_2)]$ when the ratio of $[l_2/(l_1+l_2)]$ is within the range of 30–75%.

Figure 2.17 shows that the natural resonant frequency of the cantilever beam changes with the ratio of $(l_2/(l_1+l_2))$ and the sum of (l_1+l_2). It shows that the natural resonant frequency of the cantilever beam changes slightly when $l_2/(l_1+l_2)$ is within the range of 30–75%. For example, when $(l_1+l_2)=$ 440 μm and l_3 = 10 μm, the resonant frequency is 15 ± 0.5 kHz as $l_2/(l_1+l_2)$ is within the range of 0.3–0.75. The natural resonant frequency of the cantilever beam decreases with the increase of (l_1+l_2) due to the increase of the effective mass and the decrease of the stiffness of the cantilever beam.

2.2.6 Dynamic Switching Time

The switching time is obtained using Eq. (2.21) when $y = g_0$. Substituting Eqs. (2.22), (2.24), and (2.25) into Eq. (2.21), the dynamic response equation is obtained as

$$y'' + \frac{\omega_0}{Q}y' + \omega_0^2 y = \frac{\varepsilon_0 h l_2 V^2}{2(g_0 - y)^2 m} \tag{2.30}$$

The equation governs the simple 1D nonlinear model and can be solved with a nonlinear simultaneous differential equation solver – *Mathematica* [20].

Table 2.4 Parameters of the cantilever beam for the dynamics simulations

Parameters	Values	Parameters	Values
l_1 (μm)	275	ρ_{Si} (kg/m^3)	2330
l_2 (μm)	165	ρ_{Au} (kg/m^3)	19,320
l_3 (μm)	10	ρ_{Al}(kg/m^3)	2700
w_1 (μm)	2.5	E_{Si} (GPa)	162
w_2 (μm)	5.0	E_{Al} (GPa)	70
w_m (μm)	0.6	E_{Au} (GPa)	80
g_{Si} (μm)	6.0	Mass, m(ng)	107 (Al); 264 (Au)
d_{Si} (μm)	4.0	Stiffness k(N/m)	0.86 (Al); 0.88 (Au)
h (μm)	35	Frequency ω_0 (kHz)	14.3 (Al); 9.2 (Au)

Assume that the cantilever beam has parameter values as shown in Table 2.4. Figure 2.18 presents the time-domain response for the cantilever beam coated with gold (Au) and aluminum (Al) for different Q factors. The applied voltage is 40 V. There is a substantial improvement from $Q = 0.5$ to 2, but little improvement when Q is above 2. The beam coated with Al responds faster than coated with Au since mass of the Al is smaller than the mass of the Au. Figure 2.19 shows the switching time depends significantly on the applied voltage. The calculated switching time is 23 μs when the applied voltage is 40 V and 43 μs when the applied voltage is 30 V as $Q = 1$ and the coating metal is Al. The corresponding contact force values are 77.6, 106, and 139 μN for 30, 35, and 40 V, respectively.

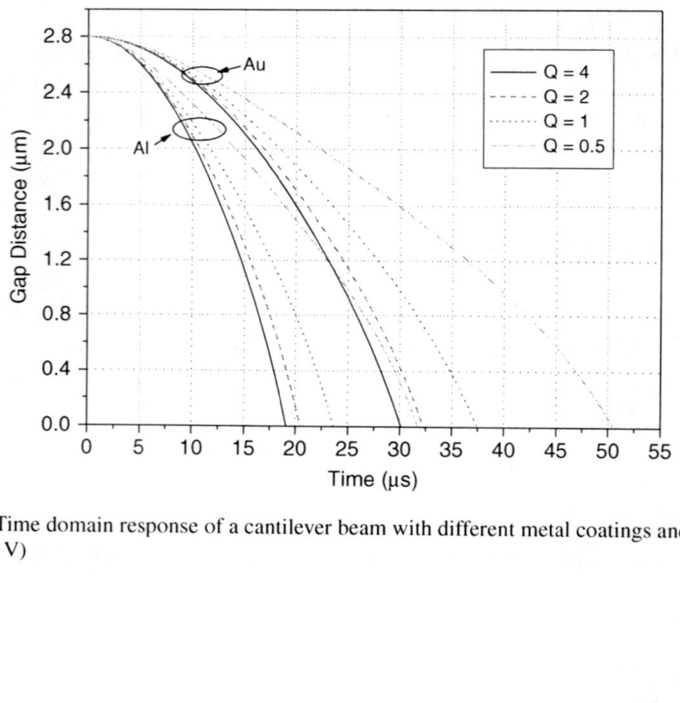

Fig. 2.18 Time domain response of a cantilever beam with different metal coatings and Q-factors ($V_{bias} = 40$ V)

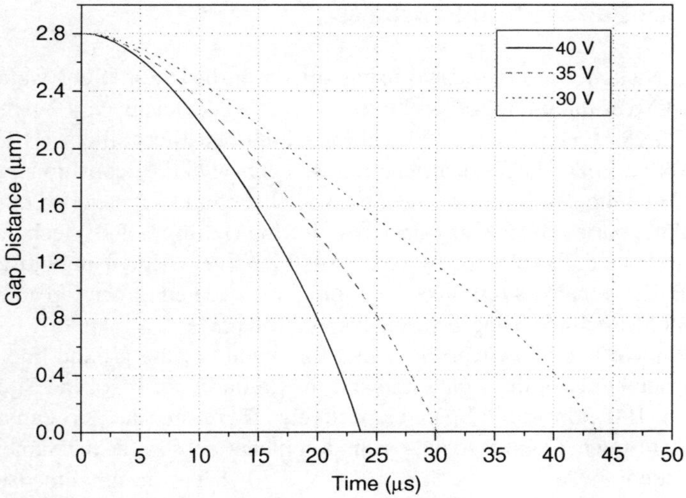

Fig. 2.19 Time domain response for the cantilever beam with different applied voltages ($Q = 1$)

2.2.7 Dynamic Release Time

The nonlinear dynamic analysis equation can also be used to model the release mechanism of the switch when$f_{ext}(t) = 0$. The restoring force ($= kg_0$) is 3.8–4.6 μN and parameter values of the switch are shown in Table 2.4. Figure 2.20 presents the release response for the cantilever beam coated with Al and $Q = 0.5$, 1, 2, and 4, respectively. When $Q = 2$ and 4, the beam oscillates and takes a longer time to stabilize. When $Q = 0.5$ and 1, the beam takes a shorter time to stabilize.

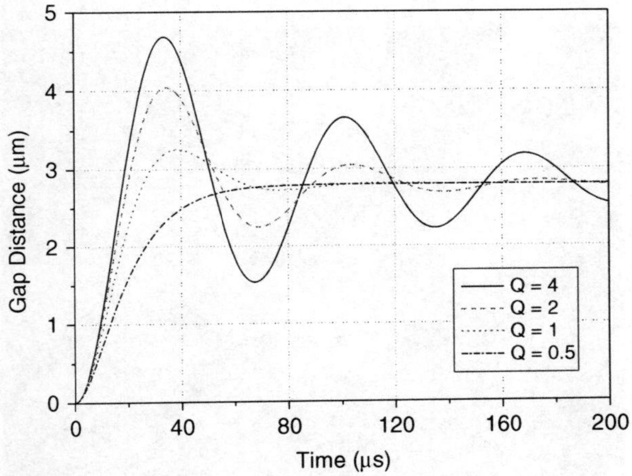

Fig. 2.20 Simulation results of release time for different Q-factors

2.3 Device Fabrication Processes

The lateral switches are fabricated using silicon-on-insulator (SOI) wafers, which include a 35-μm-thick low-resistivity silicon (LRSi) device layer, a 2-μm-thick silicon dioxide (SiO_2) layer and a 500-μm-thick high-resistivity silicon (HRSi) handle layer (>4000 Ω cm). The fabrication process begins with the deposition and patterning of SiO_2 on the device layer. Then, the switch structures are etched in the device layer until the buried oxide using deep reactive ion etching (DRIE) technique. Next, the cantilever beam is released by removing the buried oxide using buffered oxide etchant (BOE). Finally, a thin layer of aluminum is coated on the top and sidewalls of the switch structures using E-beam evaporation [21].

For a typical lateral switch, the signal line width, S, the ground line width, G, and the space between the signal line and the ground line, W, of the Si-core CPW port are 66, 100–300, and 67 μm, respectively. Therefore, the port can accommodate 150 μm-pitch ground–signal–ground coplanar probes; at the same time, the characteristic impedance of the Si-core CPW is 50 Ω. The design dimensions of the electrostatic actuator are $l_1 = 210$–275 μm, $l_2 = 165$–220 μm, $l_3 = 10$–30 μm, $w_1 =$ 2–3 μm, $w_2 = 5$–15 μm, $g_{Si} = 4$–6 μm, and $d_{Si} =$ 3–4 μm. There are four main concerns in selecting these dimensions for the switch design. First, to keep a small device area, generally <1 mm^2, a short cantilever beam is required. Second, good RF performance needs a short and wide cantilever beam and a large gap distance between the cantilever beam and the fixed electrode. Third, to obtain low actuation voltage and fast switching speed, a long and narrow cantilever beam and a small gap distance are necessary. Fourth, w_1, w_2, g_{Si}, and d_{Si} should be large enough for easy fabrication. The cantilever beam has a length of 450–500 μm. The entire device including the two ports has a length of 700 μm. The length ratio $[l_2/(l_1+l_2)]_{min}$ is 30–75% for low threshold voltage and low insertion loss.

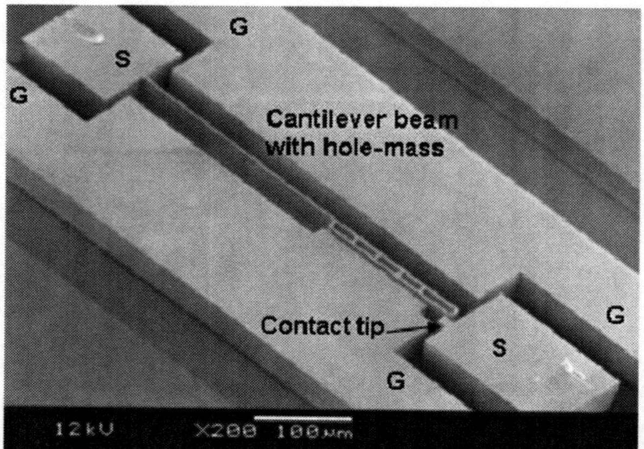

Fig. 2.21 SEM image of a lateral switch with the hole mass (G: ground, S: signal). Copyright/used with permission of/courtesy of Institute of Physics and IOP Publishing Limited

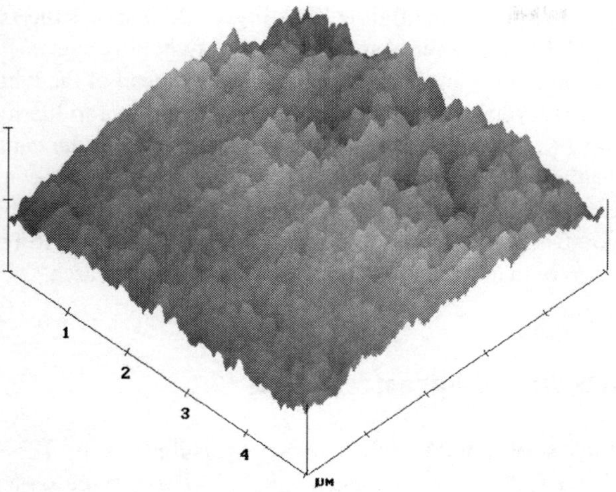

Fig. 2.22 AFM micrograph showing the surface roughness of the sidewall of a cantilever beam coated with Al. Roughness = 250 Å. Copyright/used with permission of/courtesy of Institute of Physics and IOP Publishing Limited

Figure 2.21 shows the SEM image of a lateral switch. The switch has a size of 400 μm × 700 μm in area. The actuator dimensions are $l_1 = 275$ μm, $l_2 = 165$ μm, $l_3 = 10$ μm, $w_1 = 2.4$ μm, $w_2 = 15$ μm, $g_{Si} = 6$ μm, and $d_{Si} = 4$ μm. A hole mass structure is employed in the cantilever beam of the switch with the width of 15 μm. Every hole is 10-μm wide and 30-μm long. Figure 2.22 is an AFM micrograph showing the surface roughness of a cantilever beam with Al coating. The surface roughness is 250 Å, which is much smaller than Al skin depth up to 100 GHz. Therefore, only a small percent of the total current sees this roughness. The power

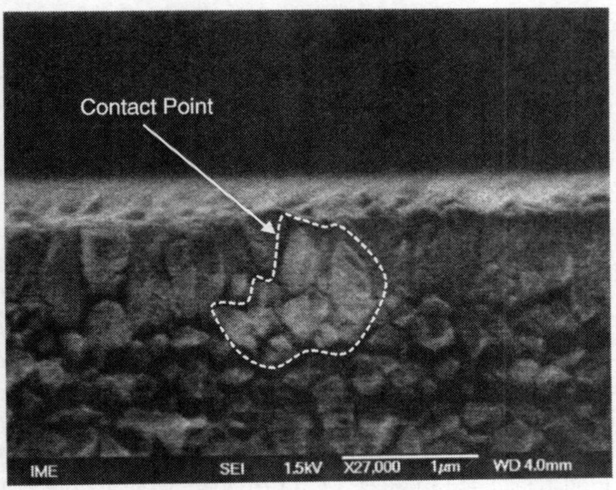

Fig. 2.23 SEM image of the contact point on the cantilever beam. Copyright/used with permission of/courtesy of Institute of Physics and IOPPublishingLimited

loss due to this roughness is negligible [22]. Figure 2.23 shows the zoomed view of the contact part of the cantilever beam after tens of switching cycles. The small point contact is at the top of the cantilever beam sidewall, instead of the whole depth. That is because the cantilever beam is not absolutely straight due to the limitation of the fabrication process. The metal deposited at the upper side of the cantilever beam is slightly thicker than that at the bottom. Therefore, the gap between the cantilever beam and the contact bump on the upper part is slightly narrower than the bottom. The contact point is about 1 μm $\times$ 0.7 μm in area. It is easy to realize effective contact since the contact force focuses on this small contact area.

2.4 Lateral Switch Characterization

The RF responses of lateral switches are measured using HP 8510C Vector Network Analyzer (VNA) with tungsten – tip 150 μm – pitch Cascade Microtech ground–signal–ground coplanar probes. The system is calibrated using short-open-load-through (SOLT) on-wafer calibration technique. All tests are performed in the standard room environment without any device package. Before RF testing, each switch is actuated tens of cycles to make the contact surfaces adapt to each other and to make a constant contact resistance. Table 2.5 lists the actuator dimensions and fitted circuit parameters of all switches discussed in this section. $g_{\mathrm{Si}} = 6$ μm, $d_{\mathrm{Si}} = 4$ μm. Except switches H and I, the other switches (switches A–G) are coated with 1.2-μm-thick Al. The second column is the key parameter to differentiate each switch. R is the total resistance ($R = R_l + R_c$).

2.4.1 The Single-Beam Switch and the Double-Beam Switch

Figure 2.24 shows measured and fitted S-parameters of a single-beam lateral switch (switch A). The design dimensions of the cantilever beam of the lateral switch are $l_1 = 275$ μm, $l_2 = 165$ μm, $l_3 = 10$ μm, $w_1 = 2.4$ μm, $w_2 = 5$ μm (solid mass), $g_{\mathrm{Si}} = 6$ μm, and $d_{\mathrm{Si}} = 4$ μm. About 1.2-μm-thick Al is deposited. The insertion loss of the switch is 0.37 dB, the return loss is 23 dB, and the isolation is 27 dB at 10 GHz. The equivalent circuit model values are fitted using the measured S-parameters in ADS. Table 2.6 shows the comparison between the fitted and calculated values of the equivalent circuit models of switch A. The fitted inductance, L, is close to the calculated value. The fitted parasitic capacitance, C_{gc}, at the on-state and C_{go} at the off-state are 28 and 20 fF, respectively, which are a little larger than the calculated values 22.9 and 15.6 fF due to the fringing field capacitance. Therefore, the fringing field capacitance is about 5.1 fF at the on-state and 4.4 fF at the off-state. The calculated DC resistance of the cantilever beam, R_l, is 0.4 Ω, while the fitted total resistance, R, is 2.6 Ω. Therefore, the contact resistance is about 2.2 Ω. The fitted total resistance increases with frequency, as shown in Fig. 2.25. The resistance is 2.6 Ω at DC and 4.2 Ω at 10 GHz. The contact resistance may also

Table 2.5 Cantilever beam parameters and fitted circuit values of switches at 10 GHz

No.	Para-meters	$L(\mu m)$	$L_1(\mu m)$	$L_2(\mu m)$	$L_3(\mu m)$	$w_1(\mu m)$	$w_2(\mu m)$	$R(\Omega)$	$L(pH)$	$C_{gc}(fF)$	$C_{go}(fF)$	$C_s(fF)$
A	1-beam	450	275	165	10	2.4	5	4.2	148	28	20	6.5
B	2.beam	450	275	165	10	2.4	5	2.4	60.4	67	56	7
	w_1/w_2 (μm) (single-beam switch with separated fixed electrode)											
C	2.0/5.0	447	210	215	22	2.0	5	5.4	149	22.6	16	13
D	2.4/5.0	447	210	215	22	2.4	5	5.1	140	22.6	16	13.2
E	2.4/15	447	210	215	22	2.4	15	4.2	130	22.6	16	13.5
	L_2 (μm) (double-beam switch)											
F	235	450	187	235	28	2.4	5	2.8	58	70	63	7
G	350	450	72	350	23	2.4	5	3.4	55.6	77	70	7
	t_{Al} (μm) (single-beam switch)											
H	0.8	450	275	165	10	2.4	5	5.0	182	56	50	5.5
I	1.5	450	275	165	10	2.4	5	1.8	147	34	24	5.8

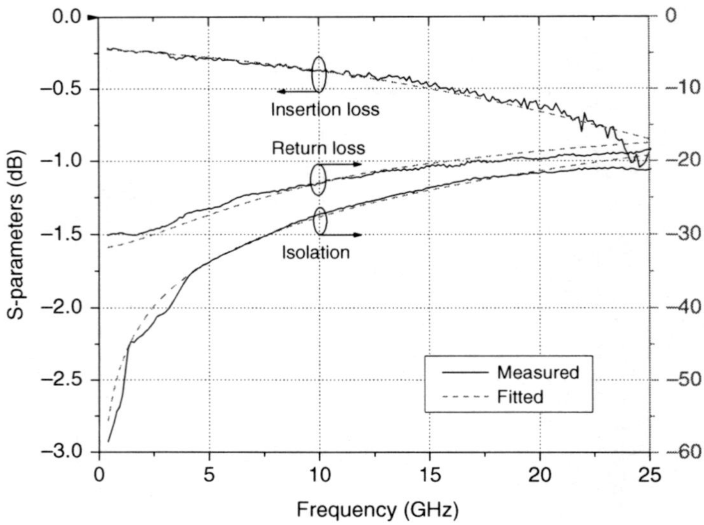

Fig. 2.24 Comparison between measured and fitted *S*-parameters of the single-beam switch (switch A)

Table 2.6 Comparison between fitted and calculated circuit values of switch A

	R_l+R_c @ DC (Ω)	R_l +R_c @ 10 GHz (Ω)	L(pH)	C_{gc}(fF)	$C_{go,}$(fF)	C_s (fF)
Fitted value	2.6	4.2	148	28	20	6.7
Calculated value	0.5	1.0	150	22.9	15.6	–

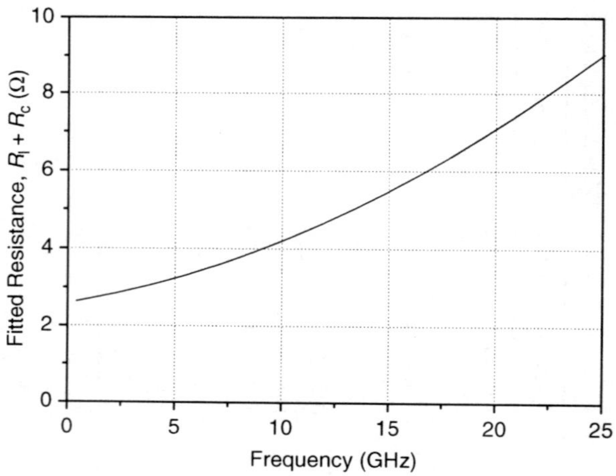

Fig. 2.25 Plot of the fitted resistance, R_l+ R_c, of switch A with frequency

increase with frequency. Third, the potential of the two ground lines is not same at the high frequency due to the asymmetrical structure of the switch, which results in more loss at high frequency. In this section, if without specific clarification, the total resistance of the switches is set to increase with frequency in the equivalent circuit model.

Figure 2.26 shows the SEM image of a double-beam switch. The switch has a size of 800 μm × 700 μm. A solid mass structure is employed in the cantilever beam. The actuator design dimensions and fitted circuit elements at 10 GHz are listed in Table 2.5 (switch B). Figure 2.27 shows the comparison between the measured and the fitted *S*-parameters of the double-beam switch. The fitted *S*-parameters agree with the measured results well. Figure 2.28 compares measured *S*-parameters between the single-beam switch (switch A) and the double-beam switch. The double-beam switch has a lower insertion loss than the single-beam switch by 0.1 dB from 0.05 to 25 GHz. The return loss of the double-beam switch is larger than the single-beam switch from 0.05 to 17 GHz and lower than the latter when the frequency is above 17 GHz. The isolation loss of two switches is close to each other. The fitted total resistance, *R*, of the double-beam switch is 2.4 Ω at 10 GHz, which is close to half of the single-beam switch. The fitted inductance, *L*, of the double-beam switch is 60.4 nH, which is also close to half of the single-beam switch. Lower inductance results in lower insertion loss at high frequencies for the double-beam switch. The parasitic capacitance at the on-state, C_{gc}, and at the open state, C_{go}, of the double beam switch is nearly two times of the single-beam switch, which results in higher return loss at low frequencies for the double-beam switch. The open capacitance, C_S, of the double-beam switch is slightly larger than the single-beam switch. Therefore, the equivalent circuit model of the double-beam switch, as shown in Fig. 2.8, is verified to be valid. The double-beam switch can provide lower insertion

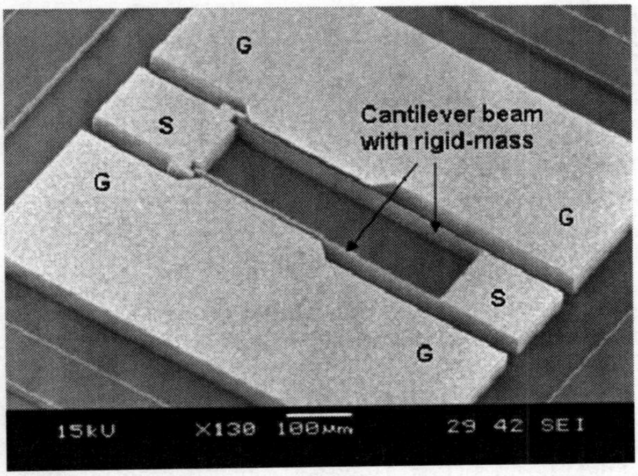

Fig. 2.26 SEM image of a double-beam lateral switch with the solid mass. Copyright/used with permission of/courtesy of Institute of Physics and IOP Publishing Limited

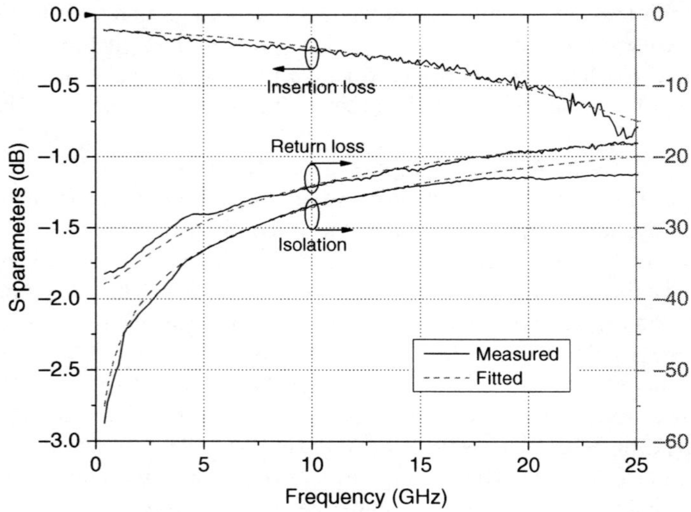

Fig. 2.27 Comparison between measured and fitted *S*-parameters of the double-beam switch (switch B)

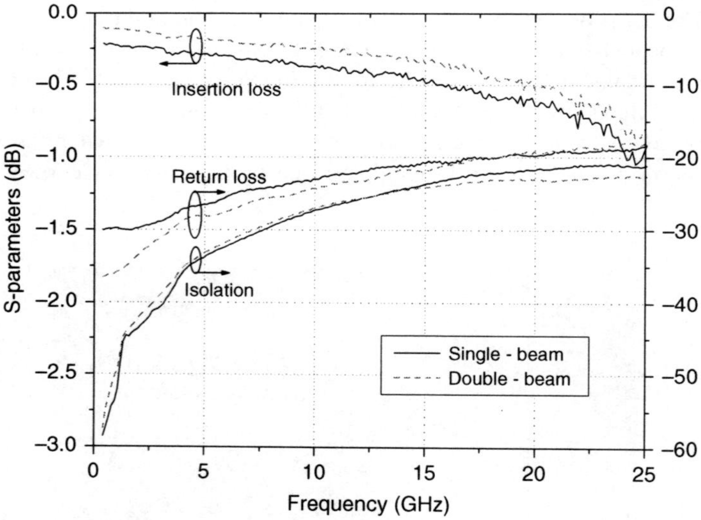

Fig. 2.28 Comparison of measured *S*-parameters between the single-beam switch (switch A) and the double-beam switch (switch B). Copyright/used with permission of/courtesy of Institute of Physics and IOP Publishing Limited

loss than the single-beam switch. However, when two beams in the double-beam switch are not symmetrical, the double-beam switch performances may deteriorate and become worse than the single-beam switch.

2.4.2 Cantilever Beam Design Effect

The cantilever beam is the main part of the signal line in the lateral switch. Therefore, the design parameters (beam width, mass shape, and length) of the cantilever beam have a significant role in determining the RF performance of the lateral switch.

Figures 2.29, 2.30, and 2.31 show the comparison between the measured and fitted *S*-parameters of switches C, D, and E, respectively. In switch C, $w_1 = 2.0$ μm, $w_2 = 5$ μm; in switch D, $w_1 = 2.4$ μm, $w_2 = 5$ μm; and in switch E, $w_1 = 2.4$ μm, $w_2 = 15$ μm. Other structure parameters are the same. In these switches, fixed electrode is separated from the ground line, as shown in Fig. 2.38. All switches have insertion loss below 0.8 dB, return loss of above 15 dB, and isolation of above 17 dB up to 20 GHz. The fitted *S*-parameters agree well with the measured results. Table 2.5 shows that both the total resistance, *R*, and the inductance, *L*, decrease with the beam width, either w_1 or w_2. The parasitic capacitance at the on-state, C_{gc}, and at the off-state, C_{go}, of these three switches are the same. The open capacitance, C_s, increases slightly with the beam width. Therefore, the beam width mainly affects the beam resistance and the inductance, as predicted in Section 2.1.2.

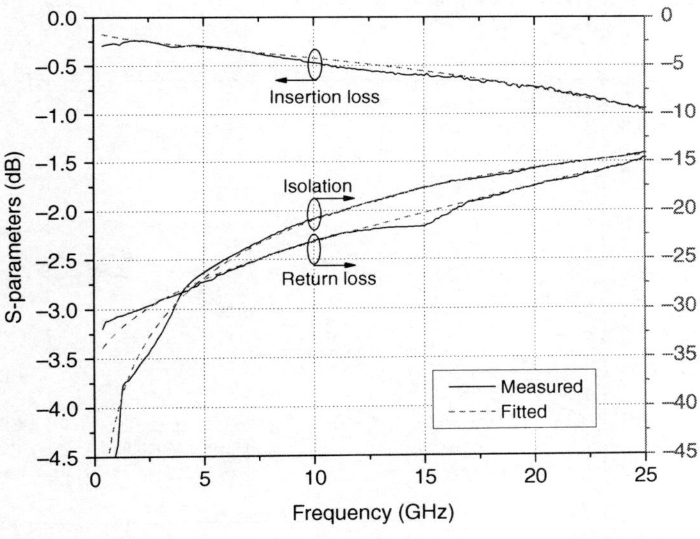

Fig. 2.29 Comparison between measured and fitted *S*-parameters of a single-beam switch with the solid mass (switch C, $w_1 = 2.0$ μm, $w_2 = 5$ μm)

Figure 2.32 shows the comparison of the measured results between switch C ($w_1 = 2.0$ μm) and switch D ($w_1 = 2.4$ μm). It is found that the switch with $w_1 = 2.4$ μm has slightly lower insertion loss and higher return loss than the switch with $w_1 = 2.0$ μm. This is because narrower cantilever beam results in larger beam resistance, R_l, and inductance, *L*, than the wider cantilever beam. The isolations loss of the two switches change marginally.

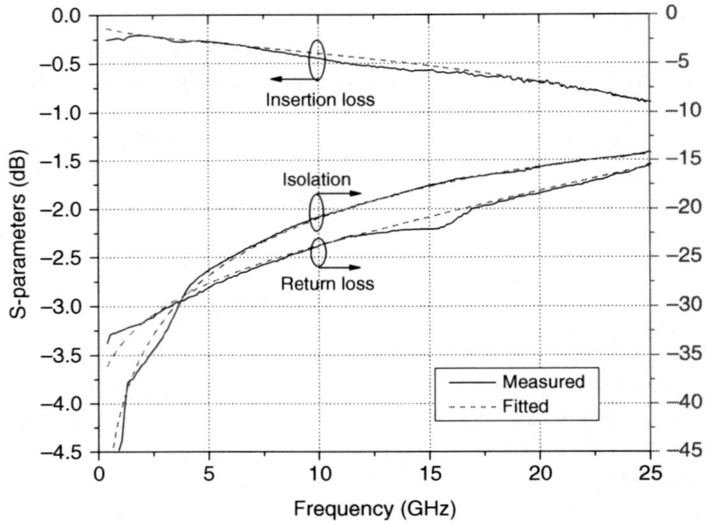

Fig. 2.30 Comparison between measured and fitted S-parameters of a single-beam switch with the solid mass (switch D, $w_1 = 2.4$ μm, $w_2 = 5$ μm)

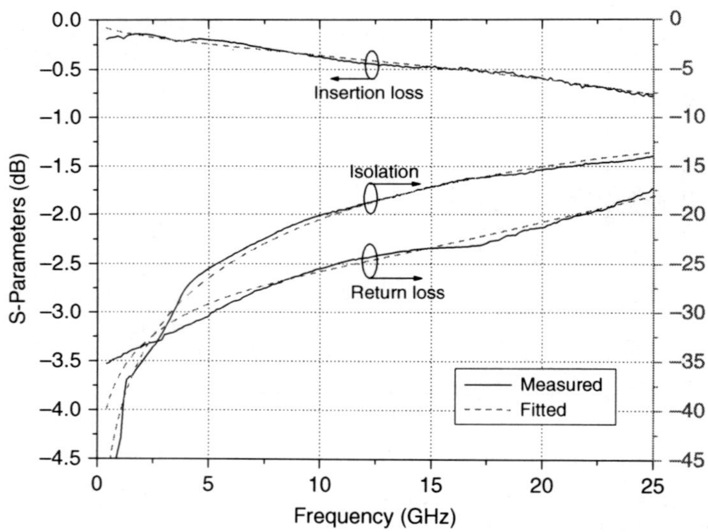

Fig. 2.31 Comparison between measured and fitted S-parameters of a single-beam switch with the hole mass (switch E, $w_1 = 2.4$ μm, $w_2 = 15$ μm)

Figure 2.33 shows the comparison of the measured S-parameters of switch D (solid mass, $w_2 = 5$ μm) and switch E (hole mass, $w_2 = 15$ μm). The insertion loss of the hole mass switch is lower than the solid mass switch by 0.1 dB. The return loss of the hole mass switch is higher than the solid mass switch by 2.5 dB.

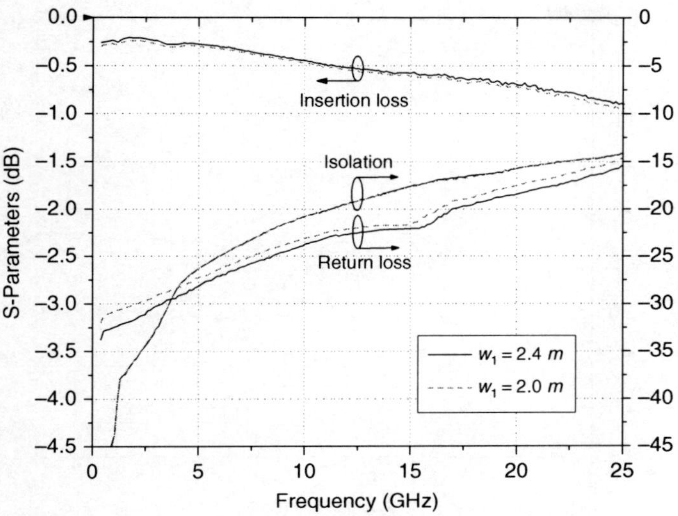

Fig. 2.32 Comparison of measured *S*-parameters between switch C ($w_1 = 2.0$ μm) and switch D ($w_1 = 2.4$ μm)

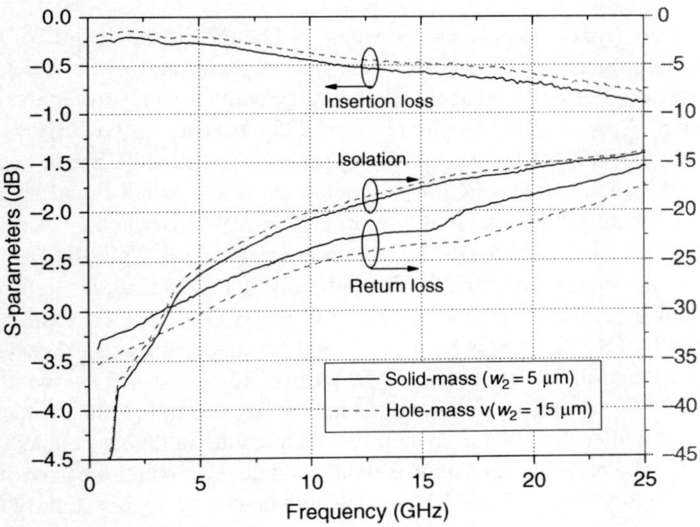

Fig. 2.33 Comparison of measured *S*-parameters between switch D (solid-mass, $w_2 = 5$ μm) and switch E (hole-mass, $w_2 = 15$ μm). Copyright/used with permission of/courtesy of Institute of Physics and IOP Publishing Limited

This is because the hole mass offers lower beam resistance $R_l$and inductance L compared to the solid mass. The isolation of the hole mass switch is slightly lower than the solid mass switch since wider mass structure result in slightly larger open capacitances, C_s.

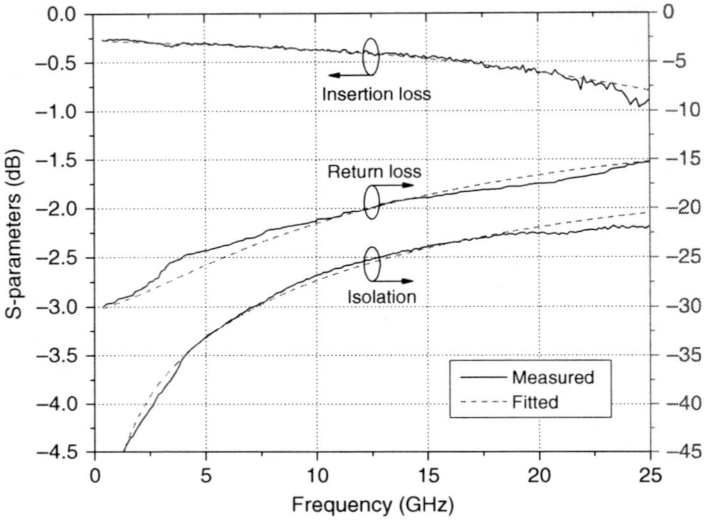

Fig. 2.34 Comparison between measured and fitted *S*-parameters of switch F ($l_2 = 235$ μm)

When sum (l_1+l_2) and l_3 are constant, a change in the length of the fixed electrode, l_2, affects the RF performance due to the effect of the shunt coupling capacitance, C_g. Figures 2.34 and 2.35 shows the comparison between the measured and the fitted *S*-parameters of switch F ($l_2 = 235$ μm) and switch G ($l_2 = 350$ μm), respectively. For both switches, sum (l_1+l_2) and l_3 are 422 and 25 μm, respectively. Figure 2.36 compares measured *S*-parameters between switch F and switch G. At 10 GHz, the insertion loss increases from 0.37 to 0.41 dB when l_2 increases from 235 to 350 μm, whereas the return loss increases from 21 to 22.7 dB. The isolations of two switches change marginally. Theoretically, the resistance, *R*, should decrease with l_2. However, the resistance of switch G is 4.3 Ω at 10 GHz, which is larger than switch F of 3.8 Ω. That is because of larger contact resistance of switch G. The threshold voltage of switch G (l_2 = 350 μm) is 30 V and that of switch F (l_2 = 235 μm) is 23 V. Therefore, when 40 V bias voltage is applied, the contact force of switch G is smaller than that of switch F, which results in higher contact resistance of switch G. The parasitic capacitance at the on-state, C_{gc}, increases from 70 to 77 fF when l_2 increases from 235 to 350 μm. The inductance decreases slightly with l_2.

2.4.3 Metal Coating Effect

The thickness of the metal coating affects the RF performance of lateral switches since it determines the switch resistance. Figure 2.37a, b compares the measured *S*-parameters of a single-beam switch with various thicknesses of Al coating at the on-state and off-state, respectively. The Al thickness is 0.8 μm for switch H, 1.2 μm

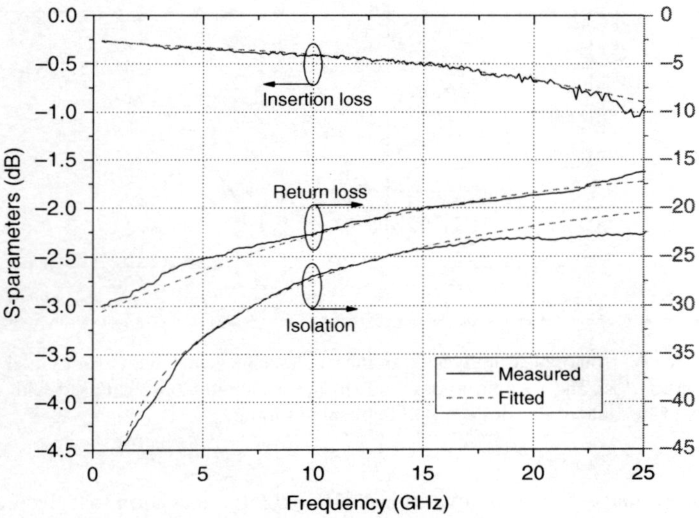

Fig. 2.35 Comparison between measured and fitted *S*-parameters of switch G ($l_2 = 350\ \mu m$)

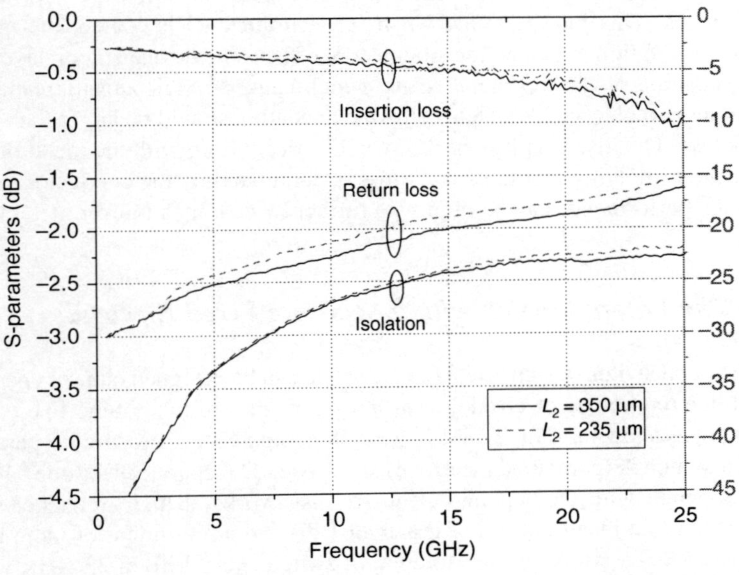

Fig. 2.36 Comparison of measured *S*-parameters between switch F ($l_2 = 235\ \mu m$) and switch G ($l_2 = 350\ \mu m$). Copyright/used with permission of/courtesy of Institute of Physics and IOP Publishing Limited

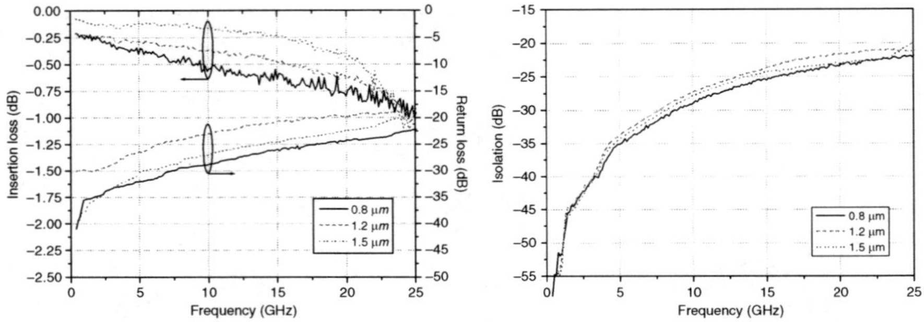

(a) The insertion loss and return loss at the on-state (b) The isolation at the off-state

Fig. 2.37 Comparison of measured results of the single-beam switch with various thick Al coating (a) the insertion loss and the return loss, and (b) the isolation. Copyright/used with permission of/courtesy of Institute of Physics and IOP Publishing Limited

for switch A, and 1.5 μm for switch I. At 10 GHz, the insertion loss decreases from 0.62 to 0.2 dB when the Al thickness increases from 0.8 to 1.5 μm. The main reason is the reduction in the total resistance from 5 to 1.8 Ω due to thicker metal coating. It is noted that the return loss of switch H with 0.8-μm-thick Al coating is larger than the other two switches with thicker metal coating. This is in contrast to the fact that the return loss increases when the metal coating increases from 1.2 to 1.5 μm. It is found that a shunt conductance G (3.5×10^{-4} S) is needed to add in the circuit model of switch H to fit the measured S-parameters, but is unnecessary for the other two switches. This indicates that when 0.8-μm-thick Al is coated, the metal on sidewalls is too thin to isolate the silicon core effect. The device silicon layer works as the substrate of the top metal strips, which causes larger parasitic capacitance C_{gc} and more dielectric loss. Hence, the metal coating should be larger than 0.8 μm for low loss. The insertion loss of the switch is decreasing with the metal thickness significantly implying that the switch loss is dominated by the conductor loss. The switch RF performance can be improved further by coating a thicker metal layer.

2.4.4 The Lateral Switch with a Separate Fixed Electrode

For the lateral switch design, the fixed electrode can be designed either as part of the ground line (combined electrode) or as a separate part of the ground line (separate electrode), as shown in Fig. 2.38. Figure 2.39 compares the measured S-parameters between switch A (combined electrode) and switch D (separate electrode). It shows that the switch with the separate electrode also works at high frequencies. Both switches have an insertion loss of less than 1 dB and a return loss of more than 15 dB up to 25 GHz. However, the isolation of switch D is 14 dB at 25 GHz, whereas the isolation of switch A is 21 dB at 25 GHz. The isolation of switch D is lower than switch A by about 7 dB from 0.05 to 25 GHz. This is because switch D has larger open capacitance, C_s, as shown in Table 2.5. The open capacitance of switch

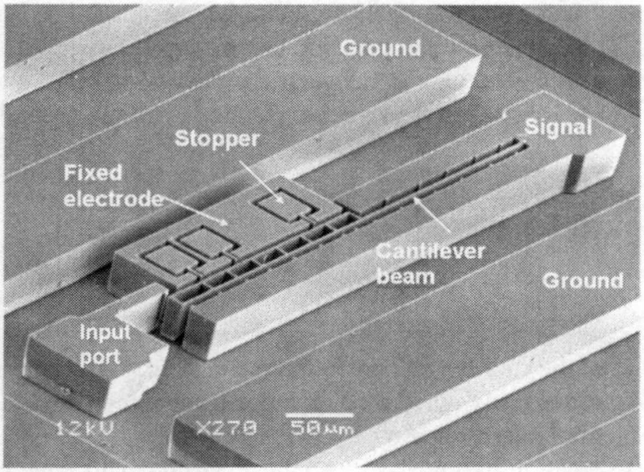

Fig. 2.38 SEM images of a lateral switch with separate fixed electrode. Copyright/used with permission of/courtesy of Springer

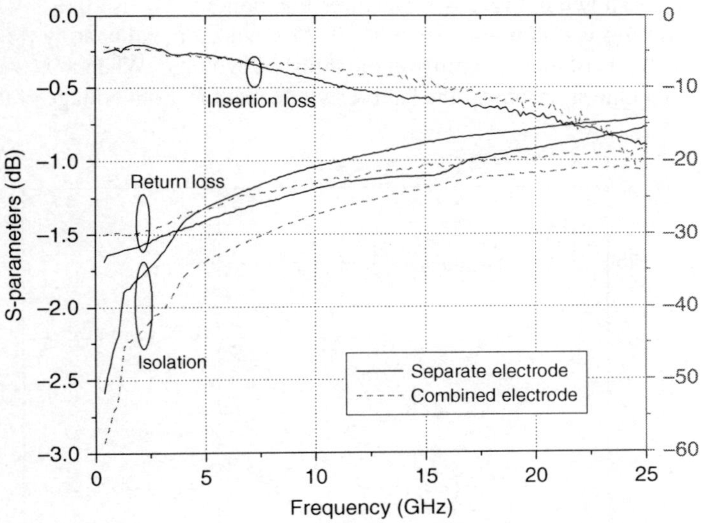

Fig. 2.39 Comparison of measured *S*-parameters between switch D (separate electrode) and switch A (combined electrode)

D is 13.2 fF, whereas the open capacitance of switch A is 6.5 fF. From Fig. 2.38, it is found that except the end of the cantilever beam, part of the signal line also faces the input port in switch D, which increases the open capacitance. Therefore, the isolation of switch D can be improved by using the cantilever beam alone as the signal line of the switch. It is noted that the parasitic capacitance, C_{gc}, of switch D of 22.6 fF is less than switch A of 28 fF, although switch D has longer electrode than switch A. This is because the parasitic capacitance of switch D is contributed by two capacitors in series. One is between the cantilever beam and the separate electrode. The other is between the separate electrode and the ground line.

2.5 Mechanical Measurements

2.5.1 Static behavior

Since the pull-in of the cantilever beam is sharp and sudden, the measurement of the pull-in voltage can be easily performed at wafer level using the standard electrical test equipment with a microscope.

Figure 2.40 shows the comparison of the measured and calculated threshold voltages, V_{th}, of the switch with various original gap distances, g_0, where $l_1 =$ 220 μm, $l_2 =$ 210 μm, $l_3 =$ 10 μm, $w_1 =$ 2.4 μm, $w_2 =$ 5 μm, and $w_{Al} =$ 0 μm. The threshold voltage increases with g_0. A 0.5-μm increase in g_0 will increase the threshold voltage by about 2.5 V. The effect of $(l_2/(l_1 + l_2))$ ratio on the threshold voltage, V_{th}, is shown in Fig. 2.41. The threshold voltage, V_{th}, is 20 ± 1 V when the $(l_2/(l_1 + l_2))$ ratio is within the range of 30–75% before metal coating. Figure 2.41 also shows effects of metal coating on the threshold voltage. When a 0.63-μm-thick Al is coated on the sidewalls of the lateral switch, the threshold voltage of the switch

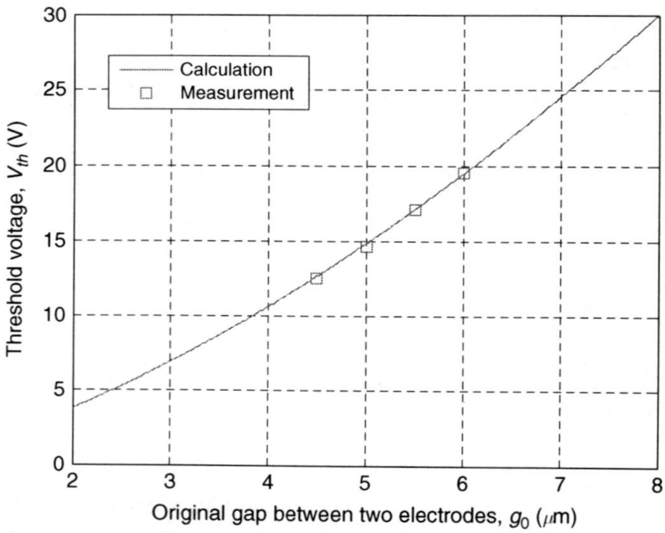

Fig. 2.40 Comparison of measured and calculated threshold voltage V_{th} of the lateral switch with various g_0 ($l_1 = 220$ μm, $l_2 = 210$ μm, $l_3 = 10$ μm, $w_1 = 2.4$ μm, $w_2 = 5$ μm, $w_{Al} = 0$ μm). Copyright/used with permission of/courtesy of Institute of Physics and IOP Publishing Limited

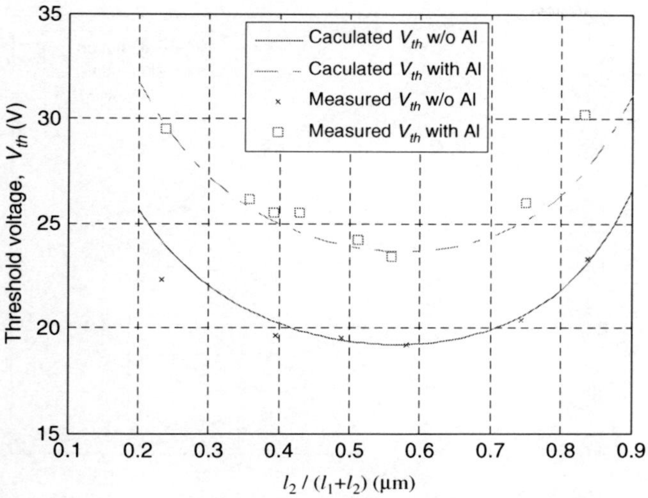

Fig. 2.41 Measured and calculated V_{th} with various $(l_2/(l_1 + l_2))$ ratio with and without (w/o) Al coating ($l_1 + l_2 = 430$ μm, $l_3 = 10$ μm, $w_1 = 2.4$ μm, $w_2 = 5$ μm, $g_{Si} = 6$ μm). Copyright/used with permission of/courtesy of Institute of Physics and IOP Publishing Limited

increases by about 5 V. However, the measured results of the switch with 0.63-μm-thick Al on the sidewalls do not fit so well to the calculation results compared to the switch without Al coating. This may be because residual stress is introduced by evaporated Al.

Figure 2.42 shows the comparison between the measured, calculated, and simulated displacement results of the free end of a switch with 0.63-μm Al coated on the sidewalls. The measurement procedures of the displacement of the free end of the cantilever beam are as follows: First, the lateral switch is fixed on a probe station, which has a microscope above it. A camera is connected to the microscope. Then, two micro-probes connected to an external power source are put on the two electrodes. After a bias voltage is applied, the camera takes a photo of the contact part once the cantilever beam becomes stable. By comparing the photo at some specified voltage with the photo at 0 V, the displacement of the free end of the cantilever beam is calculated. It can be noted that the displacement of the free end of the cantilever beam increases with the applied bias voltage. When the bias voltage increases to 23.3 V, the cantilever beam is attracted to touch the contact tip rapidly. Therefore, the threshold voltage of this switch is 23.3 V. The measurement result is consistent with the calculated and the simulated results.

2.5.2 Dynamic Behavior

A lateral switch with $l_1 = 275$ μm, $l_2 = 165$ μm, $l_3 = 10$ μm, $w_1 = 2.4$ μm, $w_2 = 5$ μm, $w_{Al} = 0.63$ μm, $g_{Si} = 6$ μm, and $d_{Si} = 4$ μm is tested and the results

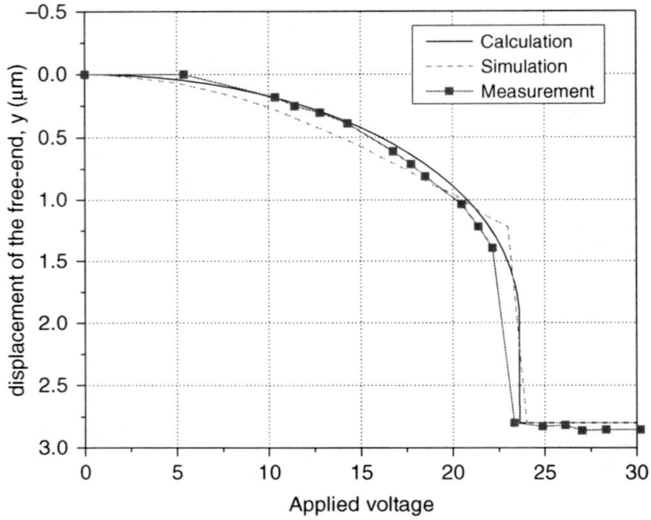

Fig. 2.42 Comparison of measured, calculated, and simulated displacement of the free end of the cantilever beam with 0.63-μm-thick Al on sidewalls ($l_1 = 275$ μm, $l_2 = 165$ μm, $l_3 = 10$ μm, $w_1 = 2.4$ μm, $w_2 = 5$ μm, $w_{Al} = 0.63$ μm, $g_0 = 4.8$ μm, $d_0 = 2.8$ μm). Copyright/used with permission of/courtesy of Institute of Physics and IOP Publishing Limited

are illustrated in Fig. 2.43. The driving voltage is 35 V. The upper waveform in Fig. 2.43a, b is the square-wave driving signal while the lower waveform is the responding waveform of the current through the switch on the series resistor. The switching time is 35 μs and the release time is 36 μs.

2.5.3 Reliability of the Lateral Switch

Resistive switches usually fail due to an excessive increase in electrical contact resistance. Generally failure occurs when the contact resistance is greater than 5 Ω, even though the cantilever beam remains deflecting and making contact. The study includes both cold and hot switching lifetimes. The cold switching refers to opening and closing the switch with zero RF signal level through the contact. The hot switching refers to opening and closing the switch with a specified RF signal level through the contact [23]. It is known that the lifetime of hot switching is shorter compared to the lifetime of cold switching due to larger heat dissipation. The testing results show that the cold switching lifetime of the lateral switch exceeds million switching cycles as shown in Fig. 2.44. The insertion loss and the return loss of a double-beam switch deteriorate by 0.1 and 1.5 dB at 10 GHz, respectively, after 1 million cold switching cycles. The isolation changes marginally. The hot switching lifetimes of these switches are quite poor. Generally, after tens of hot switching cycles, the switch fails due to stiction of the cantilever beam to the contact bump.

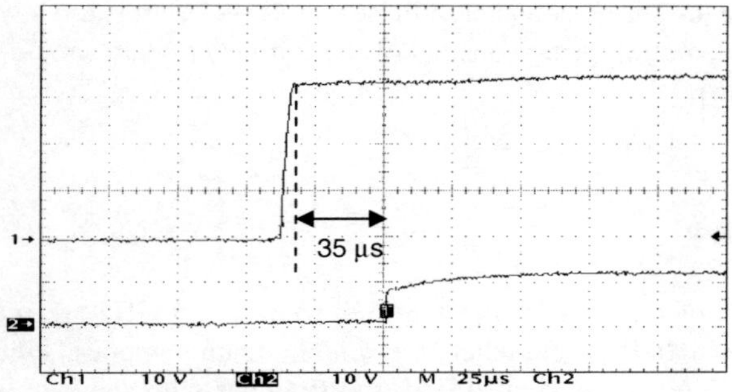

(a) the on-state of the switch.

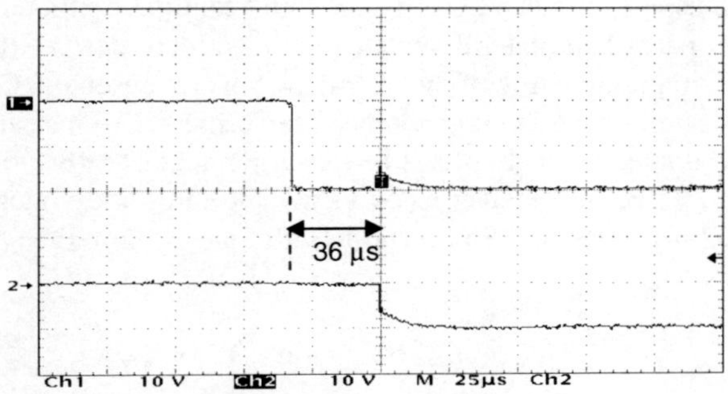

(b) the on-state of the switch.

Fig. 2.43 The experimental results of switching time of a lateral switch: (**a**) the on-state of the switch and (**b**) the off-state of the switch

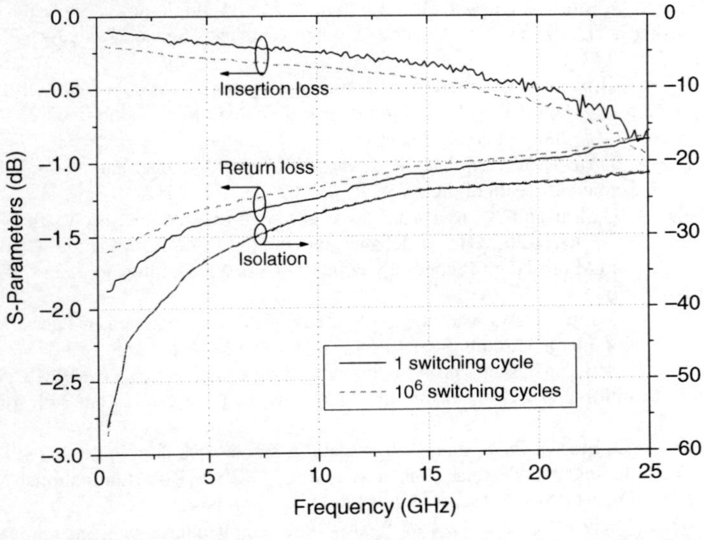

Fig. 2.44 Measured *S*-parameters of switch B with different switching cycles. Copyright/used with permission of/courtesy of Institute of Physics and IOP Publishing Limited

This is because the spring constant of these cantilever beam is too low (~1 N/m). To solve this problem, the spring constant of the cantilever beam should be designed to be >10 N/m [16].

2.6 Summary

Different types of lateral switches from 50 MHz to 25 GHz are presented in this chapter. These include single-beam and double-beam switches, switches with part of the ground line as a fixed electrode, and switches with separate fixed electrode. All lateral switches are designed using Si-core CPW and an electrostatic cantilever actuator. A high aspect ratio cantilever beam with beam mass structure is employed as the actuation part of the lateral switch. The electronic design, the mechanical design, and the circuit modeling of the lateral switch are presented. Comprehensive modeling and design of the lateral switches are verified. The measurements show that the optimized lateral switches have low insertion loss (<1 dB), high return loss, and isolation (>20 dB) at 50 MHz to 25 GHz. The threshold voltage is less than 25 V. The switching speed is 35 μs. The RF lifetime is more than 1 million cold switching cycles.

References

1. A. Q. Liu,M. Tang,A. Agarwal and A. Alphones, "Low-loss lateral micromachined switches for high frequency applications,"J.Micromech. Microeng. Vol. 15,2005,pp. 157–167.
2. M. Tang, A. Q. Liu, A. Agarwal, Z. S. Liu and C. Lu, "A single-pole double-throw (SPDT) circuit using lateral metal-contact micromachined switches," Sensor. Actuator.: Phys. A Vol. 121, 2005,pp. 187–196
3. M. Tang, A. Q. Liu and J. Oberhammer, "A silicon-on-glass single-pole double-throw (SPDT) switching circuit integrated with silicon-core metal-coated transmission line," J. Micromech. Microeng. Vol. 18, 2008, 095024 (9 pp.).
4. A. Q. Liu, A. B. Yu, M. F. Karim and M. Tang, "RF MEMS Switches and integrated switching circuits," J. Semicond. Technol. Sci. Vol. 7(3), 2007, pp. 166–176.
5. M. Tang, A. Q. Liu and A. Agarwal, "Lateral single-pole-double-throw (SPDT) MEMS switches for 50 MHz-20 GHz wideband applications," in Asia-Pacific Conference of Transducers and Micro-Nano Technology 2006 (APCOT 2006), Singapore, 25–28 June 2006, Paper No. A-26.
6. M. Tang, A. Q. Liu, A. Agarwal and Q. X. Zhang, "A new approach of lateral RF MEMS switch," Analog Integr. Circuits Signal Process. Vol. 40, 2004, pp. 165–173.
7. A. Q. Liu, W. Palei, M. Tang and A. Alphones, "Microstrip lateral RF MEMS switch integrated with multi-step CPW transition," Microwave Opt. Technol. Lett. Vol. 44(1), 2005, pp. 93–95.
8. M. Tang, A. Q. Liu, W. Palei, and A. Agarwal, "A single-pole-double-throw (SPDT) circuit using deep etching lateral metal-contact switches," IEEE MTT-S International Microwave Symposium Digest (IMS 2004), 6–11 June 2004, pp. 581–584.
9. M. Tang, A. Agarwal and A. Q. Liu, "A low loss lateral micromachined relay using substrate transfer fabrication process," in International Conference on Material for Advanced Technologies 2005 (ICMAT 2005).

10. M. Tang, A. Agarwal, Z. X. Shen and A. Q. Liu, "A high reliability and low loss lateral RF micromachined relay", in Asia-Pacific Conference of Transducers and Micro-Nano Technology (APCOT 2004), 4–7 July 2004, pp. 1042–1047.
11. W. Palei, M. Tang and A. Q. Liu, "RF MEMS switch integrated with an enhanced back to back microstrip to CPW transition," in Asia-Pacific Conference of Transducers and Micro-Nano Technology (APCOT 2004), 4–7 July 2004, pp. 479–482.
12. M. Tang, A. Agarwal, J. Li, Q. X. Zhang, P. Win, J. M. Huang and A. Q. Liu, "An approach of lateral RF MEMS switch for high performance," in Design, Test, Integration, and Packaging of MEMS/MOEMS 2003 (DTIP 2003), 5–7 May 2003, pp. 99–102.
13. M. Tang, A. Agarwal, P. Win, Q. X. Zhang, J. Li and A. Q. Liu, "A new lateral RF switch using SOI-deep-etching fabrication process," Int. J. Comput. Eng. Sci. Vol. 4(2), 2003, pp. 369–372.
14. X. M. Zhang, F. S. Chau, C. Quan and A. Q. Liu, "A study of the static characteristics of a torsional micromirror," Sensor. Actuator.: Phys. A Vol. 1–2, 2001, pp. 73–81.
15. Ansoft Corporation, Ansoft Maxwell HFSS, Release 8.0.25, 2001, www.ansoft.com
16. D. M. Pozar, Microwave Engineering, Addison-Wesley Publishing Company, Reading, MA, 1990.
17. G. B. Rebeiz, "RF MEMS Theory, Design, and Technology," John Wiley & Sons, Inc., Hoboken, NJ, 2003.
18. J. M. Gere and S. P. Timoshenko, Mechanics of Material, 4th edn., PWS Publishing Company, Boston, MA, 1997.
19. S. Rao Singiresu, Mechanical Vibrations, Addison-Wesley Publishing Company, Inc., Reading, MA, 1995.
20. Wolfram Research, Mathematica, v5.0, 2003.
21. A. A. Ayon, N. J. Kolias and N. C. MacDonald, "Tunable, micromachined parallel-plate transmission lines," In IEEE/cornell Conf. on advanced concepts in high speed semiconductor devices and circuits, NY, USA, 7–9 Aug 1995, pp. 201–208.
22. C. L. Holloway and E. F. Kuester, "Power loss associated with conducting and superconducting rough interfaces," IEEE Trans. Microwave Theor. Tech. Vol. 48(10), 2000, pp. 1601–1610.
23. J. J. Yao, "RF MEMS from a device perspective," J. Micromech. Microeng. Vol. 10, 2000, pp. R9–R38.

Chapter 3
Capacitive Shunt Switches

The focus of this chapter is capacitive shunt switches. The reliability of the capacitive switch is mainly determined by stiction between the dielectric layer and the metal layer, which is due to charge injection and charge trapping in the dielectric layer. Power handling is another important concern for the applications of the capacitive switch. For the capacitive shunt switch, it is easier to handle less than 1 W RF power for billions of cycles. A novel DC-contact capacitive shunt switch with large down/up-state capacitance ratio and broadband is developed to avoid the down-state capacitance degradation problem.

3.1 Broadband DC-Contact Capacitive Switch

3.1.1 Electromagnetic Design and Modeling

Figure 3.1 shows the schematic diagram of the DC-contact capacitive switch. In this design, the dielectric layer is fabricated on the CPW ground planes instead of the CPW center conductor. The metal bridge, the dielectric layers, and the CPW ground planes consist of two metal–insulator–metal (MIM) capacitors in shunt connection. This newly designed switch is referred to as dielectric-on-ground (DOG) switch. It not only is innovative in terms of its unique design but also helps in solving the capacitance degradation problem. The conventional capacitive switch is referred to as dielectric-on-center conductor (DOC) switch.

Figure 3.2 shows the equivalent circuit of the switch. In the up-state position, the switch capacitance $C_{\mathrm{u}}^{\mathrm{DOG}}$ is dominated by the parallel plate capacitance C_{pp} between the metal bridge and the CPW center conductor; it is expressed as

$$C_{\mathrm{u}}^{\mathrm{DOG}} \cong C_{\mathrm{pp}} = \frac{\varepsilon_0 A_1}{g + t_{\mathrm{d}}/\varepsilon_{\mathrm{r}}} \cong \frac{\varepsilon_0 A_1}{g} \tag{3.1}$$

Aibin Yu and Ai Qun Liu

A.Q. Liu, *RF MEMS Switches and Integrated Switching Circuits*,
MEMS Reference Shelf 5, DOI 10.1007/978-0-387-46262-2_3,

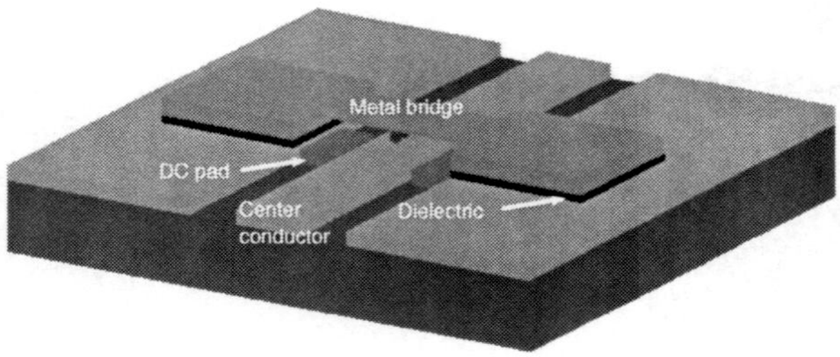

Fig. 3.1 Schematic diagram of the DOG capacitive switch

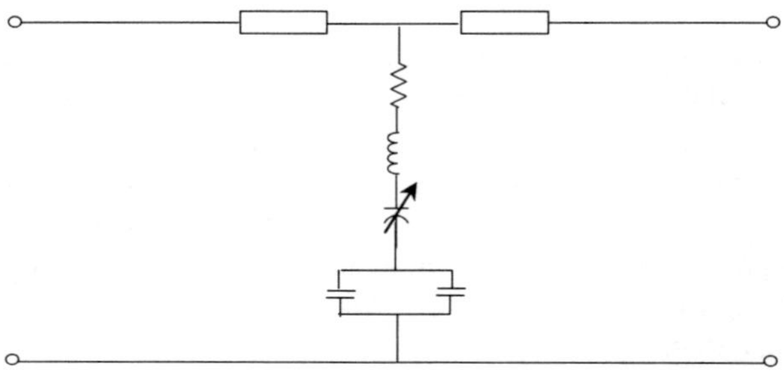

Fig. 3.2 Equivalent circuit of DOG switch

where A_1 is the overlap area between the metal bridge and the CPW center conductor and $A_1 = wW$, where w is the width of the metal bridge and W is the width of the CPW center conductor.

In the down-state position, the DC contact between the metal bridge and the CPW center conductor results in an R_sL model in series with MIM capacitors. Therefore, the down-state capacitance of the switch C_d^{DOG} is dominated by the capacitance of the MIM capacitor C_{MIM} and is expressed as

$$C_d^{DOG} = C_{MIM} = \frac{\varepsilon_0 \varepsilon_r A_2}{t_d} \tag{3.2}$$

where $A_2 = 2wW_d$ is the total area of the MIM capacitor and W_d is the width of the MIM capacitor.

Therefore, the capacitance ratio of this DOG switch is

$$r = \frac{C_d^{DOG}}{C_u^{DOG}} = \frac{\varepsilon_r g}{t_d} \cdot \frac{A_2}{A_1} = R\frac{A_2}{A_1} \tag{3.3}$$

where $R = (\varepsilon_r g)/t_d$ is the ideal capacitance ratio of the DOC switch.

According to Eq. (3.3), high capacitance ratio of the DOG switch can be obtained by increasing A_2/A_1, which gives a new degree of freedom to optimize the switch design. As the down-state capacitance is determined by the MIM capacitor, the deformation of the metal bridge and the surface roughness do not affect on the down-state capacitance value. As a result, the down-state capacitance degradation problem is avoided [1–4]. This is one of the significant advantages of this DOG switch.

Based on the circuit model as shown in Fig. 3.2, assuming that the transmission line loss and the inductance of the switch are negligible, the insertion loss of the switch is expressed as

$$S_{21} = \frac{2Z_b}{2Z_b + Z_0} \tag{3.4}$$

where Z_b is the impedance of the metal bridge. When the inductance of the metal bridge is ignored, Z_b is

$$Z_b = R_b + \frac{j}{\omega C_u} \tag{3.5}$$

Substituting Eq. (3.5) into Eq. (3.4), we get

$$|S_{21}|^2 = \frac{4\left(R_b^2 + 1/(\omega C_u)^2\right)}{(2R_b + Z_0)^2 + 4/(\omega C_u)^2} \tag{3.6}$$

where R_b is the resistance of the metal bridge, ω is the angular frequency, and C_u is the up-state capacitance of the switch which can be expressed as

$$C_u = \frac{\varepsilon_0 A}{g_0 + t_d/\varepsilon_r} + C_f \tag{3.7}$$

where C_f is the fringing field capacitance which is approximately 30% of the parallel plate capacitance between the metal bridge and the CPW center conductor. The item t_d/ε_r is ignored since it is much smaller than g_0.

The capacitive switch is shown in Fig. 3.1 with design dimensions listed in Table 3.1, normally $R_b < 10\ \Omega$ and $C_u < 100$ fF, then $1/\omega C_u >> R_b$ till 100 GHz.

Table 3.1 Design parameters of the DOG capacitive switch

Design parameters	Values
Width of the metal bridge, w (μm)	100
Length of the metal bridge, l (μm)	400
Area of MIM capacitor, A_2 (μm^2)	200 × 200
Width of the center conductor of CPW, S (μm)	200
Slot width of CPW, W (μm)	100
Initial height of the metal bridge, g_0 (μm)	2
Thickness of the metal bridge, t_b (μm)	0.6
Space between bridge anchor and lower DC pad, l_1 (μm)	10
Width of the lower DC pad, d (μm)	80
Thickness of the dielectric layer, t_d (μm)	0.15
Dielectric constant of SiN	7

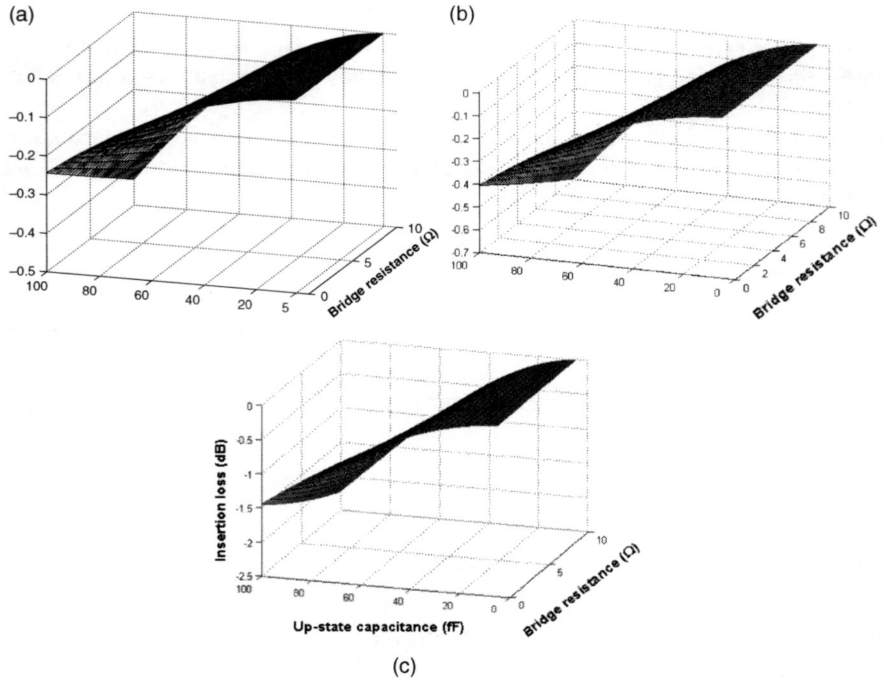

Fig. 3.3 Insertion loss changes with up-state capacitance and resistance **(a)** 5 GHz; **(b)** 20 GHz; and **(c)** 40 GHz

Figure 3.3 shows how the insertion loss S_{21} (in dB) changes with R_b and C_u at 5, 20, and 40 GHz. It is observed that the insertion loss increases with C_u and R_b. When the impedance of the metal bridge in the up-state is 10 times lower than R_b, the S_{21} is no longer sensitive to the variation in R_b and can be kept smaller than 0.2 dB.

In the down-state position of the switch, the RF performance can also be analyzed in the same way as that of the DOC switch. However, it should be noted that in the DOG switch, as the contact between the metal bridge and the CPW center conductor becomes DC contact, the down-state resistance R_s includes both the metal bridge resistance and the contact resistance, which value is around 0.5–2 Ω for most of the design and limits the highest isolation to 22–34 dB. The contact resistance depends on the size of the contact area, the mechanical force applied, and the quality of the metal-to-metal contact.

The impedance of the metal bridge at the down-state Z_b is expressed as

$$Z_b = R_s + j\left(\omega L - \frac{1}{\omega C_d^{DOG}}\right) \tag{3.8}$$

Figure 3.4a shows that the isolation changes with the down-state resistance R_s and the frequency. The down-state capacitance and inductance are assumed to be 31

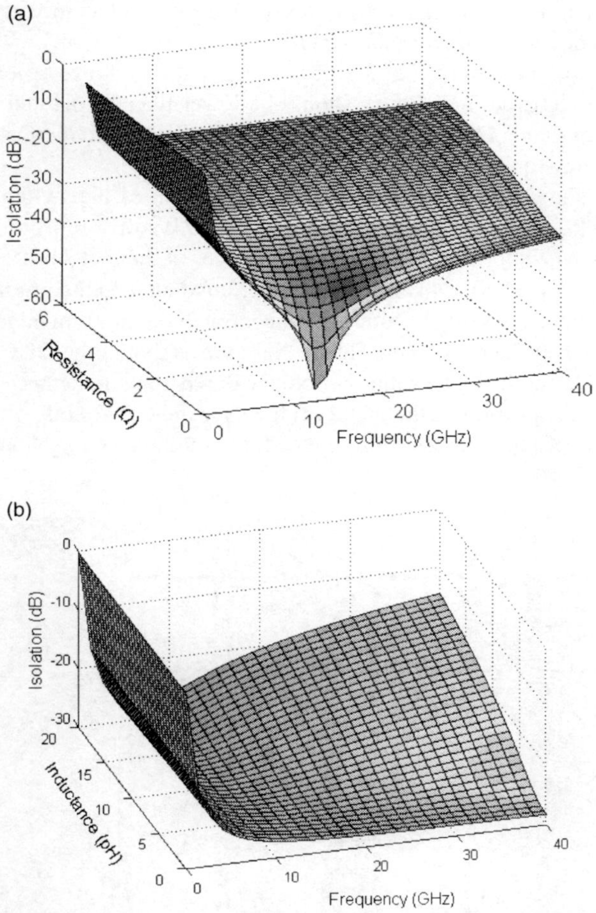

Fig. 3.4 Variation of down-state isolation: (**a**) frequency versus resistance and (**b**) frequency versus down-state inductance [5]. Copyright/used with permission of/courtesy of Elsevier B.V

pF and 5 pH, respectively. The peak of the isolation can be seen clearly at the *LC* resonant frequency $f_0 = 12.4$ GHz when $R_s < 0.5\ \Omega$. The highest isolation can be expressed as

$$S_{21} = \frac{2R_s}{Z_0} \tag{3.9}$$

When $R_s > 0.8\ \Omega$, the isolation changes a little with frequency $f > 10$ GHz and its value is mainly determined by the inductance of the switch.

Figure 3.4b shows the isolation changes with the inductance L and the frequency f. The down-state capacitance and the resistance are assumed to be 31 pF and 1 Ω, respectively. The isolation increases quickly when the inductance is larger than 5 pH and the frequency is higher than 10 GHz.

Combining (a) and (b) of Fig. 3.4, it is noted that to achieve at least 20 dB isolation until 40 GHz, the inductance should be lower than 10 pH and the resistance should be lower than 2 Ω. Smaller inductance is also helpful to increase the isolation in high frequency range.

Figure 3.5 shows the simulated current distribution of the switch in the down-state using high-frequency structure simulator (HFSS) software from Ansoft. When the metal bridge is in the down-state, the front edges of the metal bridge that are over the CPW slot act as short paths from the center conductor to the ground. Therefore, the current in the metal bridge is mainly concentrated on the front edges of the metal bridge that are over the CPW slot. This is the same as that in the DOC shunt switch. This implies that for the DOG shunt switch, the down-state inductance is also mainly determined by the portion of the metal bridge over the CPW slots. Therefore, when the dimensions of the CPW slot are reduced, the inductance of the switch can also be reduced.

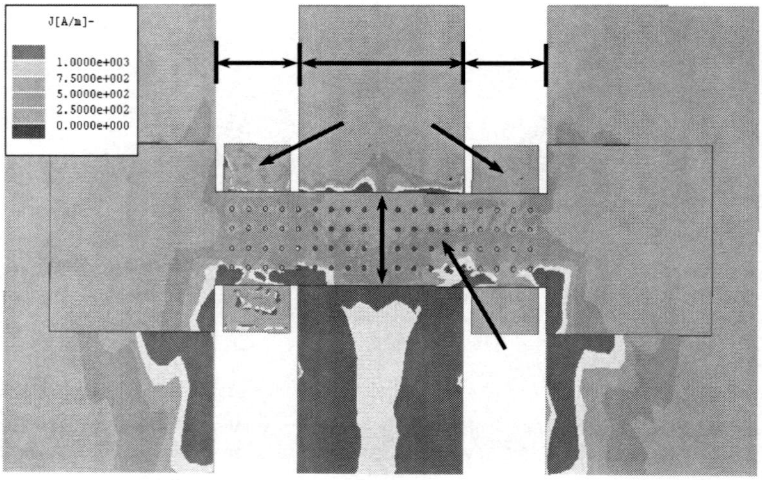

Fig. 3.5 Simulated current distribution of the DOG switch at the down-state: width of bridge $w = 100$ μm; width of CPW = 100/200/100 μm

3.1.2 Mechanical Design

Figure 3.1 shows the schematic cross section of the DOG switch. The lower DC pads are positioned under the two edges of the metal bridge. The effective spring constant of the metal bridge can be expressed as

$$k_{\text{eff}} = k' + k'' \tag{3.10}$$

where k' is the stiffness of the bridge which accounts for the material characteristics, such as Young's modulus E and the moment of inertia I. k'' is the stiffness of the bridge which is due to the residual stress σ_0 within the metal bridge [6].

Suppose that ξ is the force per unit length, the displacement y of point P can be thought as the total deflection caused by a number of small concentrate force $\xi \cdot ds$ on a small portion of the beam, where ds is a small portion of the beam which is with distance s from the support. As the structure is symmetrical, only half of the force effect is calculated and the entire deflection y is twice as large. Therefore, as

$$\begin{cases} EI\frac{d^2y_1}{dx^2} = M_A + R_A x, \quad x \le s \\ x = 0,\ y_1 = 0 \\ x = 0,\ \frac{dy_1}{dx} = 0 \end{cases} \tag{3.11}$$

and

$$\begin{cases} EI\frac{d^2y_2}{dx^2} = M_A + R_A x - \xi ds(x - s), \quad x \ge a \\ x = l,\ y_2 = 0 \\ x = l,\ \frac{dy_2}{dx} = 0 \end{cases} \tag{3.12}$$

and with the continuous condition

$$\begin{cases} y_1(s) = y_2(s) \\ \frac{dy_1(s)}{dx} = \frac{dy_2(s)}{dx} \end{cases} \tag{3.13}$$

where l is the length of the metal bridge, M_A (N m) and R_A (N) are the reaction moment and the vertical reaction at the left end, respectively. $I = wt^3/12$ is the moment of inertia for a rectangular cross section, where w and t are the width and thickness of the metal bridge, respectively; the deflection function can be expressed as

$$y_1 = \frac{M_A x^2}{2EI} + \frac{R_A x^3}{6EI}, \quad x \le s \tag{3.14}$$

$$y_1 = \frac{M_A x^2}{2EI} + \frac{R_A x^3}{6EI}, \quad x \le s \tag{3.15}$$

$$y_2 = \frac{M_A x^2}{2EI} + \frac{R_A x^3}{6EI} - \frac{\xi \cdot ds(x - s)^3}{6EI}, \quad x \ge s \tag{3.16}$$

$$M_A = -\frac{\xi \cdot ds}{l^2} s\,(l - s)^2 \tag{3.17}$$

$$R_A = \frac{\xi \cdot ds}{l^3}\,(l - s)^2\,(l + 2s) \tag{3.18}$$

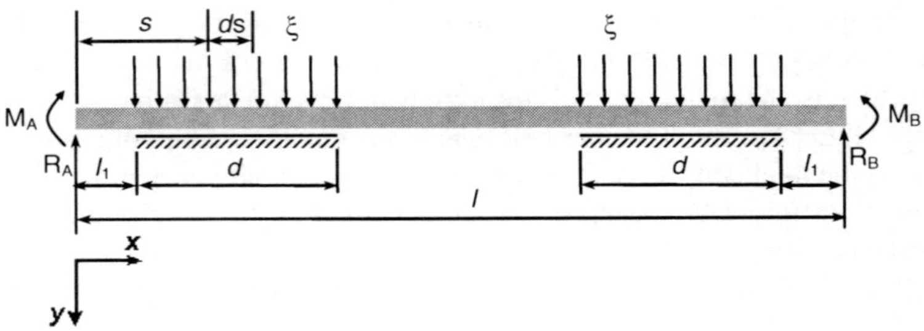

Fig. 3.6 Fixed–fixed beam with distributed load at the ends of the beam

For the deflection of the center part of the beam, the deflection of the beam at the center is used to determine the spring constant. Combining Eqs. (3.14), (3.15), (3.16), (3.17), and (3.18), when the load is distributed across the width of the electrode as shown in Fig. 3.6, the deflection at the center point of the beam ($x = l/2$) can be derived as

$$\begin{aligned} y &= 2\int_{l_1}^{l_1+d} y_2(l/2) \\ &= \frac{2\xi}{EI}\int_{l_1}^{l_1+d}\left(\frac{4s^3 - 3ls^2}{48}\right)ds \\ &= \frac{\xi d}{24EI}\left\{(2l_1 + d)\left[(l_1 + d)^2 + l_1^2\right] - l\left[(l_1 + d)^2 + l_1\,(l_1 + d) + l_1^2\right]\right\} \end{aligned} \tag{3.19}$$

The spring constant k' at the center can be found to be

$$k' = -\frac{2\xi d}{y} = 4Ewt^3\frac{1}{l\left[(l_1 + d)^2 + l_1\,(l_1 + d) + l_1^2\right] - (2l_1 + d)\left[(l_1 + d)^2 + l_1^2\right]} \tag{3.20}$$

The part of the spring constant which is caused by the biaxial residual stress within the beam is derived from modeling the beam as a stretched wire. It is noted that this model only applies for tensile stress. The force associated with biaxial residual stress, σ_0, can be expressed as

$$S = \sigma_0(1 - \upsilon)tw \tag{3.21}$$

where ν is Poisson's ratio.

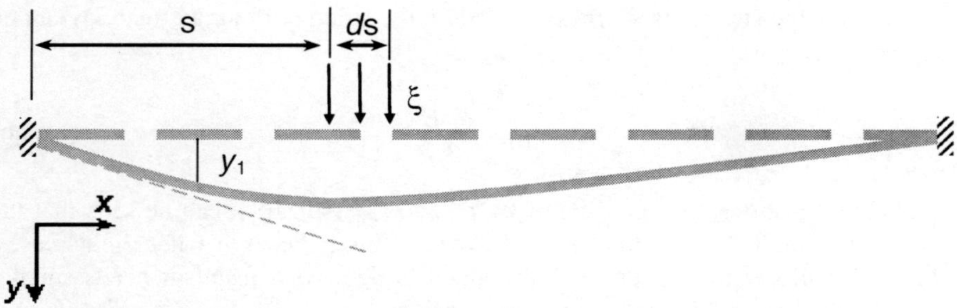

Fig. 3.7 Deflection of the fixed–fixed beam

For the load on the beam as shown in Fig. 3.7, the deflection that results from the residual stress at x can be expressed as

$$y_1 = \frac{\xi \cdot ds\,(l-s)}{Sl}x, \quad x \le s \tag{3.22}$$

$$y_2 = \frac{\xi a \cdot ds}{Sl}(l-x), \quad x \ge s \tag{3.23}$$

Therefore, the total deflection for the distributed load that caused by the residual stress at the center point of the beam can be expressed as

$$y = 2\int_{l_1}^{l_1+d} y_2(l/2) = \frac{\xi d}{2S}(2l_1 + d) \tag{3.24}$$

From Eq. (3.24), the spring constant k'' that due to the residual stress can be expressed as

$$k'' = \frac{2\xi d}{y} = \frac{4S}{2l_1 + d} = 4\sigma_0(1-\nu)wt\frac{1}{2l_1 + d} \tag{3.25}$$

When $l_1 = \frac{1}{3}l$ and $d = \frac{1}{6}l$, from Eqs. (3.20) and (3.25), then

$$k' = 32Ew\left(\frac{t}{l}\right)^3\left(\frac{27}{49}\right) \tag{3.26}$$

$$k'' = 8\sigma_0(1-\upsilon)w\left(\frac{t}{l}\right)\left(\frac{3}{5}\right) \tag{3.27}$$

As a result, the force equation of the metal bridge can be written as

$$\frac{\varepsilon_0 w(2d)}{2(g_0 - y)^2}V^2 - k_{\text{eff}}y - k_{\text{D}}y^3 = 0 \tag{3.28}$$

where k_D is the stretching stiffness. Based on the strain stiffening effect, k_D can be expressed as

$$k_D = \frac{\pi^4 Ewt}{8\, l^3} \tag{3.29}$$

Figure 3.8 shows the variation of k', k'', and k_{eff} with l_1. It can be seen that the longer the length l_1, the smaller the stiffness. This is because when l_1 increases, the load is closer to the center of the metal bridge. As a result, it needs smaller electrostatic force to drive down the metal bridge.

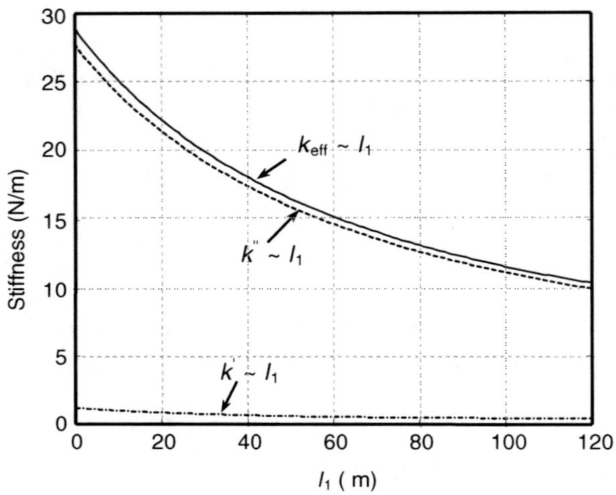

Fig. 3.8 Variation of stiffness versus different edge distances l_1

Figure 3.9 shows the calculated metal bridge height versus the applied voltage for different l_1s. As expected, the pull-down voltage decreases when l_1 increases. When $l_1 = 10\ \mu$m, the pull-down voltage is 19.2 V with 27 MPa stress in the metal bridge.

3.1.3 Experimental Results and Discussions

The DOG switch is fabricated using the surface micromachining process. Figure 3.10 shows the SEM image of the fabricated DOG switch. In this design, the driven-down DC pad is separated with the grounds, which can reduce the possibility of stiction when the metal bridge is driven down. It also reduces the breakdown of the dielectric layer.

Figure 3.11a shows the measurement results for the DOG switch. In the up-state position of the bridge, the extracted up-state capacitance C_{ue}^{DOG} is 90 fF at 10 GHz while the calculated up-state capacitance C_u^{DOG} is 80 fF. The reason for the

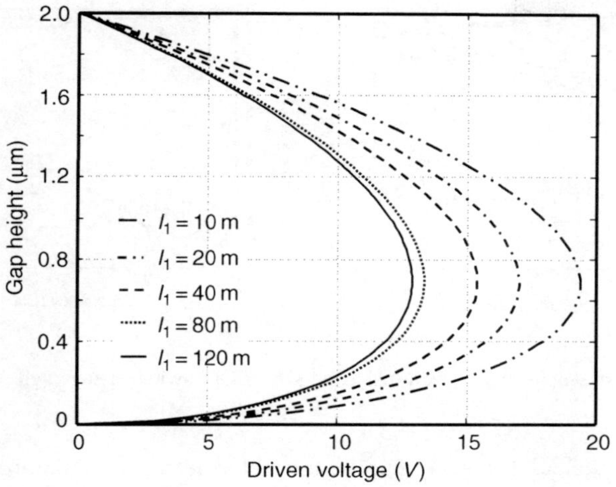

Fig. 3.9 Metal bridge displacement versus applied voltage

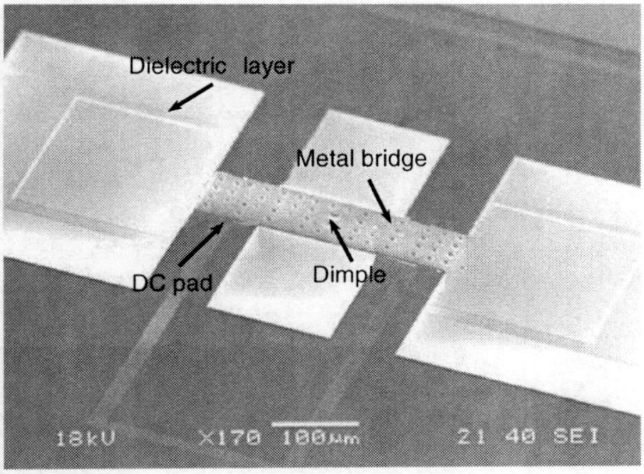

Fig. 3.10 SEM image of DOG switch on high-resistivity silicon substrate

difference between C_{ue} and C_{ud} is due to the fringing field effect. The insertion loss is 0.9 dB at 20 GHz and 2 dB at 40 dB. According to Eq. (3.6) and Fig. 3.3, when the up-state capacitance is 90 fF, insertion loss is 0.7 and 1.9 dB at 20 and 40 GHz, respectively.

Figure 3.12b shows the down-state RF performance. In the down-state position, the isolation is higher than 10 dB from 2 to 40 GHz and the highest isolation is 28 dB at 6.5 GHz. The extracted down-state capacitance C_{de}^{DOG} is 30 pF and the calculated down-state capacitance C_{d}^{DOG} is 33 pF. The small difference between the extracted

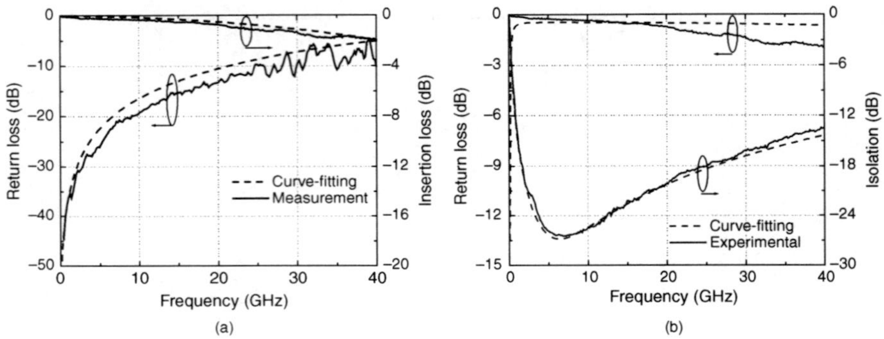

Fig. 3.11 Measurement and curve-fitted results of DOG switch using high-resistivity silicon substrate

value and the calculated value can be attributed to the fabrication and measurement error. Therefore, the down-state capacitance degradation can be avoided for the DOG switch. The experimental capacitance ratio is 330, which is much larger than that of the DOC switch. This is the highest capacitance ratio obtained so far

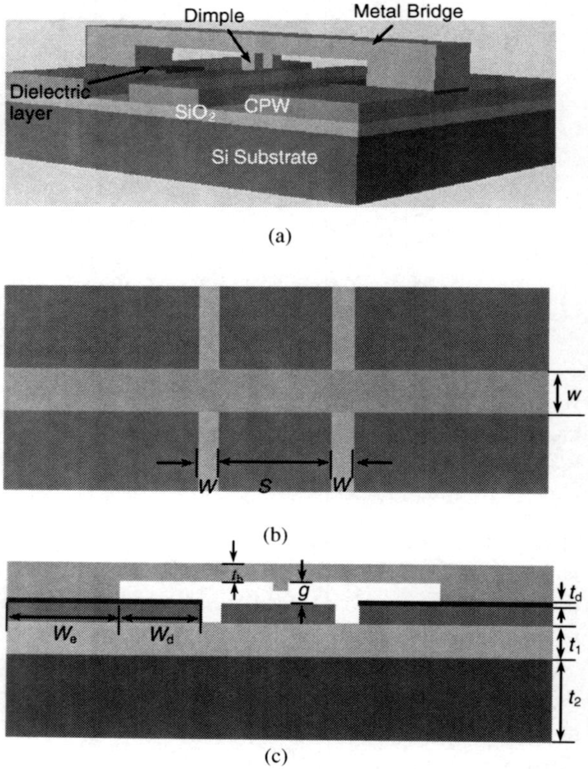

Fig. 3.12 Schematic diagram of the DOG capacitive switch: (**a**) 3D view; (**b**) top view; and (**c**) cross section [5]. Copyright/used with permission of/courtesy of Elsevier B.V

except the one using very high dielectric constant materials for the dielectric layer. The inductance can be extracted from the down-state capacitance and the resonant frequency as 28 pH. The series resistance can be extracted from the isolation at the resonant frequency as 1.2 Ω.

3.2 DOG Switch on Low-Resistivity Silicon Substrate

Another problem reported in most capacitance shunt switches is that these switches are fabricated on low loss substrates, such as high-resistivity silicon wafer, glass, or quartz wafers, or to use wafer transfer technique to reduce substrate loss. The shortcoming of the high-resistivity silicon, glass, or quartz wafers is high cost. The use of wafer transfer technique adds complexity to the fabrication process.

It is interesting to note that since the MIM capacitor is defined on the ground planes of the CPW transmission line, the width of the CPW slots can be reduced to 20 μm or narrower while the characteristic impedance Z_0 of the CPW transmission line can still be maintained at 50 Ω by properly selecting the width of the center conductor. The narrow slots and the center conductor do not have problem in obtaining large down-state capacitance since the MIM capacitor is fabricated on the ground planes. Therefore, the switch can be fabricated on a low-resistivity silicon substrate. This is another advantage of the DOG switch [5, 7].

When a low-resistivity silicon substrate is used, a thick isolation layer (silicon dioxide or polyimide) is coated on the surface of low-resistivity silicon substrate to reduce the substrate loss. When the slot width of the CPW transmission line is narrower than the thickness of the isolation layer, the insertion loss of the CPW transmission line is comparable to that of the CPW transmission line fabricated on high-resistivity silicon substrate [8]. With this strategy, the slot width of the CPW line is designed to be smaller than 20 μm.

Consequently, the center conductor of the CPW is approximately 50-μm wide in order to obtain 50-Ω characteristic impedance. For a typical capacitive switch, the down-state capacitance is limited by the width of the center conductor. However, when the dielectric layer is shifted onto the ground planes of CPW line, this limitation vanishes.

Figure 3.12 shows schematic drawing of the DOG switch fabricated on a low-resistivity silicon substrate. The dimples are used to avoid the stiction between the metal bridge and the low DC pad. All the design parameters are listed in Table 3.2. Most of the dimensions are similar to the DOG switch fabricated on high-resistivity silicon substrate, as listed in Table 3.1.

When a thick layer of SiO_2 or polyimide is coated on the silicon substrate, a 50-Ω CPW transmission line can be designed based on the following equations [9]:

$$Z_0 = \frac{30\pi}{\sqrt{\varepsilon_{\text{eff}}}} \frac{K(k_0')}{K(k_0)} \tag{3.30}$$

Table 3.2 Design parameters of DOG capacitive switch on low-resistivity silicon substrate

Design parameters	Values
Width of the metal bridge, w (μm)	150
Area of MIM capacitor, A_2 (μm^2)	150 × 250
Width of the center conductor of CPW, S (μm)	50
Slot width of CPW, W (μm)	10
Initial height of the metal bridge, g_0 (μm)	2
Thickness of the metal bridge, t_b (μm)	1.2
Height between the bridge and CPW ground, W_d (μm)	165
Thickness of the dielectric layer, t_d (μm)	0.15
Dielectric constant of SiN	7

where

$$\varepsilon_{\mathrm{eff}} = 1 + \frac{\varepsilon_{\mathrm{r1}} - 1}{2} \frac{K(k_1)}{K(k_1')} \frac{K(k_0')}{K(k_0)} + \frac{\varepsilon_{\mathrm{r2}} - \varepsilon_{\mathrm{r1}}}{2} \frac{K(k_2)}{K(k_2')} \frac{K(k_0')}{K(k_0)} \tag{3.31}$$

$$k_0 = \frac{S}{S + 2W} \tag{3.32a}$$

$$k_1 = \frac{\sinh[\pi S/4(t_1 + t_2)]}{\sinh\{[\pi(S + 2W)]/4(t_1 + t_2)\}} \tag{3.32b}$$

$$k_2 = \frac{\sinh(\pi S/4t_2)}{\sinh\{[\pi(S + 2W)]/4t_2\}} \tag{3.32c}$$

$$k_n' = \sqrt{1 - k_n^2}, \quad n = 0, 1, 2 \tag{3.32d}$$

where K is the first kind complete elliptic integrates, $\varepsilon_{\mathrm{eff}}$ is the effective dielectric constant, Z_0 is the characteristic impedance, $\varepsilon_{\mathrm{r1}}$ is the dielectric constant of silicon and is equal to 11.9, and $\varepsilon_{\mathrm{r2}}$ is the dielectric constant of SiO_2 and is equal to 4.1. t_1 and t_2 are the thickness of the silicon substrate and SiO_2, respectively.

Figure 3.13a, b is the calculated characteristic impedance and the effective dielectric constant of the CPW transmission line fabricated on the low-resistivity silicon substrate with an 18-μm-thick SiO_2 isolation layer. The dielectric constants of the silicon and SiO_2 are 11.9 and 4.6, respectively.

Based on Eqs. (3.30), (3.31), and (3.32) and Fig. 3.13, the dimensions of the CPW transmission line are $W/S/W$ = 10/50/10 μm. An 18-μm-thick SiO_2 layer is deposited on the substrate as an isolation layer.

The deposition is processed using NOVELLUS PECVD machine and the deposition parameters are listed in Table 3.3. The residual stress is controlled to be below 40 MPa. The SEM image of the DOG switch is shown in Fig. 3.14.

Figure 3.15 shows the simulated current distribution of the switch in the down-state using the HFSS software from Ansoft Co. The current is mainly concentrated on the front edge of the metal bridge portion that is over the CPW slots in the down state. This implies that the down-state inductance is also mainly determined by the

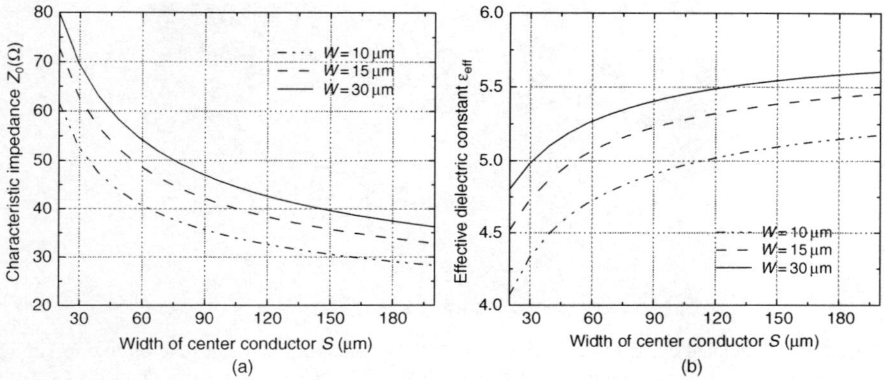

Fig. 3.13 Calculated ϵ_{eff} and Z_0 of a CPW on low-resistivity silicon. (The thickness of SiO_2 layer is 18 μm)

Table 3.3 USG deposition parameters

Parameters	Value
N_2O gas (sccm)	300
SiH_4 gas (sccm)	9500
Carrier gas N_2 (sccm)	50
RF power (W)	800
Pressure (Torr)	2.2
Temperature (°C)	400
Deposition time (min)	36

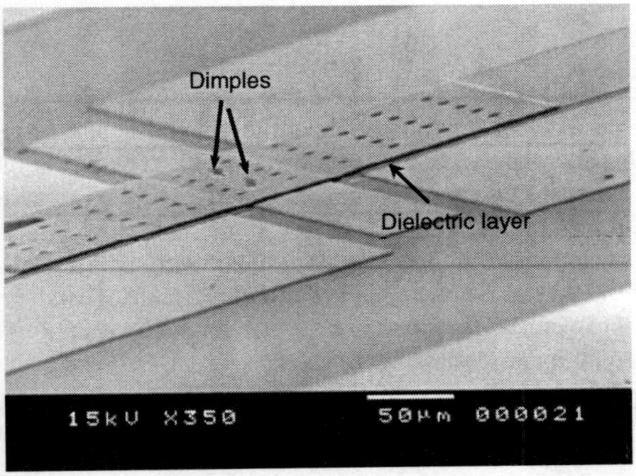

Fig. 3.14 SEM image of DOG switch on low-resistivity silicon substrate [5]. Copyright/used with permission of/courtesy of Elsevier B.V

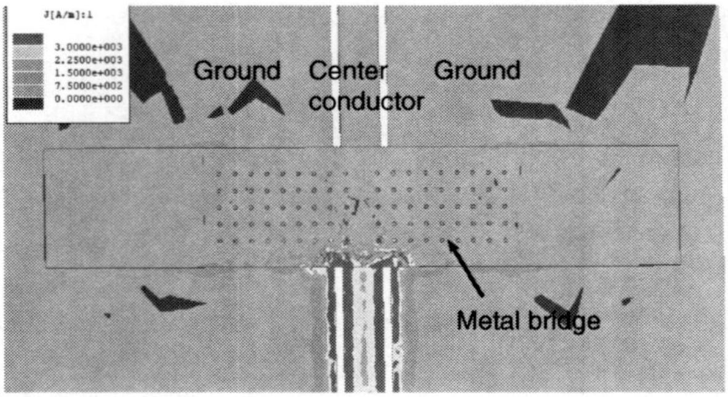

Fig. 3.15 Simulated current distribution of the DOG switch at the down state on low-resistivity silicon substrate (width of bridge $w = 100$ μm; CPW $W/S/W = 10/50/10$ μm)

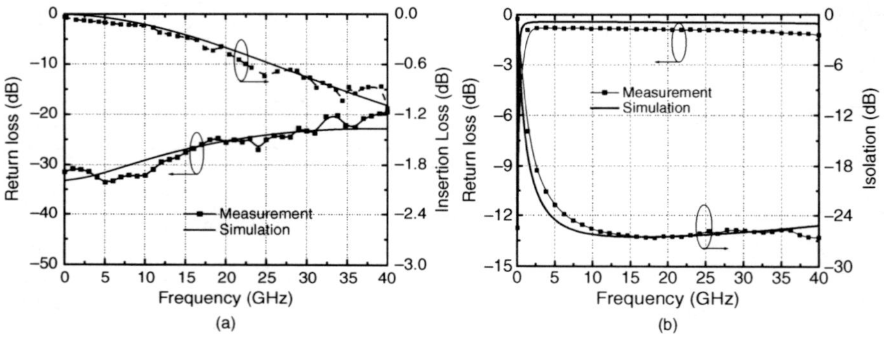

Fig. 3.16 Measurement and curve-fitted results: (**a**) up state and (**b**) down state. Copyright/used with permission of/courtesy of Elsevier B.V

portion of the metal bridge over the CPW slots. Therefore, when the dimensions of the CPW slot are reduced, the inductance of the switch can also be reduced.

Figure 3.16 a shows the measured and the curve-fitted results of the switch in the up state. The extracted up-state capacitance is 30 fF and the insertion loss is lower than 1.2 dB up to 40 GHz. Figure 3.17 shows the measured insertion loss of the CPW transmission lines (with 700-μm long). It is observed that the loss varies from 0.15 to 0.3 dB when the frequency is lower than 40 GHz. Therefore, the insertion loss of the DOG switch on the low-resistivity silicon wafer is the same as that of the DOG switch on high-resistivity silicon wafer.

When comparing the insertion losses of the two DOG switches, the DOG switch on low-resistivity silicon wafer has smaller insertion loss, because its up-state capacitance is smaller compared to that of the DOG switch on high-resistivity silicon wafer.

Figure 3.16b shows the measured and the curve-fitted results in the down-state position. It is observed that the isolation is 13 dB at 1 GHz, 26 dB at 20 GHz,

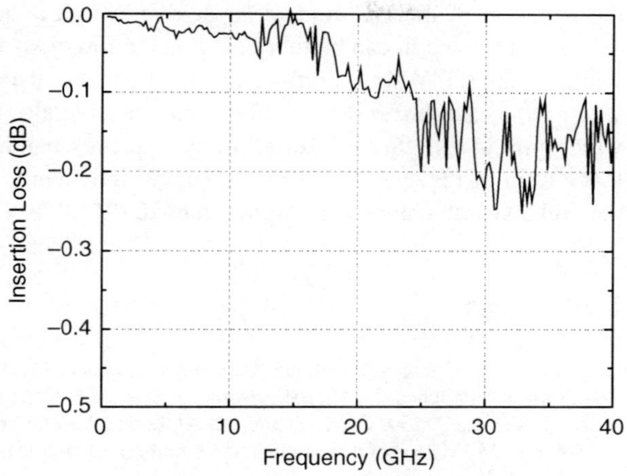

Fig. 3.17 Insertion loss of CPW transmission line

and 27 dB at 40 GHz. The extracted down-state capacitance is 30 pF at 1 GHz. The designed down-state capacitance is 31 pF. The small difference between the extracted value and the calculated value can be attributed to the fabrication and measurement error. Therefore, it can be concluded that the down-state capacitance degradation can be avoided in the design of the DOG switch. In addition, based on the experimental results, the capacitance ratio of the DOG switch is 1000, which is much larger than that of the DOC switch of 150 at most. The extracted down-state resistance and inductance are 1.3 Ω and 2 pH, respectively. The RF performances and the extracted RLC values for this DOG switch are listed in Table 3.4.

Table 3.4 RF characteristics of the DOG switch on low-resistivity silicon wafer

	Up state			Down state		
	1 GHz	20 GHz	40 GHz	1 GHz	20 GHz	40 GHz
S_{11} (dB)	30	29	19	1.2	0.9	0.8
S_{21} (dB)	0.2	0.5	1.2	13	26	27

3.3 Summary

In this chapter, design of the CPW transmission line and the DC-contact capacitive switches are presented. For the DC-contact capacitive switch, the dielectric layer is patterned on the ground planes of the CPW transmission line. Therefore, the down-state capacitance of this switch is not related to the overlap area between the metal bridge and the center conductor. The issue of the down-state capacitance degradation is solved and the down-state capacitance of 30 pF is obtained. As a result, the isolation of the DC-contact capacitive switch is higher than 15 dB from 2 to 40 GHz. The switch can be fabricated on the low-resistivity silicon wafer where the width of the CPW slot is only 18 μm. Therefore, using the low-resistivity silicon wafer reduces not only the cost of the substrate but also the inductance of the switch, which provides higher isolation in the frequency band above 10 GHz. The capacitance ratio of the switch is 1000 and small inductance implies that the isolation of the DOG switch can be kept higher than 20 dB till 40 GHz.

References

1. Z. J. Yao, S. Chen, S. Eshelman, D. Denniston and C. Goldsmith, "Micromachined low-loss microwave switches," J. Microelectromech. Syst. Vol. 8(2), 1999, pp. 129–134.
2. A. B. Yu, A. Q. Liu, Q. X. Zhang, A. Alphones, L. Zhu and S. A. Peter, "Improvement of isolation MEMS capacitive switch via membrane planarization," Sensor. Actuator. A: Phys. Vol. 119, 2005, pp. 206–213.
3. J. B. Muldavn and G. M. Rebeiz, "High-isolation inductively-tuned X-band MEMS shunt switches," IEEE MTT-S Int. Microwave Symp. Digest, Boston, MA, USA, June 2000 169–172.
4. A. B. Yu, A. Q. liu, Q. X. Zhang and H. M. Hosseini, "Effects of surface roughness on electromagnetic characteristics of capacitive switches," J. Micromech. Microeng. Vol. 16, 2006, pp. 2157–2166.
5. A. B. Yu, A. Q. Liu, Q. X. Zhang and H. M. Hosseini, "Micromachined DC contact switch on low-resistivity silicon substrate," Sensor. Actuator. A Vol. 127, 2006, pp. 24–30.
6. J.-M. Huang, K. M. Liew, C. H. Wong, S. Rajendran, M. J. Tan and A. Q. Liu, "Mechanical design and optimization of capacitive micromachined switch," Sensor. Actuator. A: Phys., Vol. 93(3), 2001, pp. 273–285.
7. Q. X. Zhang, A. B. Yu and A. Q. Liu, "Metal contact capacitive switch on low-resistivity silicon wafer," Transducers'05, Seoul, Korea, June 5–9, 2005.
8. J. N. Burghartz and D. C. Edelstein, "Spiral inductors and transmission lines on silicon technology using copper-damascene interconnects and low-loss substrates," IEEE Trans. Microwave Theory Tech., Vol. 45, 1997, pp. 1961–1968.
9. R. N. Simons, Coplanar Waveguide Circuits, Components, and Systems, Wiley, New York, NY, 2001.

Chapter 4
Coplanar Waveguide Transmission Line

The objective of this chapter is to investigate silicon-core (Si-core) metal-coated coplanar waveguide transmission (CPW) line [1]. The Si-core CPW is designed, fabricated, and experimented. Then, attenuation related to different substrate materials, core materials, and fabrication processes are analyzed. The experimental results show that the Si-core CPW supports quasi-TEM mode propagation up to 25 GHz with attenuation less than 4 dB/cm.

4.1 Design of the Si-Core CPW

The geometry of the Si-core CPW is similar to the conventional thin film CPW as shown in Fig. 4.1. It consists of three parallel plate waveguides. In this structure, each waveguide of the Si-core CPW is formed from a single-crystal silicon plate with thickness of 35–75 μm which is coated with a thin layer of metal on the top and sidewalls. RF signal can then propagate along the metal on the top surface and also along sidewalls of the transmission line. The core material can either be low-resistivity silicon ($\rho \leq 10\ \Omega$ cm) or high-resistivity silicon ($\rho \geq 1000\ \Omega$ cm). The substrate can be either Si on glass (SiOG) or high-resistivity silicon in silicon on insulator (SOI). The recesses formed in the substrate under waveguides serve three purposes: (1) to help the release of movable structures in the RF MEMS circuits; (2) to avoid metal connection to short circuit after metal coating; and (3) to reduce the dielectric loss of the transmission line.

The characteristic impedance of CPW depends significantly on its geometrical parameters, including the center conductor width, S, the ground width, G, and the conductor separation, W, and the substrate material properties, such as the substrate thickness, H, and the relative permittivity, ε_r. In the Si-core CPW, other process parameters also affect the characteristic impedance. These include the thickness, T, and the resistivity, ρ, of the core material, the depth of the recesses, h_r, the undercut width, w_r, the thickness of the metal on the top, t_t, and at sidewalls, t_s.

Min Tang, Muhammad Faeyz Karim and Ai Qun Liu

A.Q. Liu, *RF MEMS Switches and Integrated Switching Circuits*,
MEMS Reference Shelf 5, DOI 10.1007/978-0-387-46262-2_4,

Fig. 4.1 Schematic cross-sectional view of the CPW. Copyright/used with permission of/courtesy of Institute of Physics and IOP Publishing Limited

(a) Conventional CPW

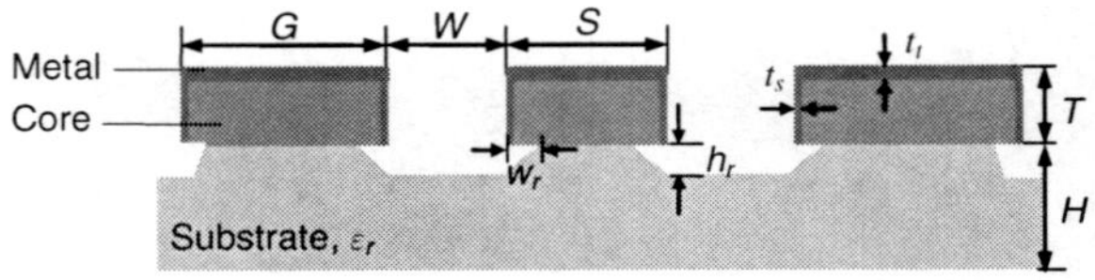

(b) Si-core CPW

The design of the Si-core CPW involves three steps. First, based on the process requirements, some parameter values are determined such that the thickness of the glass substrate $H = 500\ \mu\text{m}$, the permittivity of the glass substrate $\varepsilon_r = 4.6$, the thickness of the silicon core $T = 50$–$80\ \mu\text{m}$, and the recess parameters $w_r = 1.6h_r$ to $1.8h_r$. To accommodate 150-μm-pitch ground–signal–ground coplanar probes, the distance between the two ground lines ($S + 2W$) is 200 μm and the ground line width, G, falls within the range of 100–400 μm. Second, the characteristic impedance, Z_0, for the normal finite CPW [2] can be expressed as

$$Z_0 = \frac{40\pi}{\sqrt{\varepsilon_{\text{eff}}}} \frac{K(k_1)}{K(k_1')} \tag{4.1a}$$

where

$$\varepsilon_{\text{eff}} = 1 + \frac{\varepsilon_r - 1}{2} \frac{K(k_2')K(k_1)}{K(k_2)K(k_1')} \left[1 + \frac{T}{2W} \frac{K(k_1)}{K(k_1')}\right] + \frac{T}{2W} \frac{K(k_1)}{K(k_1')} \left[2 + \frac{T}{2W} \frac{K(k_1)}{K(k_1')}\right] \tag{4.1b}$$

$$k_1 = \frac{S + 2W + 2G}{S + 2W} \sqrt{\frac{(S + 2W)^2 - S^2}{(S + 2W + 2G)^2 - S^2}} \tag{4.1c}$$

$$k_2 = \frac{\sinh\left[\frac{\pi(S+2W+2G)}{4H}\right]}{\sinh\left[\frac{\pi(S+2W)}{4H}\right]} \sqrt{\frac{\sinh\left[\frac{\pi(S+2W)}{4H}\right] - \sinh\left(\frac{\pi S}{4H}\right)}{\sinh\left[\frac{\pi(S+2W+2G)}{4H}\right] - \sinh\left(\frac{\pi S}{4H}\right)}} \tag{4.1d}$$

$$k_n' = \sqrt{1 - k_n^2}, \quad n = 1, 2 \tag{4.1e}$$

where K is the complete elliptic integrals of the first kind and ε_{eff} is the effective relative permittivity. The center conductor width, S, and the conductor separation, W, can be used to determine the required characteristic impedance. The dimension values of the center conductor width, S, the ground width, G, and the conductor separation, W, can be determined using three-dimensional (3D) finite-element method (FEM) simulation.

Figure 4.2a shows the electric field (E-field) distribution of the Si-core CPW at 20 GHz. The electric field radiates from the center signal conductor to the two ground conductors. The dominant mode of the transmission line is quasi-transverse

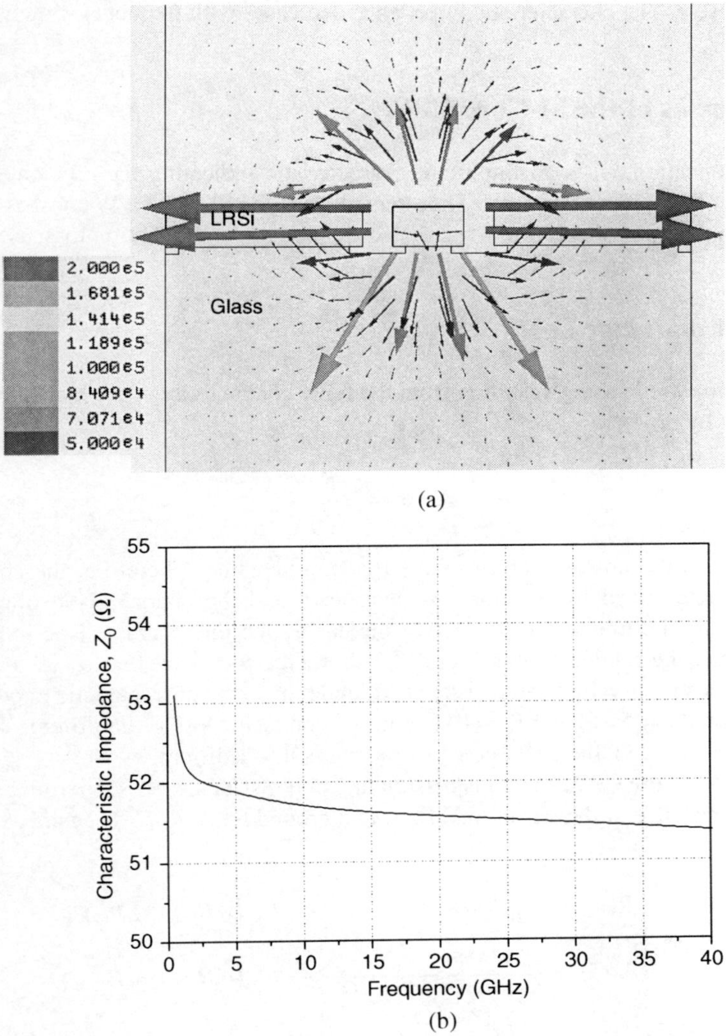

Fig. 4.2 (**a**) Vector E-field distribution and (**b**) characteristic impedance of the Si-core CPW ($S/W/G = 110/45/300$ μm, $h_r = 12$ μm, $w_r = 20$ μm). Copyright/used with permission of/courtesy of Institute of Physics and IOP Publishing Limited

electromagnetic (TEM) mode, which is similar to the conventional thin film CPW transmission line. It is found that there is strong penetration of the E-field into the substrate and the silicon core. Therefore, the material properties of the substrate and the core have significant effects on the performance of the Si-core CPW. The simulated characteristic impedance versus frequency of a Si-core CPW is shown in Fig. 4.2b. The characteristic impedance of this Si-core CPW is 51.3–52 Ω when the dimension values of CPW *S*/*W*/*G* are equal to 110/45/300 μm, $h_r = 12$ μm, $w_r = 20$ μm, $T = 50$ μm, $H = 500$ μm, $\varepsilon_r = 4.6$, $t_t = 1$ μm, and $t_s = 0.5$ μm, respectively. The characteristic impedance decreases with frequency slowly.

4.2 Losses of the Si-Core CPW

The attenuation, α, is an important characteristic indicating the efficiency of the coplanar lines' transmit power. The attenuation of the Si-core CPW consists of three losses, namely conductor loss, α_c, dielectric loss, α_d, and radiation loss, α_r.

4.2.1 Conductor Loss

The conductor loss, α_c, resulting from the finite conductivity of the metal, in Np/cm is given by

$$\alpha_c = \frac{R}{2Z_0} \tag{4.2}$$

where R is the unit resistance of the transmission line. Therefore, the conductor loss is determined by the line unit resistance and the characteristic impedance. Decreasing the line unit resistance or increasing the line characteristic impedance can reduce the conductor loss. Figure 4.3 shows the conductor loss, α_c, as a function of the transmission line unit resistance, R, under different characteristic impedances, Z_0. When $Z_0 = 50$ Ω and $R = 40$ Ω/cm, the conductor loss is 3.5 dB/cm. When Z_0 decreases to 42 Ω, the conductor loss increases to 4.1 dB/cm.

In the Si-core CPW, the transmission line unit resistance, R, is determined by the metal resistance of the signal line, $R_{s,m}$, and ground lines, $R_{g,m}$. They are expressed as

$$R_{s,m} = \begin{cases} R_{s,dc} = \frac{1}{\sigma(2t_sT + t_tS)} & (\text{if } R_s(f) \le R_{s,dc}) \\ R_s(f) = \frac{1}{\sigma\delta_s[2T(1-e^{-t_s/\delta_s}) + S(1-e^{-t_t/\delta_s})]} & (\text{if } R_s(f) > R_{s,dc}) \end{cases} \tag{4.3a}$$

$$R_{g,m} = \begin{cases} R_{g,dc} = \frac{1}{\sigma(2t_sT + t_tG)} & (\text{if } R_g(f) \le R_{g,dc}) \\ R_g(f) = \frac{1}{\sigma\delta_s[2T(1-e^{-t_s/\delta_s}) + G(1-e^{-t_t/\delta_s})]} & (\text{if } R_g(f) > R_{g,dc}) \end{cases} \tag{4.3b}$$

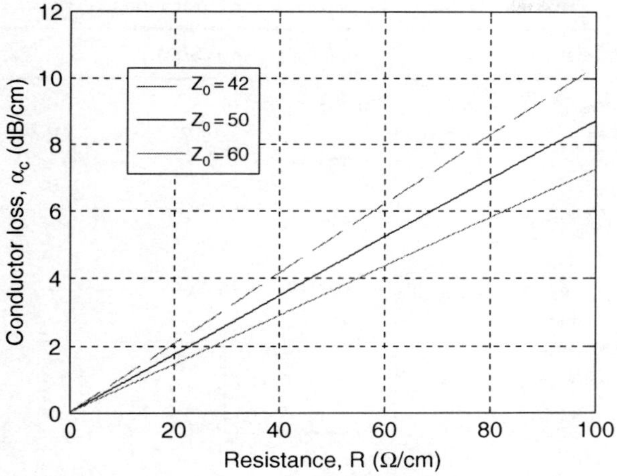

Fig. 4.3 Conductor loss with respect to different line unit series resistances and characteristic impedances

where T and S are the thickness and the width of the signal line; t_s and t_t are the metal thickness on the top and sidewalls of the transmission line, as shown in Fig. 4.1b; σ is the conductivity of the metal and δ_s is the skin depth, which is determined by Eq. (2.5). At low-frequency range, the resistance is determined by the DC resistance, $R_{s,dc}$ and $R_{g,dc}$. However, when the frequency increases, the current is crowded outside the surface of the conductor. The line resistance is determined by the $R_s(f)$ and $R_g(f)$ at high-frequency range.

4.2.2 Dielectric Loss

The dielectric loss, α_d, is due to the displacement current in the guiding medium of the transmission line. The dielectric loss, α_d, in Np/cm is given by [3]

$$\alpha_d = \frac{1}{2} G Z_0 = \frac{\omega\sqrt{\mu_0 \varepsilon_0 \varepsilon_r}\tan\delta}{2} \times \frac{1}{100} \tag{4.4a}$$

where $\tan\delta$ is the loss tangent of the guiding medium. The loss tangent specifies the lossy nature of the dielectric material and is defined as

$$\tan\delta = \frac{\sigma_d}{\omega \varepsilon_0 \varepsilon_r} \tag{4.4b}$$

where σ_d is the electrical conductivity of the dielectric material. Table 4.1 provides the material properties of four dielectric materials used in this study and their dielectric loss at 20 GHz. The dielectric loss is air, glass, high-resistivity silicon (HRSi), and low-resistivity silicon (LRSi).

Table 4.1 Dielectric properties of the fabrication materials

Material	ε_r	tan δ	σ_d (S/m)	α_d (dB/cm)
Glass	4.6	0.003	–	0.12
HRSi	11.9	–	0.05	0.24
LRSi	11.9	–	100	474.30
Air	1.0	0	–	0

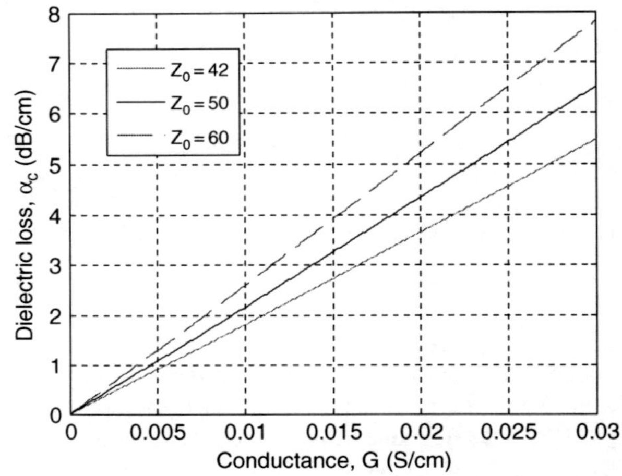

Fig. 4.4 Conductor loss with respect to different line unit shunt conductances and characteristic impedances

Figure 4.4 shows the dielectric loss, α_d, as a function of the unit shunt conductance, G, with the characteristic impedance, Z_0, as a parameter. The dielectric loss increases with the unit conductance and the characteristic impedance. When $Z_0 = 50\ \Omega$ and $G = 0.005$ S/cm, the dielectric loss is 1.08 dB/cm. When G increases to 0.02 S/cm, the dielectric loss increases to 4.34 dB/cm.

4.2.3 Radiation Loss

In addition to the conductor loss and the dielectric loss, radiation from unwanted parasitic modes and coupling of power to surface waves contribute to radiation loss, α_r. The parasitic mode in the CPW is the odd mode with anti-phase voltages in the two slots, which is excited at the discontinuities. Radiation from this mode can be minimized by maintaining the symmetrical structure of the circuits and avoiding its excitation by using air bridges or bond wires at periodic intervals. For the CPW with a finite substrate thickness, the surface wave modes of the substrate are also responsible for the leakage. Surface waves are the guided modes of a dielectric slab with finite thickness. Excitation of surface modes depends on the polarizations and symmetries of the transmission line.

Figure 4.5 shows the radiation loss, α_r, as a function of frequency, f, with the substrate permittivity, ε_r, as a parameter. In this scenario, $H = 500$ μm, $S = 110$ μm, $W = 45$ μm, and $G = 300$ μm. It shows that the radiation loss is substantially affected by the permittivity of the substrate at high-frequency range. The higher the permittivity, the faster the increase of radiation loss with frequency. For instance, when $\varepsilon_r = 4.6$ (glass) at 40 GHz, the radiation loss of the substrate is 0.006 dB/cm. When $\varepsilon_r = 11.9$, the radiation loss of the substrate increases to 0.046 dB/cm. However, compared with the conductor loss and the dielectric loss, the radiation loss is negligible when frequency is below 40 GHz. The radiation losses dominate for frequencies over 200 GHz for coplanar lines with dimensions of the order of a few tens of microns, as demonstrated in the experiments [4].

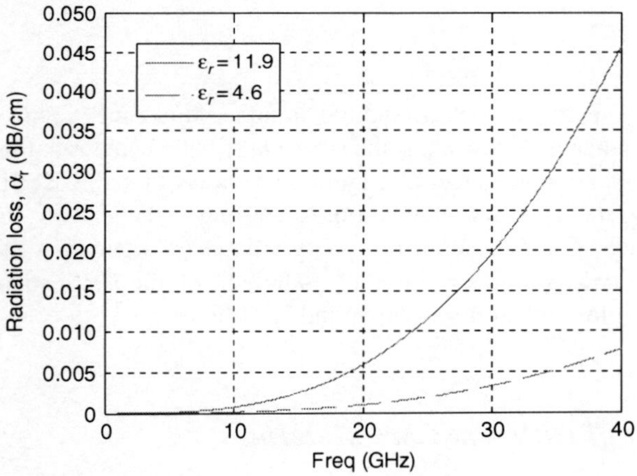

Fig. 4.5 Radiation loss curves with respect to different frequencies and substrate permittivities

4.3 Effect of Material Properties and Fabrication Processes

Factors affecting the RF properties of the Si-core CPW are studied in this section. These factors include the substrate material, the core material, and the fabrication process.

4.3.1 Effect of Different Substrate Materials

Figure 4.6 shows the simulation results of the Si-core CPW on glass and on high-resistivity silicon substrate. The core material is the low-resistivity silicon with

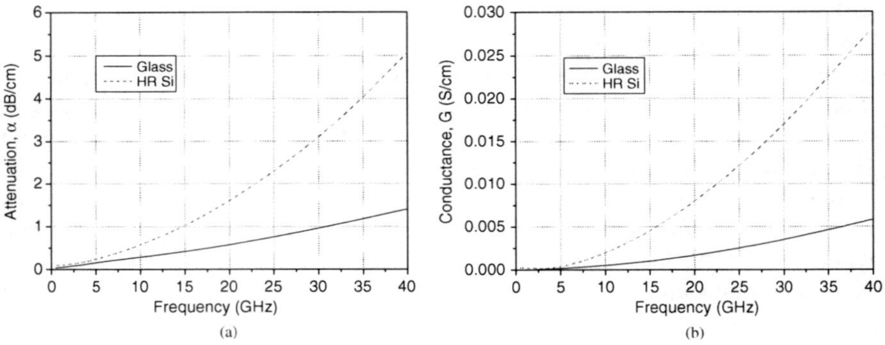

Fig. 4.6 Comparison of simulation results between the Si-core CPW on glass and HRSi substrate ($S/W/G = 110/45/300$ μm, $h_r = 12$ μm, and $w_r = 20$ μm, LRSi core): (**a**) attenuation and (**b**) shunt conductance G

$\rho = 1\ \Omega\text{cm}$. The attenuation, α, and the shunt conductance, G, are determined by S-parameters, since the substrate is the main contributory factor to the dielectric loss through the shunt conductance, G. Figure 4.6a shows that a 50-Ω Si-core CPW on the glass substrate has lower attenuation. According to the analysis in Section 4.2.2, the conductance, G, of the high-resistivity silicon is larger than the glass. When the frequency increases, the shunt conductance, G, on the HRSi substrate increases faster than the glass substrate as shown in Fig. 4.6b.

4.3.2 *Effect of Different Core Materials*

The effect of the core material is shown in Fig. 4.7. In the simulation models, the substrate is 500-μm-thick glass. $S/W/G$ are 110/45/300 μm. h_r, w_r, and T are 12, 20, and 50 μm, respectively. The characteristic impedance, Z_0, increases slightly from 48 to 50 Ω when the core material changes from the low-resistivity silicon to the high-resistivity silicon. The LRSi-core CPW has higher attenuation than the HRSi-core CPW. This difference becomes larger when the frequency increases.

The $RLGC$ parameters of the Si-core CPW transmission lines have been extracted and compared in Fig. 4.8. Figure 4.8a shows that when the core resistivity of the Si-core CPW increases, the unit inductance, L, increases by only a small amount and the unit capacitance, C, decreases by only a small amount. Therefore, the characteristic impedance which is determined by $\sqrt{L/C}$ at high frequencies (>5 GHz) increases. The unit resistance, R, of the LRSi-core CPW is slightly lower than the HRSi-core CPW when the frequency is below 3 GHz as shown in Fig. 4.8b. The shunt conductance, G, of the LRSi-core CPW is significantly larger than that of the HRSi-core CPW, especially at high frequencies. This results in higher dielectric loss

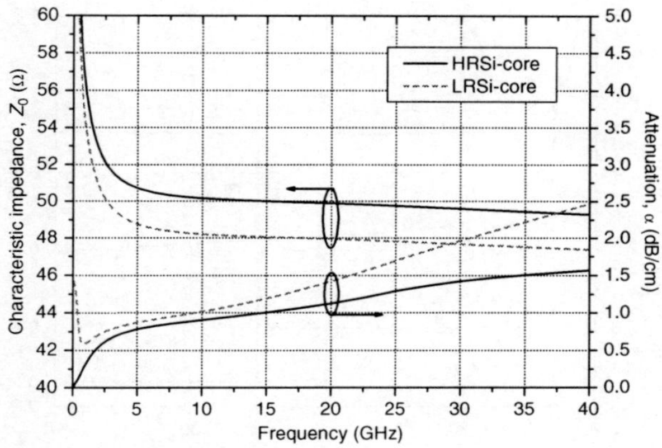

Fig. 4.7 Comparison of the RF properties of Si-core CPW with different core resistivities ($S/W/G = 110/45/300$ μm, $h_r = 12$ μm, and $w_r = 20$ μm). Copyright/used with permission of/courtesy of Institute of Physics and IOP Publishing Limited

for LRSi-core CPW than HRSi-core CPW. Therefore, the attenuation of LRSi-core CPW is larger than the HRSi-core CPW.

4.3.3 *Effect of Different Fabrication Processes*

Besides the geometrical dimensions, the types of the substrate material, and the core material, the RF properties of the Si-core CPW are also affected by the variation in fabrication process. These include the thickness of the transmission line, T, the etching depth of the glass recesses, h_r, and the undercut width, w_r, as shown in Fig. 4.9.

The characteristic impedance, Z_0, of the Si-core CPW is reduced from 55 to 44 Ω when the transmission line thickness, T, increases from 50 to 80 μm, as shown in Fig. 4.10a. This is because when the line unit capacitance, C, increases the characteristic impedance reduces. Figure 4.10b shows that the attenuation slightly decreases with an increase in the conductor thickness. This is because the surface current is distributed in the larger area, which leads to the lower conductor loss.

Figure 4.11a, b shows that when the substrate is etched more, with larger w_r and h_r, the characteristic impedance of the Si-core CPW increases from 47.6 to 51.4 Ω, and the attenuation decreases slightly. This is because when the connection of the conductor to the substrate is smaller, there is smaller wave penetration into the substrate, which leads to smaller unit capacitance and smaller dielectric loss.

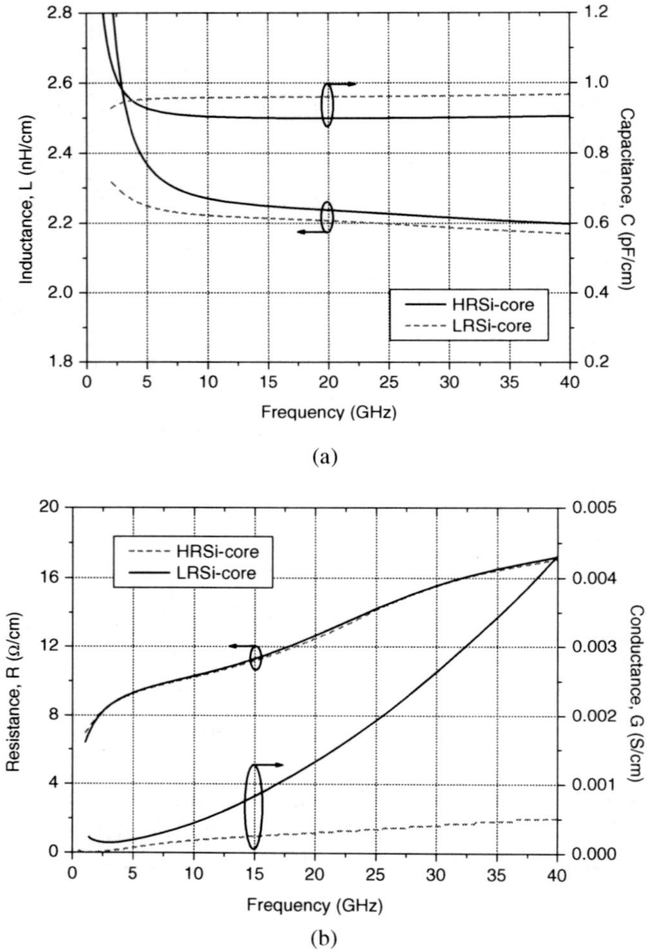

Fig. 4.8 Comparison of the distributed *RLGC* parameters of the Si-core CPW on glass substrate: (**a**) capacitance and inductance and (**b**) conductance and resistance

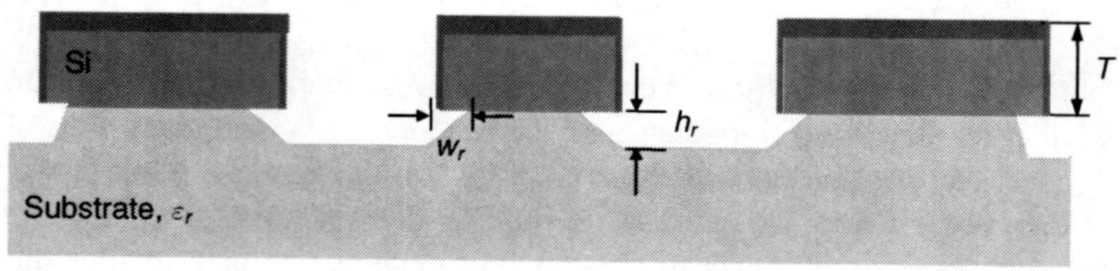

Fig. 4.9 Schematic of cross-sectional view of the Si-core CPW transmission line

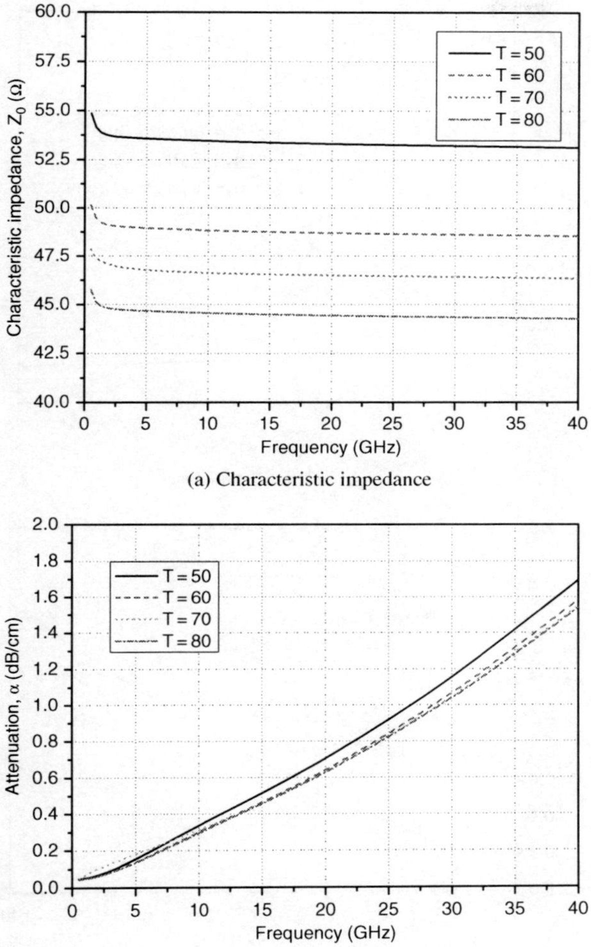

(a) Characteristic impedance

(b) Attenuation

Fig. 4.10 Simulation results of HRSi-core CPWs with various conductor thicknesses, T ($S/W/G$ = 110/45/300 μm)

The characteristic impedance of the Si-core CPW is significantly affected by the variation in the fabrication process. Therefore, the process control is important for the RF properties of the Si-core CPW, in particular the characteristic impedance, which determines the impedance matching of the RF circuits.

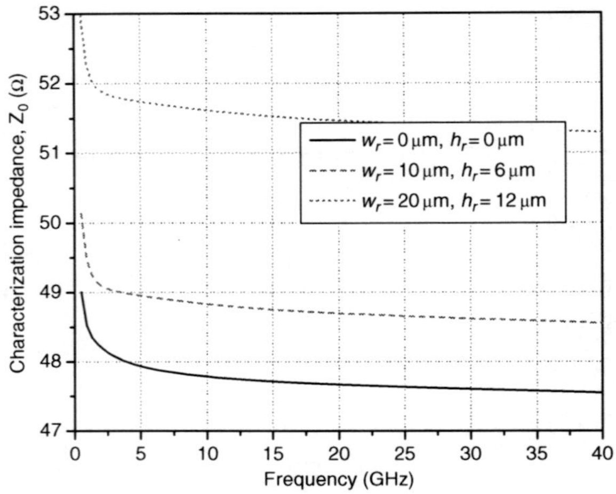

(a) Characteristic impedance Z_0

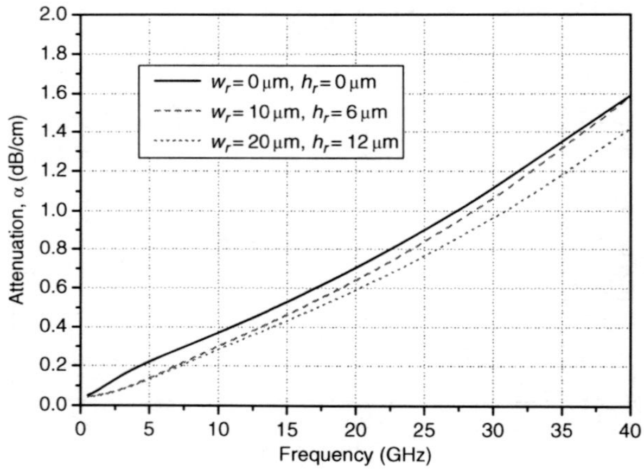

(b) Attenuation

Fig. 4.11 Simulation results of HRSi-core CPWs with various undercut dimensions ($S/W/G$ = 110/45/300 μm). Copyright/used with permission of/courtesy of Institute of Physics and IOP Publishing Limited

4.4 Experimental Results and Discussions

The RF responses of the Si-core CPW are measured using the HP 8510C Vector Network Analyzer with tungsten-tip 150-μm-pitch Cascade Microtech ground–signal–ground coplanar probes. The system is calibrated using the standard short-open-load-through (SOLT) on-wafer calibration technique. A 5-mm plastic plate is placed between the probe chuck and the sample to remove higher order modes of propagation. All experiments are performed in the room environment without any

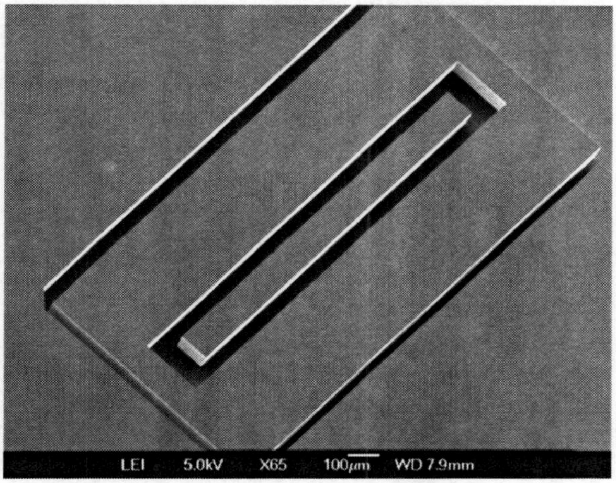

Fig. 4.12 SEM image of a 1-mm-long Si-core CPW transmission line

packaging. A SEM image of a Si-core CPW coated with 1-μm-thick gold (Au) is shown in Fig. 4.12. The thickness of the transmission line, *T*, is 62 μm and the length, *L*, is 1 mm.

4.4.1 Comparison Between Simulation Results and Measurement Results

Figure 4.13 shows the measurement results with the post-simulation results of a 1-mm-long HRSi-core CPW on glass. The *S*/*W*/*G* are 110/45/400 μm, respectively. The thickness of the transmission line is 62 μm. The glass cavity parameters, h_r and w_r, are about 6 and 10 μm, respectively. In the post-simulation model of HFSS, all geometrical parameters are set based on the experimental values. For example, the metal layer spread on glass has a thickness of 1 μm on the top and 0.3 μm on the sidewalls. Figure 4.13 shows that the simulation results are in good agreement with the measurement results. The characteristic impedance of the glass substrate, HRSi-core CPW is about 47 Ω when the frequency is above 10 GHz. The attenuation increases with frequency, which is 4 dB/cm at 25 GHz.

The unit *RLGC* parameters of simulation results and measurement results are compared in Fig. 4.14. The measured unit *RLGC* values match well with the simulated results. The unit inductance, *L*, and the unit capacitance, *C*, are kept constant from 10 to 25 GHz, which are 2.85 nH/cm and 1.3 pF/cm, respectively, as shown in Fig. 4.14a. The resistance, *R*, and shunt conductance, *G*, increase with the frequency, as shown in Fig. 4.14b. Table 4.2 compares simulated and measured attenuation due to the unit resistance, *R*, and the shunt conductance, *G*, of the HRSi-core CPW on glass at 20 GHz. The measured conductance, *G*, is 0.0017 S/cm, resulting in the

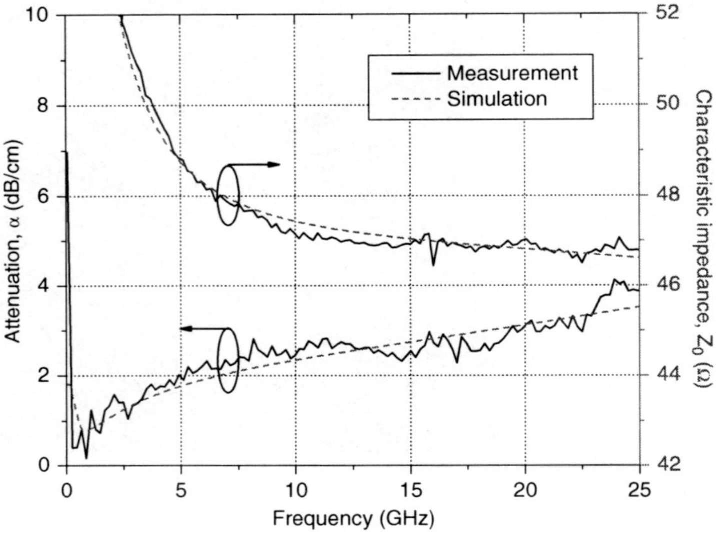

Fig. 4.13 Comparison of the measurement and the simulation results of an HRSi-core CPW on the glass substrate ($T = 62\,\mu$m, $h_r \approx 6\,\mu$m, $w_r \approx 10\,\mu$m, and 1-μm-thick Au coating). Copyright/used with permission of/courtesy of Institute of Physics and IOP Publishing Limited

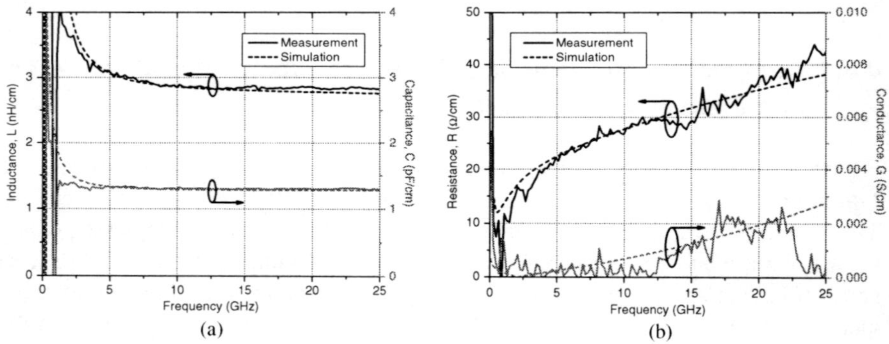

Fig. 4.14 Comparison of the measured and the simulated *RLGC* parameters of an HRSi-core CPW on a glass substrate: **(a)** *L* and *C*, and **(b)** *R* and *G*. Copyright/used with permission of/courtesy of Institute of Physics and IOP Publishing Limited

dielectric loss, α_d, of 0.34 dB/cm. The measured resistance, *R*, of 37.6 Ω/cm contributes to the conductor loss, α_c, of 3.48 dB/cm. The conductor loss is nine times higher than the dielectric loss. Therefore, the large unit resistance, *R*, is the dominant factor of the attenuation of the Si-core CPW.

By investigating the fabrication process in Chapter 5, the large unit resistance is due to low-quality metal deposition. The metal is deposited using E-beam evaporation. Generally, the E-beam evaporation has poor step coverage. The metal

Table 4.2 Comparison of the simulated and measured attenuation contributed by the unit resistance and the conductance of the glass-based HRSi-core CPW at 20 GHz

	Z_0 (Ω)	R (Ω/cm)	G (S/cm)	α_c (dB/cm)	α_d (dB/cm)	α (dB/cm)
Measurement	47.0	37.6	0.0017	3.48	0.34	3.82
Simulation	46.8	35.2	0.002	3.27	0.4	3.67

coated on sidewalls is found to be only 1/3 of the metal coated on the top surface. Therefore, only 3000–4000 Å gold can be coated on sidewalls when 1-μm-thick gold is deposited on the top of silicon structures. On the other hand, the surface current of the proposed Si-core CPW is mainly concentrated on sidewalls of the apertures, as shown in Fig. 4.2a. Therefore, more loss is caused by the thin metal coating of sidewalls. The increase of the shunt conductance, G, caused by the metal spreading on the substrate during E-beam evaporation is very small. Compared to the conductor loss, the dielectric loss is negligible.

4.4.2 Effect of Geometrical Parameters

The performance of the Si-core CPW depends on the geometrical parameters significantly, including the signal width, S, the ground width, G, and the conductor separation, W. Four 1-mm-long LRSi-core CPW with various $S/W/G$ are fabricated on glass and measured. Their $S/W/G$ parameters are summarized in Table 4.3. Note that $S{+}2W$ and G are the same for designs A, B, and C. Design D is optimized using Ansoft HFSS, whose simulated characteristic impedance is 50 Ω. The other parameters are $T \approx 50$ μm, $h_r \approx 6$ μm, and $w_r \approx 10$ μm.

Table 4.3 Geometrical parameters of four 1-mm-long Si-core CPWs

Design no.	Signal, S (μm)	Space, W (μm)	Ground, G (μm)	Sum, $S + 2W$ (μm)
A	680	10	150	700
B	300	200	150	700
C	60	320	150	700
D	110	45	400	200

The characteristic impedance, Z_0, and the attenuation, α, extracted from measured S-parameters are shown in Fig. 4.15a, b. The variation of the characteristic impedance of the Si-core CPW is larger when the frequency is below 5 GHz, indicating slow-wave mode propagation. However, when the frequency rises above 5 GHz, the characteristic impedance, Z_0, reduces gradually for all structures, indicating that the quasi-TEM mode propagation is supported above 5 GHz. In the quasi-TEM mode propagation, the characteristic impedance of the transmission line is inversely

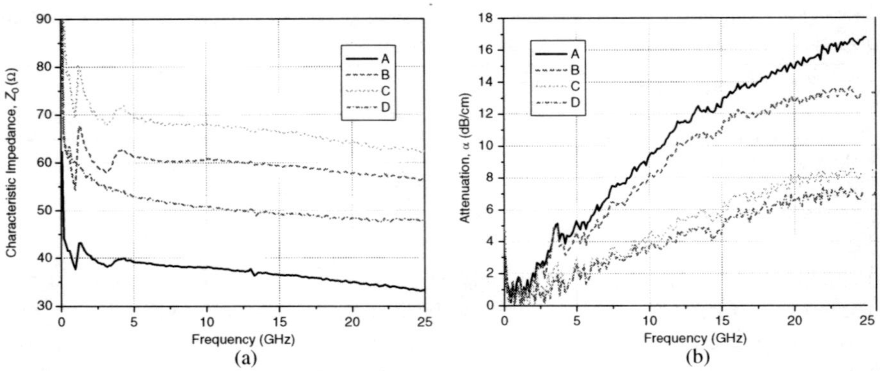

Fig. 4.15 Measured results of 1-mm-long LRSi-core CPWs on glass with different dimensions: (**a**) characteristic impedance and (**b**) attenuation

proportional to the square root of the line capacitance. When the space between the signal line and the ground line, W, is smaller, the unit inductance, L, is less and the unit capacitance, C, is larger, which lead to smaller characteristic impedance. This tendency can be observed in designs A, B, and C. At high frequency, impedance matching is important to minimize signal oscillations that are caused by the reflection of the electromagnetic wave at the impedance discontinuity. Therefore, it is most desirable to design the transmission line with the characteristic impedance of 50 Ω. The characteristic impedance of design D is about 50 Ω at 5–25 GHz. Figure 4.15b shows that the attenuation of the Si-core CPW increases with the signal line width, S. When (S+2W) is a constant, the Si-core CPW with wider signal line shows larger attenuation. That is because more electromagnetic field leaks into the substrate, e.g., dielectric loss is larger. However, design D ($S/W/G$ = 110/45/400 μm) has lower attenuation than design C ($S/W/G$ = 60/320/150 μm). That is because the sum (S+2W) of design C is larger than the glass substrate thickness of 500 μm, whereas the sum (S+2W) of design D is less than 500 μm. This leads to larger frequency dispersion and radiation loss in design C as compared to design D. Therefore, geometrical parameters of the Si-core CPW can be optimized to obtain the desired characteristic impedance and, simultaneously, suppress the unwanted radiation loss and the frequency dispersion when designing the Si-core CPW.

4.4.3 Effect of Material Properties

Both the substrate material and the core material affect the RF performance of the Si-core CPW. The effect of the substrate material is illustrated in Figs. 4.16 and 4.17. For the SOI-based LRSi-core CPW, the parameters of the CPW transmission line

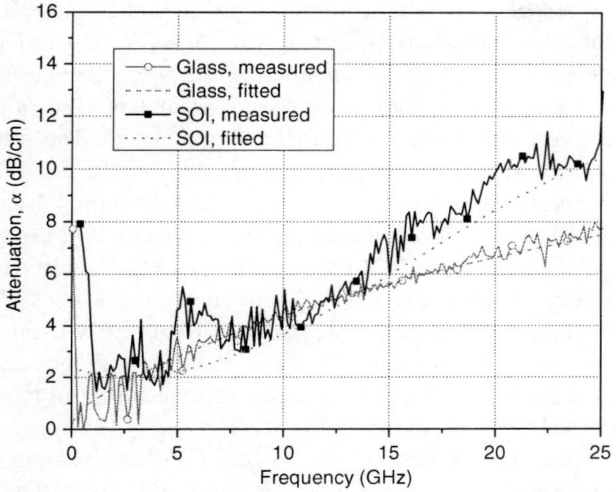

Fig. 4.16 Comparison of attenuation of LRSi-core CPW on SOI and glass substrate

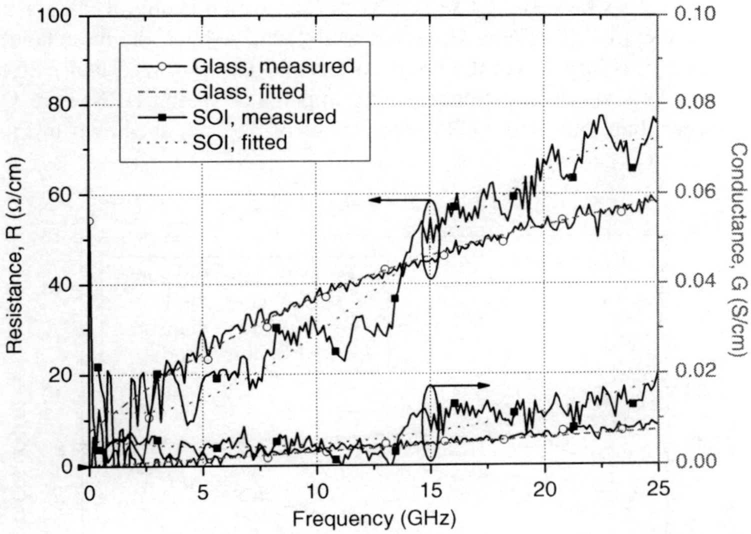

Fig. 4.17 Comparison of resistance and conductance of LRSi-core CPW on SOI and glass substrate

S/W/G are 66/67/100 μm. The substrate is high-resistivity silicon with the thickness of 500 μm. The core material is low-resistivity silicon with the thickness of 35 μm. Between the substrate and the device, silicon layer is a 2-μm-thick thermal oxide; 1.2-μm-thick Al is coated. For the glass-based LRSi-core CPW, *S/W/G* are

110/45/400 μm, respectively. The substrate is 500-μm-thick glass. The core material is 60-μm-thick low-resistivity silicon. $h_r \approx 6$ μm, $w_r \approx 10$ μm; 1-μm-thick gold is coated. The measured unit *RLGC* parameters, the characteristic impedance, and the attenuation are extracted from measured *S*-parameters. The fitted unit *RLGC* parameters are fitted from the measured *RLGC* parameters. The fitted attenuation and the fitted characteristic impedance are calculated from the fitted *RLGC* parameters, which agree well with the measured results. As in Section 4.3.1, the SOI-based CPW has larger shunt conductance than the glass-based CPW, especially at high-frequency range. Therefore, the SOI-based CPW has higher dielectric loss than the glass-based CPW, as shown in Fig. 4.17. For any frequency below 13 GHz, the resistance of the SOI-based CPW is less than the glass-based CPW due to thicker metal coating. However, when the frequency is above 13 GHz, both the resistance and the conductance of the SOI-based CPW arise abruptly, resulting in larger attenuation than the glass-based CPW, as shown in Fig. 4.16.

The effect of the core material of the Si-core CPW is illustrated in Figs. 4.18 and 4.19. The parameters of the transmission line *S/W/G* are 110/45/400 μm. The substrate is 500-μm-thick glass. $h_r \approx 6$ μm, $w_r \approx 10$ μm, and 1-μm-thick Au is deposited. The thickness of waveguides of Si-core CPWs is about 60 μm. The unit capacitance of the LRSi-core CPW is 1.33 pF/cm, which is slightly larger than the HRSi-core CPW of 1.30 pF/cm. However, as in Section 4.3.2, the inductance of the HRSi-core CPW is larger than the LRSi-core CPW. They are 2.85 and 2.26 nH/cm, respectively. As a result, the characteristic impedance of the HRSi-core CPW of 47 Ω is larger than that of the LRSi-core CPW of 41.2 Ω, as shown in Fig. 4.18.

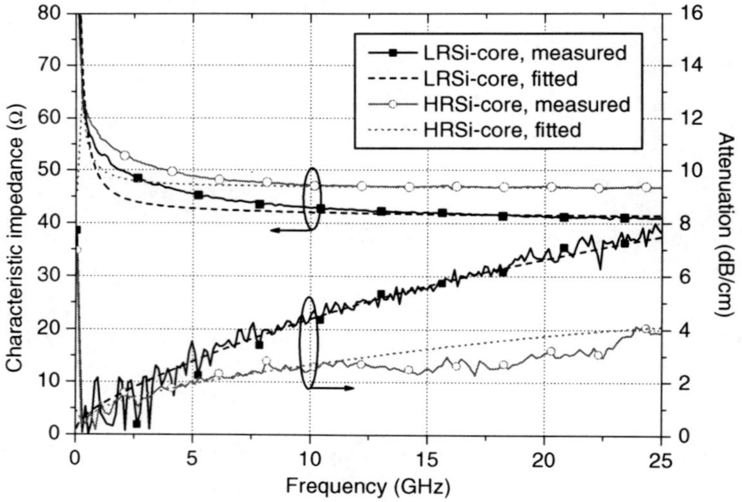

Fig. 4.18 Comparison of characteristic impedance and attenuation between the HRSi-core CPW and the LRSi-core CPW on glass. Copyright/used with permission of/courtesy of Institute of Physics and IOP Publishing Limited

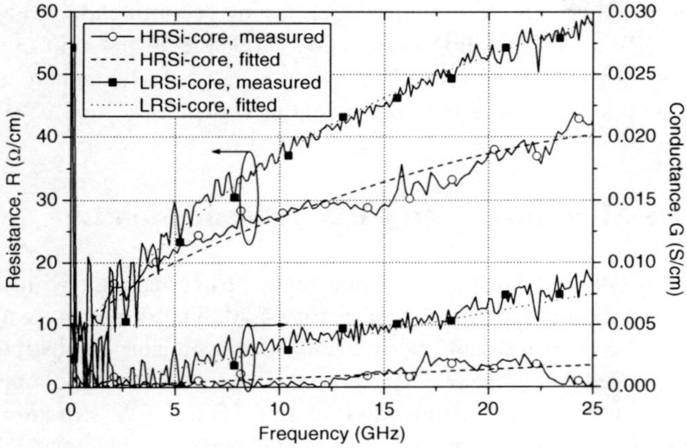

Fig. 4.19 Comparison of resistance and conductance between the HRSi-core CPW and the LRSi-core CPW on glass

The HRSi-core CPW has lower attenuation than the LRSi-core CPW, as shown in Fig. 4.19. That is because the unit shunt conductance, *G*, of the HRSi-core CPW is less than the LRSi-core CPW, resulting in lower dielectric loss. The unit resistance, *R*, of the HRSi-core CPW is significantly lower than the LRSi-core CPW, resulting in lower conductor loss.

Table 4.4 shows the comparison between the measured and fitted *RLGC* parameters, the characteristic impedance, the attenuation between the SOI-based LRSi-core CPW, the glass-based LRSi-core CPW, and the glass-based HRSi-core CPW at 20 GHz. The glass-based HRSi-core CPW has the lowest resistance, conductance, and attenuation. The conductor loss is the dominant factor of the attenuation. Thicker metal should be coated to reduce the attenuation of the Si-core CPW.

Table 4.4 Comparison of measured and fitted *RLGC* values and attenuations of Si-core CPW with different substrates and core materials at 20 GHz

	SOI-based LRSi core		Glass-based LRSi core		Glass-based HRSi core	
CPW parameters	Measured	Fitted	Measured	Fitted	Measured	Fitted
R (Ω/cm)	69.7	65.1	53	53	37.6	37.4
L (nH/cm)	4.1	4.1	2.26	2.29	2.85	2.83
G (S/cm)	0.012	0.011	0.007	0.006	0.0017	0.0014
C (pF/cm)	1.95	1.95	1.33	1.33	1.30	1.30
Z_0 (Ω)	45.9	46.2	41.2	41.5	47	46.7
α_c (dB/cm)	6.59	6.11	5.58	5.55	3.48	3.48
α_d (dB/cm)	2.39	2.37	1.25	1.08	0.34	0.28
α (dB/cm)	10.04	8.48	6.88	6.63	3.82	3.76

Based on this study, the following guidelines are recommended to achieve low-loss Si-core CPW: (i) depositing sufficiently thick metal and improving the step coverage of the metal deposition; (ii) using low-loss substrate, such as glass; and (iii) using high-resistivity silicon as the core material.

4.5 Surface-Micromachined CPW Transmission Line

A coplanar waveguide (CPW) consists of a center strip conductor with semi-infinite ground planes on either side as shown in Fig. 4.20. The dimensions of the center strip, the gap, the thickness, and permittivity of the dielectric substrate are determined by the effective dielectric constant ε_{eff}, the characteristic impedance Z_0, and the attenuation, α, of the transmission line. This CPW structure supports a quasi-TEM mode of propagation. The CPW offers several advantages over the conventional microstrip line [5–12] as (1) it simplifies fabrication, (2) easy shunt as well as series surface mounting of active and passive devices, (3) it eliminates the need for wraparound and via holes, and (4) it reduces radiation loss.

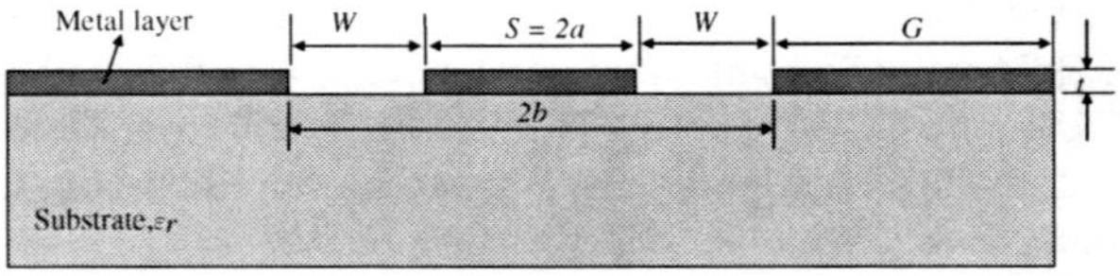

Fig. 4.20 Schematic of the coplanar waveguide (CPW) structure

4.5.1 Characteristic Impedance and Effective Dielectric Constant

The cross-sectional view of CPW is shown in Fig. 4.20. The CPW center strip conductor width S is equal to $2a$ and the distance of separation between the two ground planes is $2b$. Consequently, the slot width W is equal to $b - a$. The CPW conductors and the dielectric substrates are assumed to have perfect conductivity and relative permittivity, respectively. Hence, the structure is considered to be low loss and the dielectric substrate materials are considered to be isotropic.

In this section, the mathematical derivations of ε_{eff} and Z_0 using the conformal mapping techniques are presented. Two assumptions are made. First, the conductor thickness t is zero and, second, the magnetic walls are present along all the dielectric boundaries including the CPW slots. The CPW is then divided into several partial regions and the electric field is considered to exist successively only in the partial region. In this manner, the capacitance of each partial region is determined separately. The total capacitance is the sum of the partial capacitances [13]. The total capacitance C_{CPW} can be expressed as [14]

$$C_{\text{CPW}} = C_1 + C_{\text{air}} \tag{4.5}$$

where

$$C_1 = 2\varepsilon_0 (\varepsilon_{r1} - 1) \frac{K(k_1)}{K(k_1')} \tag{4.6a}$$

where K is complete elliptic integral of first kind and the modulus of the elliptic integrals $K(k_1)$ and $K(k_1')$ are given by

$$k_1 = \frac{\sinh(\pi S/4h_1)}{\sinh\{[\pi(S+2W)]/4h_1\}} \tag{4.6b}$$

and

$$k_1' = \sqrt{1 - k_1{}^2} \tag{4.6c}$$

where h_1 is the height of the substrate, S is the CPW central strip conductor width, and W is the slot width.

The capacitance C_{air} is given by

$$C_{\text{air}} = 2\varepsilon_0 \frac{K(k_3)}{K(k_3')} + 2\varepsilon_0 \frac{K(k_4)}{K(k_4')} \tag{4.7a}$$

where

$$k_3 = \frac{\sinh(\pi S/4h_3)}{\sinh\{[\pi(S+2W)]/4h_3\}} \tag{4.7b}$$

$$k_4 = \frac{\sinh(\pi S/4h_4)}{\sinh\{[\pi(S+2W)]/4h_4\}} \tag{4.7c}$$

$$k_3' = \sqrt{1 - k_3{}^2} \tag{4.7d}$$

$$k_4' = \sqrt{1 - k_4{}^2} \tag{4.7e}$$

Substituting $h_3 = h_4 = \infty$ into Eqs. (4.7a), (4.7b), and (4.7c), we have

$$C_{\text{air}} = 4\varepsilon_0 \frac{K(k_0)}{K(k_0')} \tag{4.8a}$$

and

$$k_3 = k_4 = k_0 = \frac{S}{S+2W} \tag{4.8b}$$

Substituting Eqs. (4.6a) and (4.8a) into Eq. (4.5), C_{CPW} is expressed as

$$C_{\mathrm{CPW}} = 2\varepsilon_0\left(\varepsilon_{\mathrm{r1}} - 1\right)\frac{K\left(k_1\right)}{K\left(k_1'\right)} + 4\varepsilon_0\frac{K\left(k_0\right)}{K\left(k_0'\right)} \tag{4.9}$$

Under the quasi-static approximation, $\varepsilon_{\mathrm{eff}}$ is defined as [14]

$$\varepsilon_{\mathrm{eff}} = \frac{C_{\mathrm{CPW}}}{C_{\mathrm{air}}} \tag{4.10}$$

Substituting Eqs. (4.8a) and (4.9) into Eq. (4.10) is expressed as

$$\varepsilon_{\mathrm{eff}} = 1 + \frac{\varepsilon_{\mathrm{r1}} - 1}{2}\frac{K(k_1)}{K(k_1')}\frac{K(k_0')}{K(k_0)} \tag{4.11}$$

The characteristics impedance Z_0 is defined as [14]

$$Z_0 = \frac{1}{cC_{\mathrm{air}}\sqrt{\varepsilon_{\mathrm{eff}}}} \tag{4.12}$$

Substituting Eq. (4.8a) into Eq. (4.12) gives

$$Z_0 = \frac{1}{4c\varepsilon_0\sqrt{\varepsilon_{\mathrm{eff}}}}\frac{K\left(k_0'\right)}{K\left(k_0\right)} \tag{4.13a}$$

where $c = 3 \times 10^8$ m/s and $\varepsilon_0 = 8.854 \times 10^{-12}$ F/m, then Eq. (4.13a) is simplified as

$$Z_0 = \frac{30\pi}{\sqrt{\varepsilon_{\mathrm{eff}}}}\frac{K\left(k_0'\right)}{K\left(k_0\right)} \tag{4.13b}$$

In Fig. 4.21, the characteristics impedance curve is plotted with different slot and conductor width values. The characteristic impedances for different dielectric substrates are also included. For the same ratio, $S/(S+W)$ of the slot conductor, the substrate with a lower dielectric constant has a higher characteristic impedance.

The vector current distribution of the coplanar waveguide is calculated using the method of moment software IE-3D as shown in Fig. 4.22. In CPW transmission line, the electric field radiates from the center signal conductor to the two ground conductors and into the substrate where the dominant mode for this kind of transmission line is quasi-transverse electromagnetic (TEM) mode, while the magnetic field lines enclose around only the center signal conductor. The electric current calculated from IE-3D is 250.3 amp/m which is dominant compared to magnetic current of 1×10^{-5} V/m. This is the reason for the strong penetration of the E-field from the signal line to ground plane and into the substrate. In Fig. 4.22, the bar indicates the different color representations in decibels for different attenuation levels. It is noticed that the signal passes with minimum attenuation of 0.1–0.3 dB from input to the output port.

The material property of the substrate has significant effects on the performance of the CPW transmission line.

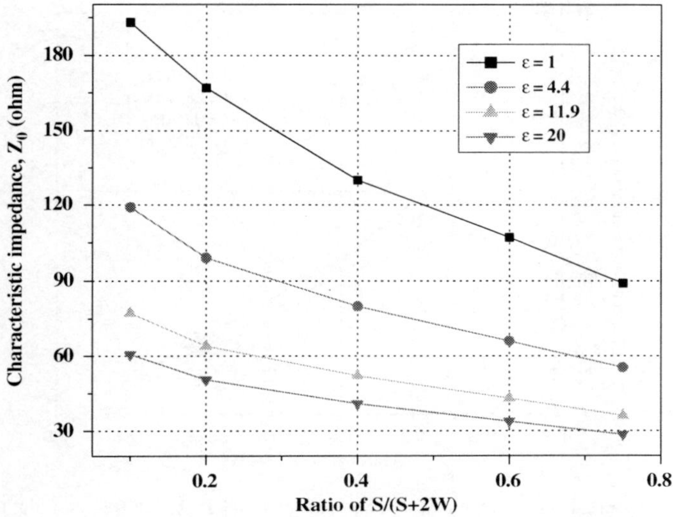

Fig. 4.21 Characteristic impedance of coplanar waveguide versus different dielectric constant of the substrate material

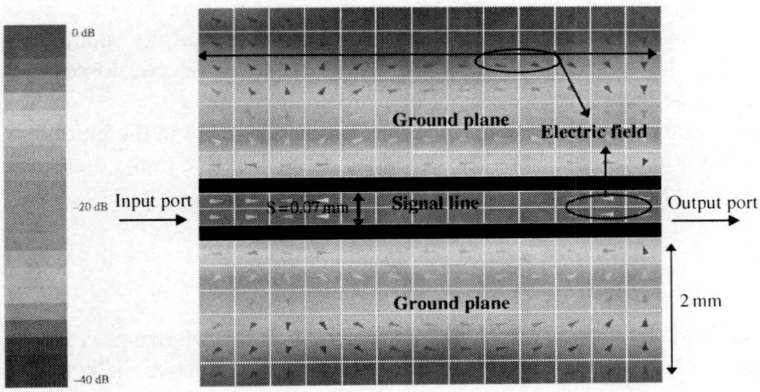

Fig. 4.22 Simulation result of the current distribution in the coplanar waveguide. Copyright/used with permission of/courtesy of IEEE

Figure 4.23 shows that for fixed $(S+2W)$ and aspect ratio $S/(S+2W)$ the Z_0 decreases as the thickness of the metal t increases for silicon substrate. For the fixed aspect ratio $S/(S+2W)$ of 0.1 and the metal thickness 0.5 μm, the characteristic impedance is 78.37 Ω. On the other hand, when the metal thickness is 10 μm the characteristic impedance is 72.54 Ω; for the fixed aspect ratio of 0.75, the characteristic impedance is 37.5 Ω for 10 μm and 38.4 Ω for 2-μm metal thickness.

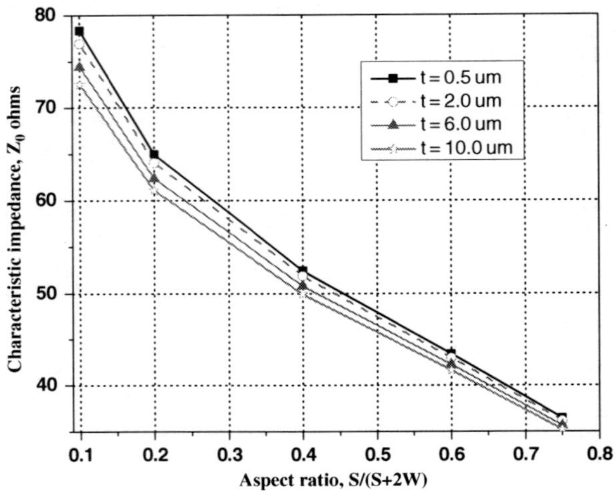

Fig. 4.23 Simulation result of characteristic impedance of CPW with different metal thicknesses

4.5.2 Losses of CPW Structure

For the CPW structure, two major factors are considered for the transmission line losses. The total CPW attenuation is expressed as the sum of the attenuation due to the dielectric losses in the substrate and that due to the conductor losses in the strip and the ground planes.

The dielectric loss, α_d is due to the displacement current in the guiding medium of the CPW transmission line. The attenuation constant due to the dielectric loss is expressed as [15]

$$\alpha_d = \frac{\pi}{\lambda_0}\frac{\varepsilon_r}{\sqrt{\varepsilon_{eff}}}q\tan\delta_e(\mathrm{N/m}) \tag{4.14}$$

where λ_0 is the free space wavelength in meters, ε_r is the relative permittivity of the substrate,tan δ_e is the dielectric loss tangent, ε_{eff} is the effective dielectric constant. The loss tangent tan δ_e remains constant with respect to the frequency for different dielectric materials. However, the dielectric loss α_d increases linearly with the frequency. The loss tangent is an important factor to be taken into consideration as there are material losses due to bulk conductivityσ. The loss tangent is defined as

$$\tan\delta_e = \frac{\sigma}{\omega\varepsilon_0\varepsilon_{eff}} \tag{4.15}$$

The material with high loss tangent results in a higher dielectric loss. The material or substrate properties are listed in Table 4.5.

The attenuation constant due to the conductor loss α_c in the central strip conductor and the ground planes of the CPW is given below. In deriving this expression,

Table 4.5 Properties of the dielectric materials

Materials	ε_r	$\tan\delta_e$
High-resistivity (HR) silicon	11.9	0.025 S × m
Low-resistivity (LR) silicon	11.9	100 S × m
Air	1.0	0.0 S × m

the thickness t of the CPW conductors is assumed to be greater than the skin depth δ in the metal. The conductor loss α_c is defined as [15]

$$\alpha_c = \frac{R_c + R_g}{2Z_0} \tag{4.16}$$

where R_c is the series resistance in ohms per unit length of the center conductor given by

$$R_c = \frac{R_s}{4S\left(1-k_0^2\right)K^2\left(k_0\right)}\left[\pi + \ln\left(\frac{4\pi S}{t}\right) - k_0 \ln\left(\frac{1+k_0}{1-k_0}\right)\right] \tag{4.17}$$

R_g is the distributed series resistance in ohms per unit length of the ground planes and is given by

$$R_g = \frac{k_0 R_s}{4S\left(1-k_0^2\right)K^2\left(k_0\right)}\left[\pi + \ln\left(\frac{4\pi(S+2W)}{t}\right) - \frac{1}{k_0}\ln\left(\frac{1+k_0}{1-k_0}\right)\right] \tag{4.18}$$

R_s is the skin effect surface resistance and is given by

$$R_s = \sqrt{\frac{2}{\omega\delta\sigma}} \tag{4.19}$$

where σ is the conductivity and δ is the skin depth. Therefore, the total attenuation α can be expressed as

$$\alpha = \alpha_c + \alpha_d \tag{4.20}$$

Based on Eq. (4.14), the dielectric loss is linearly proportional to frequency. Figure 4.24 shows the relationship between the frequency and the attenuation. The attenuation at the frequency of 20 GHz is 0.37 dB/cm.

Figure 4.25 shows the simulation results of the attenuation constant α and the characteristic impedance Z_0 when the ground plane separation (S+2W) is constant. It is noted that the attenuation decreases slowly with the increase in characteristic impedance. Minimum attenuation occurs at the characteristic impedance which is close to 50 Ω. When the characteristic impedance is raised above 50 Ω, the attenuation increases gradually.

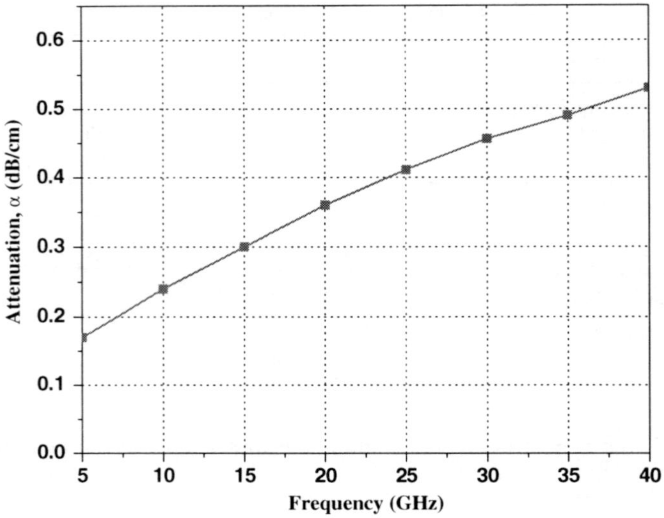

Fig. 4.24 Simulation results of attenuation constant due to conductor loss as a function of frequency of the CPW structure. Copyright/used with permission of/courtesy of IEEE

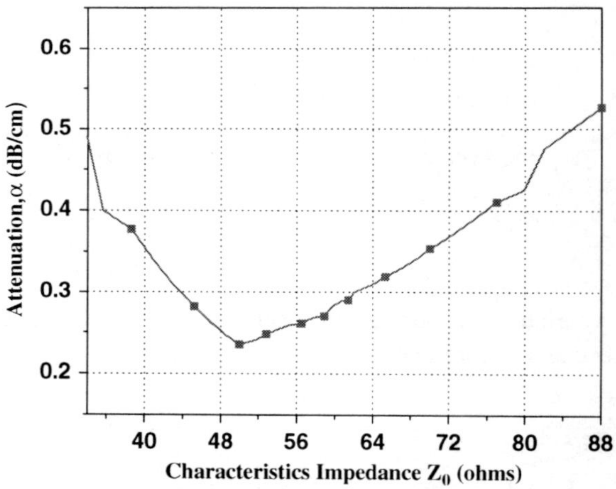

Fig. 4.25 Simulation results of coplanar waveguide attenuation constant versus characteristic impedance for an aspect ratio, $S/(S+2W)$. Copyright/used with permission of/courtesy of IEEE

4.5.3 *Effects of Material Properties*

The material properties have a significant effect on the performance of the CPW structure. They affect both the insertion loss and the RF transmission properties of the structure. Usually the insertion loss of the CPW structure should be made as

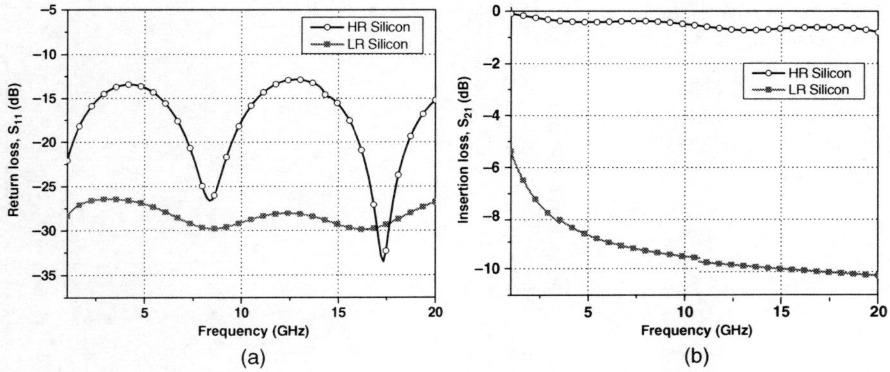

Fig. 4.26 Comparison of the simulated results of the HR silicon and LR silicon: **(a)** return loss S_{11} and **(b)** insertion loss S_{21}. Copyright/used with permission of/courtesy of IEEE

low as possible so that its influence on the performance of any microwave devices, such as filters, switches, and couplers, is negligible. The CPW structure length is 10 mm.

The simulation results of the high-resistivity (HR) silicon and low-resistivity (LR) silicon are shown in Fig. 4.26. The resistivity of HR silicon is 4000 Ω × cm, while that of LR silicon is 20 Ω × cm. The thickness of the metal layer is 2 μm The insertion losses for HR silicon and LR silicon at 5 GHz are 0.3 and 8.5 dB, respectively. Similarly, the insertion loss for the HR silicon at 15 GHz is 0.7 dB and for LR silicon is 10 dB. Therefore, it can be concluded that the CPW structure realized on LR silicon has higher attenuation compared to HR silicon. As a result, silicon with high resistivity is preferred in RF MEMS devices.

4.5.4 Effects of Metal Thickness

The effect of metal thickness t of the CPW transmission line at the fixed frequency of 20 GHz is shown in Fig. 4.27. As expected, the attenuation rises rapidly when the metal thickness becomes less than two to three times the skin depth δ Figure 4.28 shows the EM simulation results of the effect of the metal thickness on the insertion loss. The metal is gold (Au) with a conductivity of 4.1×10^7S/m. When the metal thickness is 4 and 6 μm the insertion losses are 0.2 dB at 5 GHz and 0.40 dB at 15 GHz, respectively. When the metal thickness is reduced to 2 μm, a higher insertion loss of 0.75 dB at 15 GHz is observed. Therefore, it can be concluded that the thicker the deposition layer of Au, the lower the conductor loss.

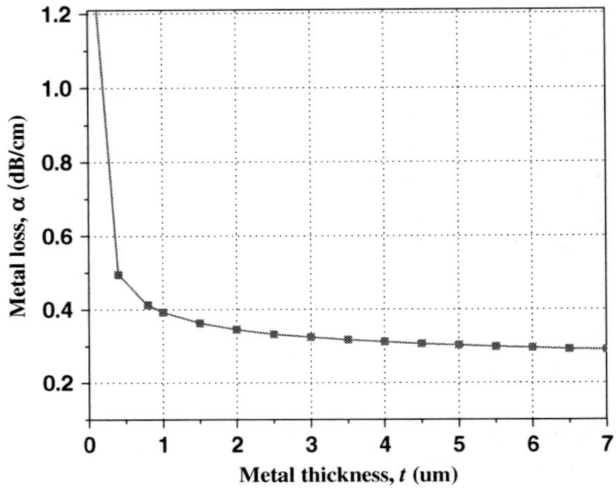

Fig. 4.27 Simulation result of the conductor loss versus different conductor thicknesses t at 20 GHz. Copyright/used with permission of/courtesy of IEEE

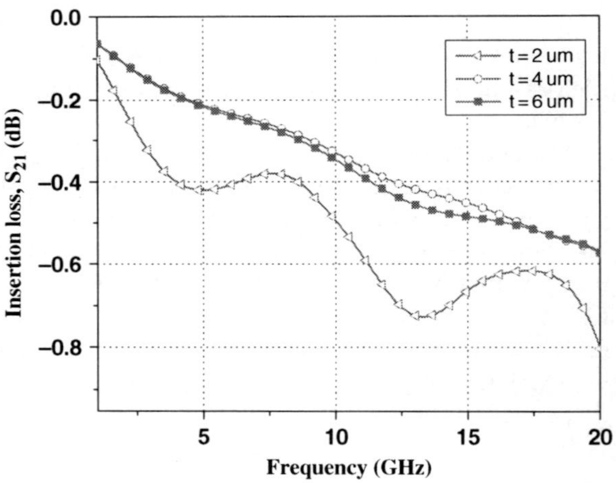

Fig. 4.28 Simulations results of the insertion loss versus frequency with different metal thicknesses as parameter. Copyright/used with permission of/courtesy of IEEE

4.5.5 Experimental Results and Discussions

The RF performance of the silicon CPW structure are measured using the HP 8510C vector network analyzer with gold tip –150 μm pitch from cascade microtech ground–signal–ground coplanar probes. The system is calibrated using standard short-open-load-through (SOLT) on wafer calibration technique. A 5-mm plastic plate is placed between the probe chuck and the sample to remove higher order modes of propagation. All experiments are performed in the room environment without packaging. SEM image of the CPW structure is shown in Fig. 4.29.

The dimensions of the CPW structure are $S = 70$ μm and $W = 115$ μm. The CPW structure length is 10 mm. It is fabricated using high-resistivity silicon

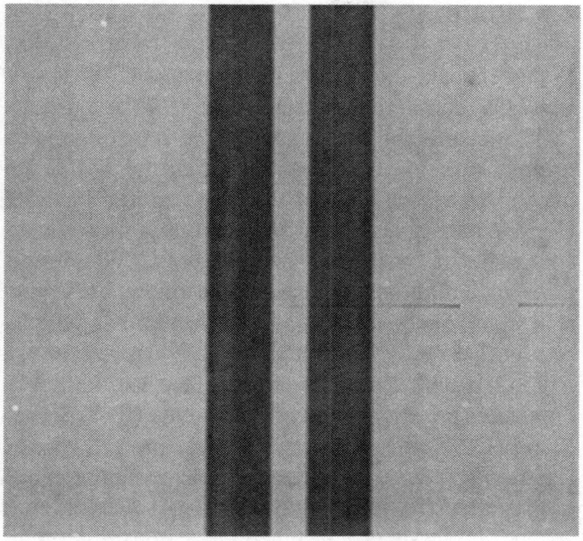

Fig. 4.29 Top view of SEM image of the CPW structure

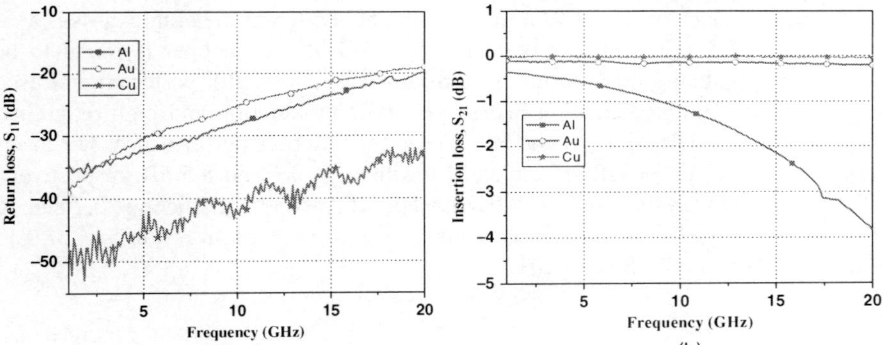

Fig. 4.30 Measurement results of RF properties of the CPW using different materials for the conductor layer: **(a)** return loss S_{11} and **(b)** insertion loss S_{21}. Copyright/used with permission of/courtesy of IEEE

substrate. The conductor layers are deposited using three different types of metal materials. The measurement results are shown in Fig. 4.30. Aluminum (Al) has a high insertion loss of 0.5 dB at 5 GHz and 2.1 dB at 15 GHz. Gold (Au) and copper (Cu) materials show excellent result in terms of insertion loss. Gold has an insertion loss of 0.17 dB at 5 GHz and 0.2 dB at 15 GHz, whereas copper is slightly better with an insertion loss of 0.08 dB at 5 GHz and 0.13 dB at 15 GHz. However, the CPW transmission line and RF MEMS switch are fabricated using gold as the conductor layer. The reason for this is that, although copper shows a better RF performance, gold is more advantageous due to its resistance to oxidation, increased reliability, and stability.

4.6 Summary

In this chapter, the Si-core metal-coated CPW is studied. The design and the full-wave EM simulation of the Si-core CPW are presented. The losses of the Si-core CPW transmission line related to the substrate material, the core material, and the process variations are analyzed using the *RLGC* model. The analysis shows that the substrate material and the core material have influence on the dielectric loss. High-resistivity and low-loss material is more favorable for the substrate and the core material for the low-loss Si-core CPW transmission line. The process variations, including the thickness of the silicon backbone and the undercut dimensions, also significantly affect the RF performance of the Si-core CPW. The experimental results also verify that the Si-core CPW supports quasi-TEM mode propagation up to 25 GHz with attenuation of less than 4 dB/cm. It is found that the conductor loss dominates the attenuation of the Si-core CPW. Some useful guidelines for low-loss Si-core CPW are as follows: (i) depositing sufficiently thick metal layer and improving the step coverage of the metal deposition; (ii) utilizing low-loss substrate, such as glass; and (iii) using high-resistivity silicon as core material.

The surface-micromachined CPW transmission line is also designed, fabricated, and experimented. The study shows that material properties play an important role in the performance of the CPW transmission line. Different metal thickness is analyzed for better insertion loss performance. Even though copper is found to be slightly better than gold in terms of performance, gold is still preferred and used in the practical implementation because of its low oxidation and high reliability. High-resistivity (HR) silicon substrate is used for better RF performance. The insertion loss at 5 GHz for HR silicon and LR silicon is 0.3 and 8.5 dB, respectively. Similarly, the insertion loss at 15 GHz for HR silicon and LR silicon is 0.7 and 10 dB, respectively. The measurement results for gold show an insertion loss of 0.17 dB at 5 GHz and 0.2 dB at 15 GHz.

References

1. M. Tang, A. Q. Liu and J. Oberhammer, "A silicon-on-glass single-pole double-throw (SPDT) switching circuit integrated with silicon-core metal-coated transmission line," J. Micromech. Microeng. Vol. 18, 2008, 095024 (9 pp)
2. B. C. Wadell, Transmission Handbook, Artech House, Inc., Norwood, MA, 1991.
3. D. M. Pozar, Microwave Engineering, Addison-Wesley, Reading, MA, 1990.
4. M. Riaziat, R. Majidi-Ahy and I.-J. Feng, "Propagation modes and dispersion characteristics of coplanar waveguides," IEEE Trans. Microwave Theory Tech. Vol. 38(3), 1990, pp. 245–251.
5. C. P. Wen, "Coplanar waveguide: a surface strip transmission line suitable for nonreciprocal gyromagnetic device applications," IEEE Trans. Microwave Theory Tech. Vol. 17(12), 1969, pp. 1087–1090.
6. J. L. B. Walker, "A survey of European activity on coplanar waveguide," 1993 IEEE MTT-S Int. Microwave Symp. Dig., Atlanta, GA, Vol. 2, pp. 693–696, June 14–18, 1993.
7. A. K. Sharma and T. Itoh (editors), Special issue on modeling and design of coplanar monolithic microwave and millimeter-wave integrated circuits, IEEE Trans. Microwave Theory Tech. Vol. 41(9), 1993.

8. T. Sporkmann, "The evolution of coplanar MMICs over the past 30 years," Microwave J. Vol. 41(7), 1998, pp. 96–111.
9. T. Sporkmann, "The current state of the art in coplanar MMICs," Microwave J. Vol. 41(8), 1998, pp. 60–74.
10. J. Browne, "Broadband amps sport coplanar waveguide," Microwaves RF Vol. 26(2), 1987, pp. 131–134.
11. J. Browne, "Coplanar waveguide supports integrated multiplier systems," Microwaves RF Vol. 28(3), 1989, pp.137–138.
12. J. Browne, "Coplanar circuits arm limiting amp with 100-dB gain," Microwaves RF Vol. 29(4), 1990, pp. 213–220.
13. C. Veyres and V. F. Hanna, "Extension of the application of conformal mapping techniques to coplanar lines with finite dimensions," Int. J. Electron. Vol. 48(1), 1980, pp. 47–56.
14. S. Gevorgian, L. J. P. Linner and E. L. Kollberg, "CAD models for shielded multilayered CPW," IEEE Trans. Microwave Theory Tech. Vol. 43(4), 1995, pp. 772–779.
15. R. E. Collin, Foundations for Microwave Engineering, McGraw-Hill, New York, NY, 1992.

Chapter 5
Single-Pole-Multi-Throw Switching Circuits

In this chapter, different types of single-pole-multiple-throw (SPMT) switching circuits are designed using a silicon-core metal-coated coplanar waveguide (CPW) transmission line and integrated lateral switches. These include the single-pole-double-three (SP2T) [1–5], single-pole-three-throw (SP3T), and single-pole-four-throw (SP4T) switching circuits [6]. However, it is to focus SP2T circuit analysis and full-wave EM simulation and experimental results discussion. The advantages of the SPMT switching circuits are low loss, high isolation, high reliability and stability, and small footprint.

5.1 SP2T Switching Circuit

Based on different circuit topologies, three different SP2T switching circuits are developed. They are the in-line SP2T switching circuit, the parallel SP2T switching circuit, and the single-beam SP2T switching circuit. As the basis of the SPMT switching circuit, the in-line SP2T switching circuit is discussed in detail. These include the design, equivalent circuit modeling, and the full-wave EM simulation.

5.1.1 Design of the In-line SP2T Switching Circuit

The schematic of the in-line SP2T switching circuit is shown in Fig. 5.1a. It consists of a T-junction with a series switch located at each of the output arms. The signal can therefore be routed to two different output ports with one switch at off state and the other switch at on state. Figure 5.1b shows the layout of the in-line SP2T switching circuit. The lateral metal-contact switch is placed at each output port. Two cantilever beams are integrated on the switching circuit. The cantilever beam is equipped with a fixed connection at the output port. The free end of the cantilever beam comes into contact with the contact bump at the T-junction upon turning on

Min Tang and Ai Qun Liu

A.Q. Liu, *RF MEMS Switches and Integrated Switching Circuits*,
MEMS Reference Shelf 5, DOI 10.1007/978-0-387-46262-2_5,

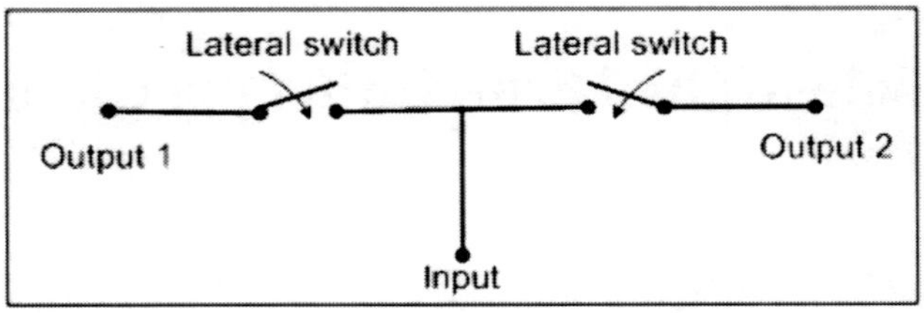

(a) Schematic diagram

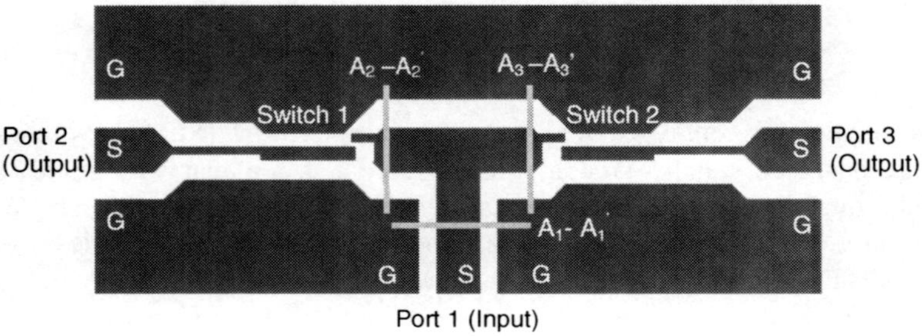

(b) Design layout

Fig. 5.1 The in-line SP2T switching circuit (**a**) schematic diagram and (**b**) design layout (G, ground; S, signal; $A_i - A_i'$, bond wires). Copyright/used with permission of/courtesy of Elsevier

the switch. Therefore, the cantilever beam does not only act as a movable electrode of the actuator but also forms part of the signal line. The ground lines beside the cantilever beam are extended toward the cantilever beam to avoid a drastic increase in the characteristic impedance of this part. The gap distance between the cantilever beam and the ground lines is 30 μm. At the free end of the cantilever beam, one ground line protuberates toward the cantilever beam to act as a fixed electrode. DC bias voltage is applied between the signal line and ground lines of the output port. When the bias voltage is applied between the signal line and the ground lines at port 2 or 3, one lateral switch is at the on state and the other is at the off state. Then, the RF signal propagates from input port to one of the output ports. Due to the discontinuity of the CPW ground plane, the RF power can be converted from the desired CPW mode to the parasitic coupled slot line mode. To equalize the potential of the ground plane, bond wires are used at the discontinuity. In Fig. 5.1b the bond wires are denoted as $A_i - A_i'$. Totally, three bond wires are used in this design.

5.1.2 Modeling and Simulation of the SP2T Switching Circuit

Since only one switch in the SP2T switching circuit is activated to the close state at any time, the equivalent circuit consists of an open stub with an open-state switch which connects between the input transmission line section and an output section with a close-state switch. The complete transmission line model is shown in

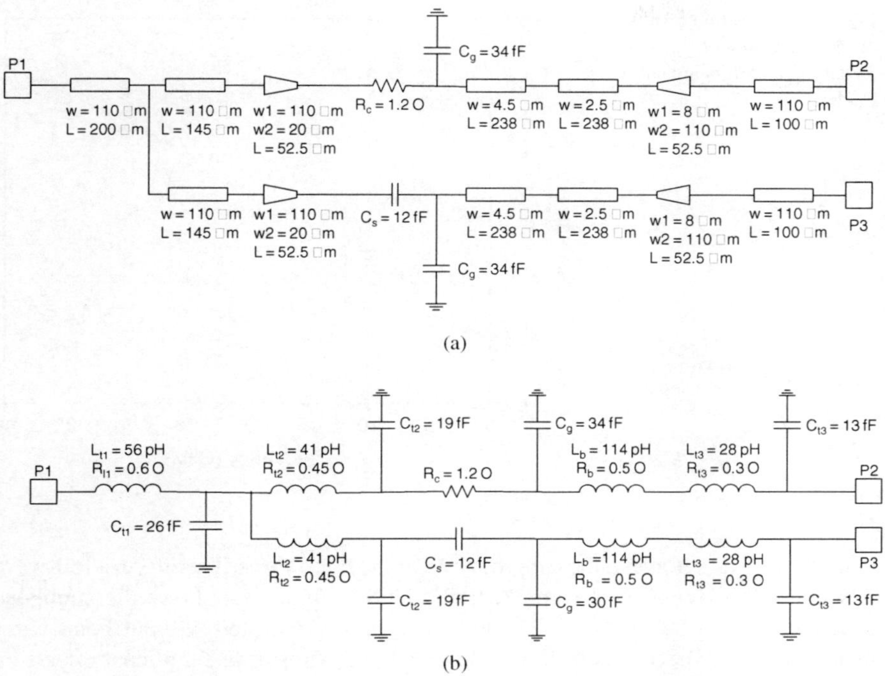

Fig. 5.2 Equivalent circuits of SP2T switching circuit: (**a**) complete transmission line model and (**b**) lumped equivalent circuit model

Fig. 5.2a, where lateral switches are modeled using a simple capacitor $C_s = 12$ fF at the open state and a contact resistor $R_c = 1.2\ \Omega$ at the close state. The parasitic capacitance between the cantilever beam and ground lines, C_g, is also considered in the model.

Figure 5.2b shows the lumped equivalent circuit model of the SP2T switching circuit. The cantilever beam is described by the beam inductance, L_b, and the beam resistance, R_b. The transmission line sections are replaced with their lumped *RLC* circuits. Since the transmission line section is quite short (<500 μm), the shunt conductance is extremely small (<10^{-4} S) and is ignored in the model. The series-L and shunt C configuration means that the insertion loss of the SP2T switching circuit exhibits a "low-pass" characteristic and the cutoff frequency depends on the length of the open stub. It is important to reduce the length of the stub by placing switches as close as possible to achieve wide bandwidth. In Fig. 5.1b, the electrical length of the stub is only 3° at 10 GHz. This is relatively small and the SP2T switching circuit can maintain an impedance match up to around 40 GHz. As shown in Fig. 5.3, the simulated return loss at the input port and the output port is higher than 13 dB at 40 GHz when one switch is closed and the other is open.

The equivalent circuit model provides a useful insight for the design of SP2T switching circuit. However, it does not take into account the parasitic coupling that

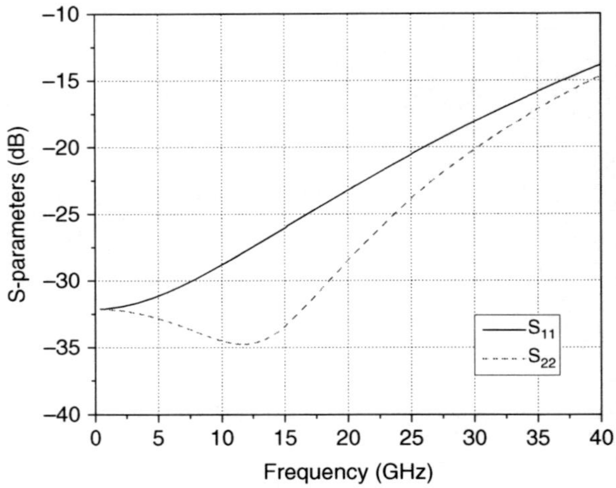

Fig. 5.3 Simulation results of the return loss of the SP2T switching circuit using circuit model

exits at the T-junction and the suppress role of the bond wires. Therefore, a full-wave EM analysis is introduced using Ansoft's HFSS. Figure 5.4 shows the simulated *S*-parameters of the in-line SP2T switching circuit with and without bond wires using Ansoft's HFSS. It is observed that the bond wires have significant effects on the RF responses at high frequencies range (>15 GHz). When the circuit is designed with three bond wires, the insertion loss is below 0.8 dB and the isolation is above 27 dB up to 30 GHz. However, when the circuit is designed without bond wires, the insertion loss is 1.3 dB and the isolation is 18 dB at 30 GHz. The surface current distribution with one switch open and the other switch closed is illustrated in Fig. 5.5. It can be seen that the current mainly flows along the sidewalls of the transmission line

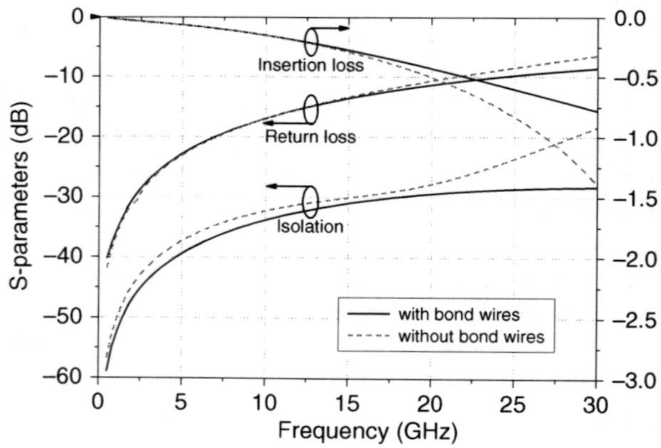

Fig. 5.4 Full-wave EM simulation results of *S*-parameters of the in-line SP2T switching circuit with and without bond wires

Fig. 5.5 Surface current distribution on the in-line SP2T switching circuit with one on-state switch and the other off-state switch

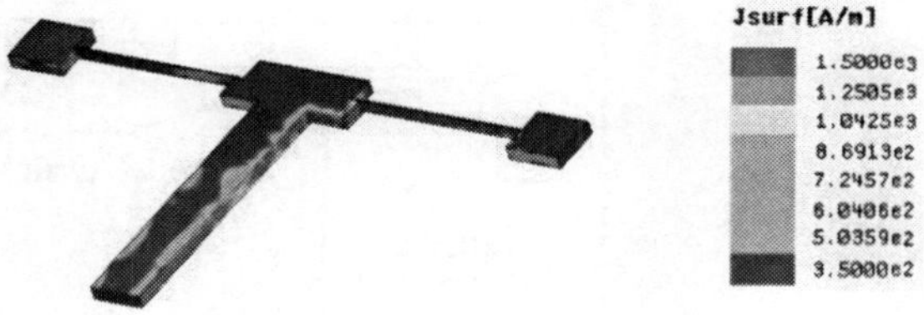

(a) The right switch is in the off-state and the left switch is in the on-state.

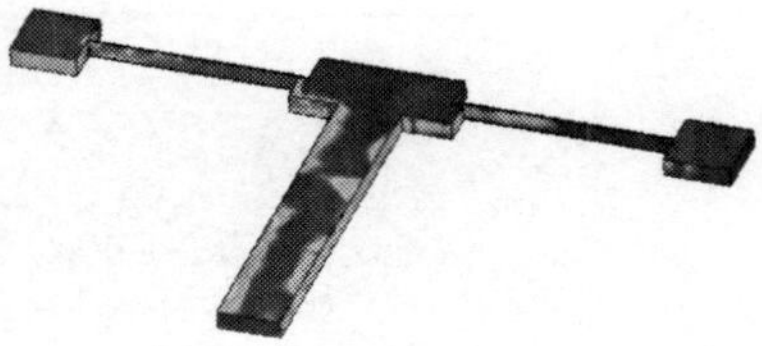

(b) The left switch is in the on-state and the right switch is in the off-state.

and the cantilever beam. When the left switch is at its open state, very little current is flowing along the left beam due to the capacitive coupling. When the left switch is at its close state, most current is flowing along the left beam. The right beam has the similar behavior as the left beam. It is noted that the current distribution patterns in Fig. 5.5a, b are not exactly symmetrical since the mesh generated by the computer is asymmetrical.

5.1.3 Different Types of SP2T Switching Circuits

Two other different designs of the SP2T switching circuits will be discussed, which improve the RF performance and compactness of the SP2T switching circuits. In the in-line SP2T switching circuit, two lateral switches are in a line. However, two lateral switches can also be parallel to each other. The parallel SP2T switching circuit is implemented with this configuration, as shown in Fig. 5.6. Two cantilever beams share one ground line. At the free end of the cantilever beams, the common ground line protuberates toward two cantilever beams, respectively, to act as the common fixed electrode of two lateral switches. When applying bias voltage between the signal line and ground lines at port 2, the cantilever beam of switch 1 moves toward the common ground line pulled by the electrostatic force until its free end hits the T-junction. Then, the RF signal propagates along the transmission line from port 1 to port 2. Little signal goes to port 3 because switch 2 is at its open state. Similarly, when switch 2 is closed, RF signal propagates along the transmission line from port 1 to port 3 and little goes to port 2. When this SP2T switching circuit is in a circuit, the high DC potential is applied on the signal line through an RF choke inductor

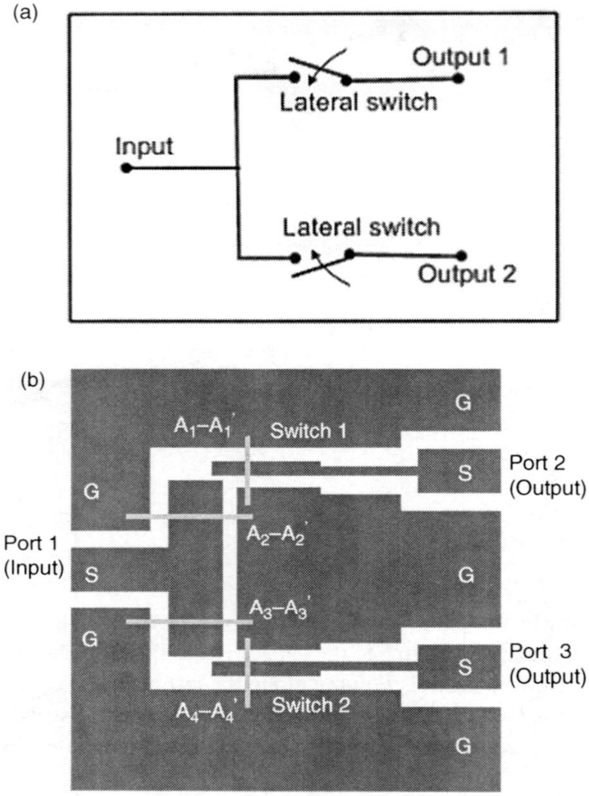

Fig. 5.6 The parallel SP2T switching circuit: (**a**) schematic diagram and (**b**) layout design (G, ground; S, signal; $A_i - A_i'$, bond wire). Copyright/used with permission of/courtesy of Institute of Physics and IOP Publishing Limited

and the DC ground potential is applied on the common ground line to actuate the corresponding switch. A DC block capacitor is inserted between the switch signal line and other circuit parts. Because of the discontinuity of the CPW ground plane, the RF power can be converted from the desired CPW mode to the parasitic coupled slot line mode. To equalize the potential of ground planes, four bond wires are used at the discontinuities of ground planes.

The first two designs of the SP2T switching circuits require two cantilever beams. The last design of the SP2T switching circuit only requires one cantilever beam to move in both directions, as shown in Fig. 5.7. Two fixed electrodes are located at either side of the cantilever beam. The ground potential of the bias voltage is applied on the cantilever beam and sufficient bias voltage is applied on one of the fixed electrode. Then, the cantilever beam moves toward this fixed electrode and hits the contact bump on the corresponding output port. As a result, the RF signal propagates along the transmission line from the input port to this output port and is blocked to

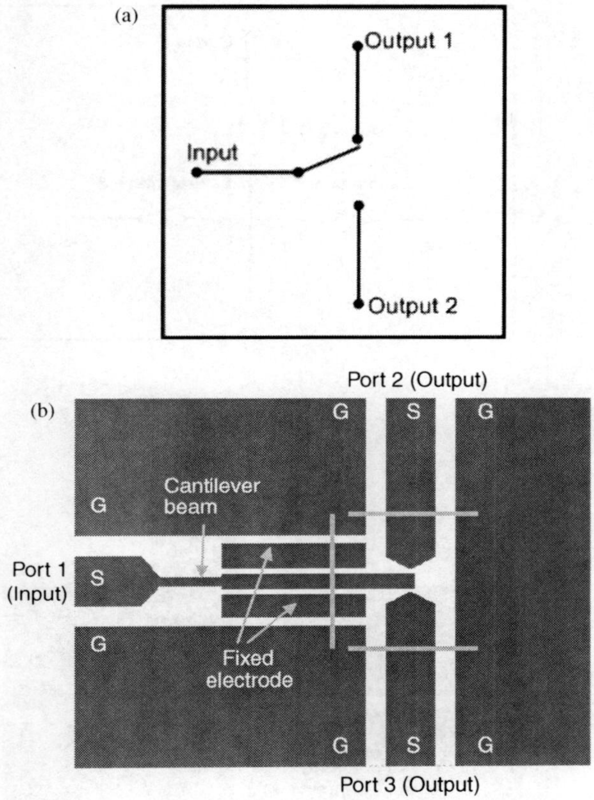

Fig. 5.7 The single-beam SP2T switching circuit: (**a**) schematic diagram and (**b**) layout design (G, ground; S, signal; $A_i - A_i'$: bond wire)

the other port. To suppress the slot line mode at high-frequency range, three bond wires are used to equalize the potential of ground plane which are denoted as $A_i - A_i'$ in Fig. 5.7b. The size of the SP2T switching circuit is reduced because only one beam is used.

5.2 Design of the SP3T Switching Circuit

Based on the design of the in-line SP2T switching circuit, a new type of SP3T switching circuit is developed as shown in Fig. 5.8a, b. Each branch is perpendicular to the adjacent branch. The bias voltage is applied between the signal line and the ground line at the output port. At any time only one switch is activated to the close state and other two switches are at the open state. Then, the RF signal propagates along the transmission line from the input port to only one of the

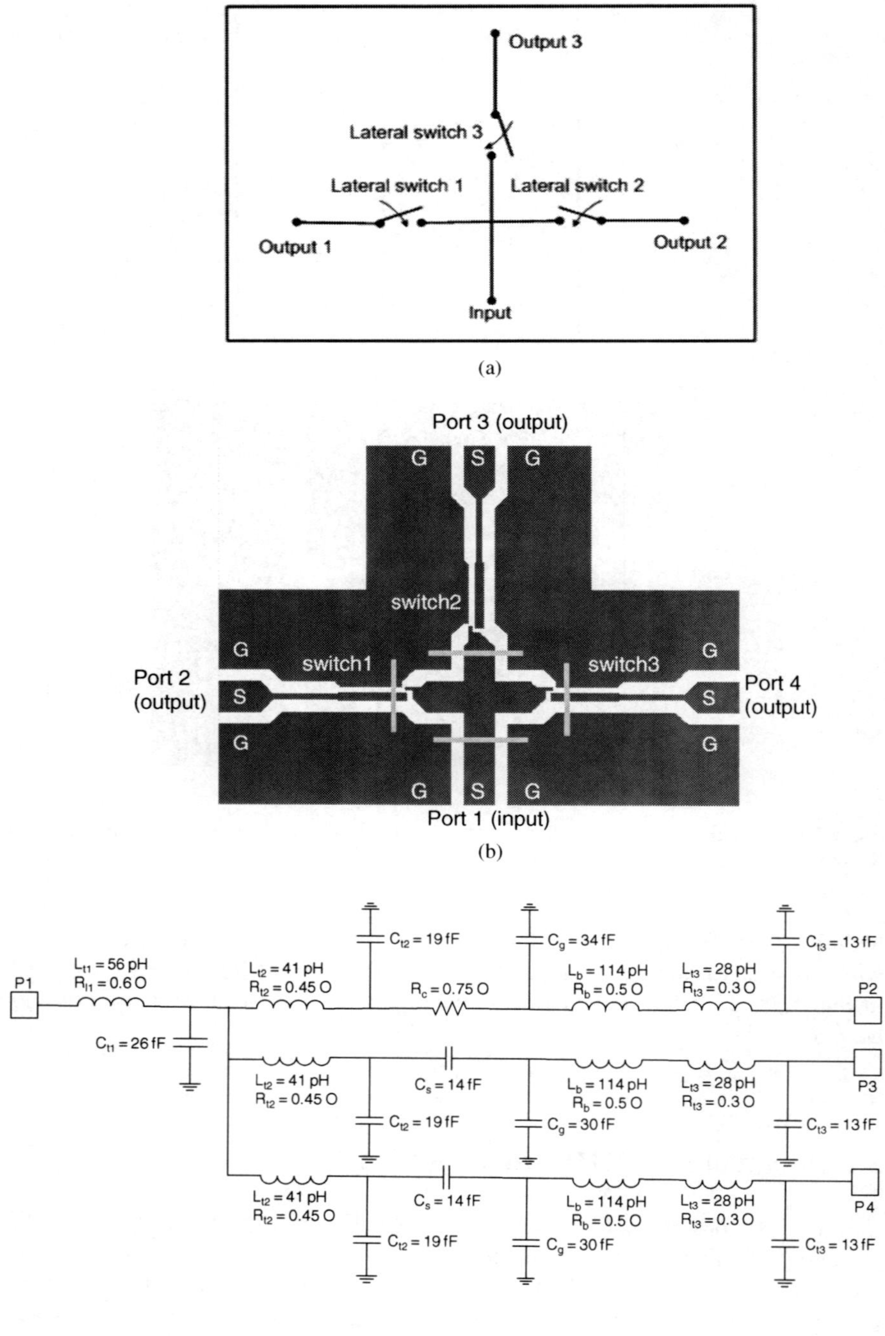

Fig. 5.8 The SP3T switching circuit: (**a**) schematic diagram; (**b**) layout design (G, ground; S, signal; $A_i - A_i'$, bond wire); and (**c**) lumped equivalent circuit

output ports. Four bond wires are required to equalize the potential of the ground planes. The lumped equivalent model is obtained by adding a branch with an open switch to the lumped equivalent model of the SP2T switching circuit, as shown in Fig. 5.8c. The SP3T switching circuit consists of two open stubs connected between the input transmission line section and output branch section with a closed switch. Since the open stubs are capacitive in nature, when there are more open stubs, higher loss is expected at high frequencies range. Therefore, the insertion loss of the SP3T switching circuit is expected to be larger than the in-line SP2T switching circuit.

It is noted that the branches to port 2 (output 1) and port 4 (output 3) are perpendicular to the input port, whereas the branch to port 3 (output 2) is in the same line with the input port. This asymmetrical structure results in different RF performances between port 2 and port 4, and port 3. The measurement results are discussed in Section 5.4.

5.3 Design of the SP4T Switching Circuit

Based on the design of the in-line SP2T switching circuit, a SP4T switching circuit is designed as a star-shaped five-port switching circuit as shown in Fig. 5.9. The angle between any two branches is 72°. Four lateral switches are located at the four output branches, respectively. Since part of the ground line is designed as the fixed electrode, the bias voltage is applied between the signal line and the ground lines of one output port to close the corresponding lateral switch. When one of the lateral switches is activated at the on-state, the remaining three switches are kept at the off-state. Then, RF signal propagates from the input port to only that particular output port and is isolated to other three output ports. The lumped equivalent circuit of the SP4T switching circuit is shown in Fig. 5.9c. The SP4T switching circuit consists of three open stubs connected between the input section and the output section with the lateral switch. The insertion loss of the SP4T switching circuit is larger than the SP3T and SP2T switching circuits because there are more open stubs connected to the active branch.

5.4 Experimental Results and Discussions

The SPMT switching circuits are fabricated on a glass wafer using a substrate transfer process. The details of the fabrication process are discussed in Chapter 9. The substrate is a 500-μm-thick glass. The device structure layer is 62-μm-thick high-resistivity silicon ($\rho > 4000\ \Omega$ cm), which is coated with 1-μm-thick gold. After the fabrication process, 1-mil-thick gold wires are bonded at the discontinuities to equalize the potential of the ground plane using the wire-bonding technique. The

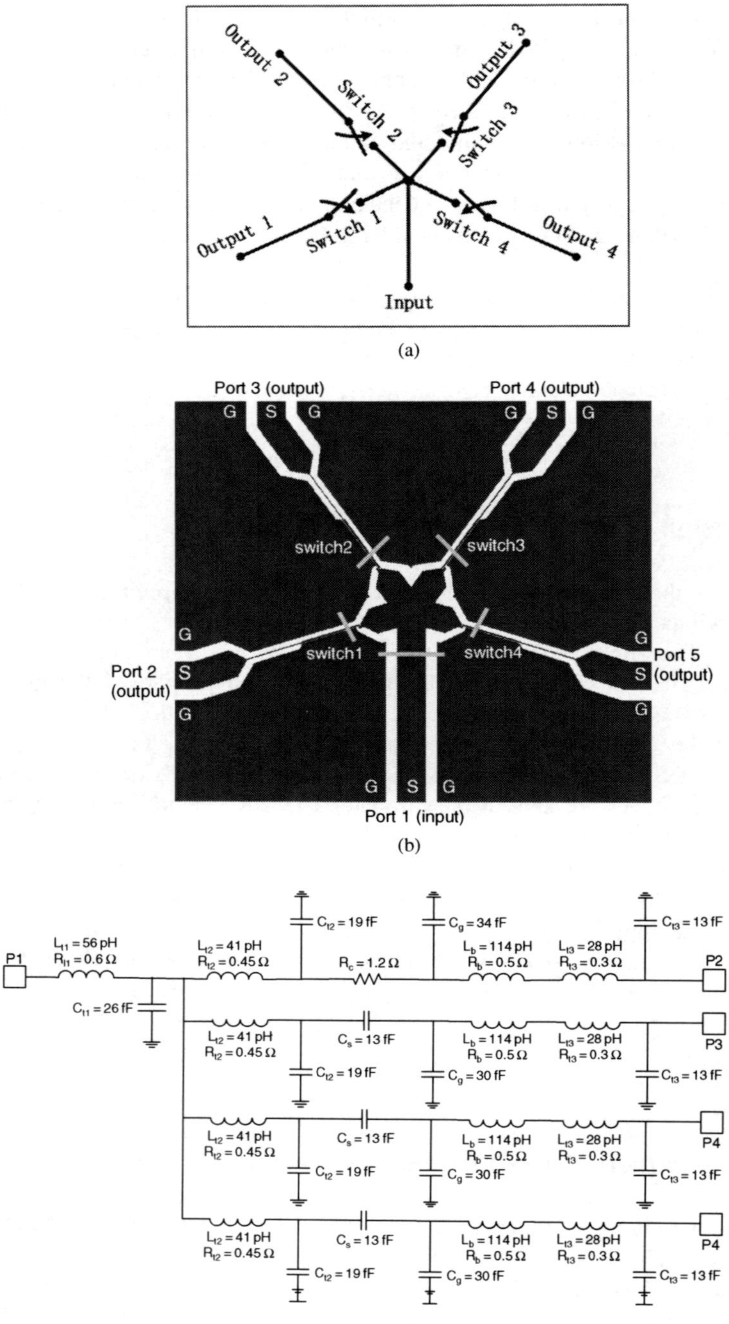

Fig. 5.9 The SP4T switching circuit: (**a**) schematic diagram; (**b**) layout design (G, ground; S, signal; $A_i - A_i'$, bond wire); and (**c**) lumped equivalent circuit. Copyright/used with permission of/courtesy of Elsevier B.V

actuator design in all switching circuits is $l_1 = 230\ \mu\text{m}$, $l_2 = 220\ \mu\text{m}$, $l_3 = 18\ \mu\text{m}$, $w_1 = 2.5\ \mu\text{m}$, $w_2 = 5\ \mu\text{m}$ (solid mass), $g_{\text{Si}} = 4\ \mu\text{m}$, and $d_{\text{Si}} = 3\ \mu\text{m}$. The insertion loss and the return loss of the SPMT switching circuits are determined by S_{21} and S_{11}, respectively, through the input and output branches that contain the lateral switch, while the switch in the other output branches is in the off state. The isolation loss of the SPMT switching circuit is characterized by S_{21} along the signal line with the switch at the off state.

5.4.1 In-line SP2T Switching Circuit

The SEM image of the SP2T switching circuit is shown in Fig. 5.10a. It occupies 1.48 mm^2 (1.85 mm × 0.8 mm) in area. The measured and fitted *S*-parameters of the SPST switch are shown in Fig. 5.10b. The SPST switch has an insertion loss of 0.7 dB, a return loss of 23 dB, and an isolation of 25 dB up to 25 GHz. The fitted lumped circuit parameters of the SPST switch are as follows: $L_b = 114$ pH, $C_{g,on} = 34$ fF, $C_{g,\text{off}} = 30$ fF, $C_s = 3.5$ fF. The fitted total switch resistance ($R_b + R_c$) increases with frequency due to the skin effect and is fitted as

$$R_b + R_c = 1.5 + \sqrt{f} \tag{5.1}$$

The measured RF results of the SP2T switching circuit with and without bond wires are shown in Fig. 5.10c, d, respectively. The bond wires are essential in suppressing the parasitic slot line mode and in keeping the desired quasi-TEM mode, especially at high frequency. The insertion loss of the in-line SP2T switching circuit is less than 1 dB up to 22 GHz with three bond wires. The return loss and the isolation are higher than 15 and 25 dB up to 22 GHz, respectively, as shown in Fig. 5.10d. The insertion loss is 1 dB at 6 GHz and increases rapidly to 5.5 dB at

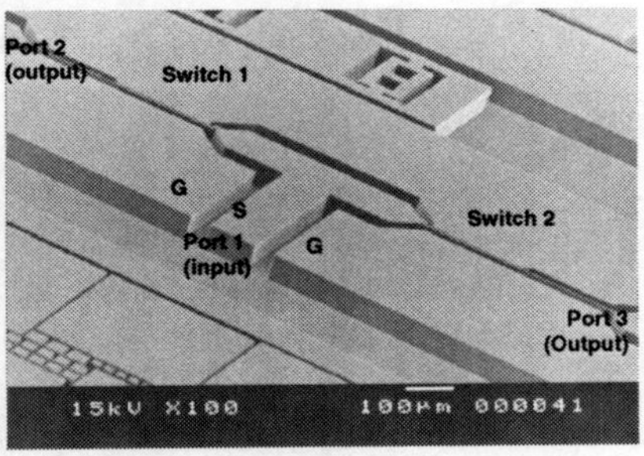

(a) SEM image

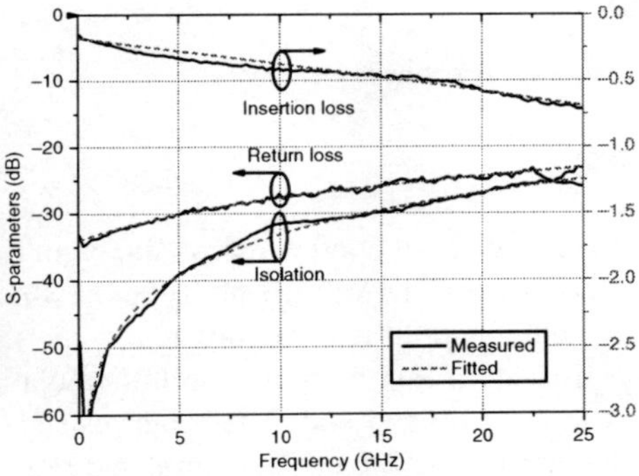

(b) SPST switch on glass

Fig. 5.10 The in-line SP2T switching circuit: (**a**) SEM image; (**b**) measured and fitted results of an SPST switch on glass; (**c**) measured results without bond wires; and (**d**) measured and fitted results with three bond wires

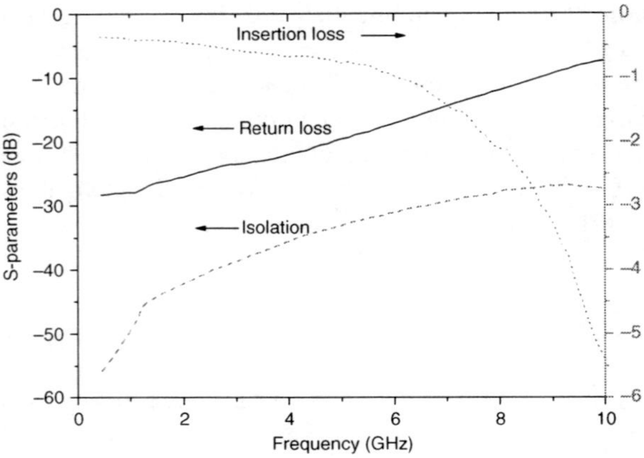

(c) Measured results of the SP2T switching circuit without bond wires

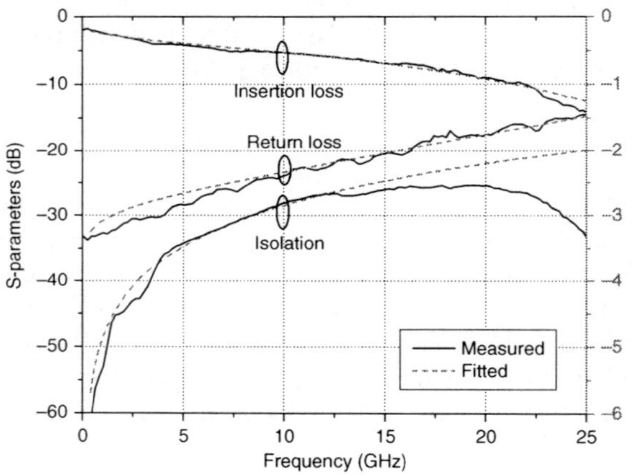

(d) Measured and fitted results of the SP2T switching circuit with three bond wires

Fig. 5.10 (Continued)

10 GHz without bond wires, as shown in Fig. 5.10c. The return loss is reduced from 17 dB at 6 GHz to 7 dB at 10 GHz, which is unfavorable compared to the SP2T switching circuit with bond wires.

Based on Fig. 5.2b, the circuit simulation results of the SP2T switching circuit is shown in Fig. 5.10d. The beam resistance, R_b, is found to be proportional to $\sqrt{f}$ due to the skin effect. The simulated return loss and insertion loss agree well with the experimental results. The simulated isolation agrees with the measured isolation up to 12 GHz. However, the measured isolation is better than the simulated results from 12 to 25 GHz because the circuit model does not describe the parasitic coupling effect at the T-junction.

5.4.2 Different Designs of the SP2T Switching Circuits

The SEM image of the parallel SP2T switching circuit with bond wires is shown in Fig. 5.11a. The circuit size is 1.3 mm^2 (1 mm $\times$ 1.3 mm) in area.

The SEM image of the single-beam SP2T switching circuit is shown in Fig. 5.11b. Since the fixed electrodes of this design are separated from the ground lines, the bias lines are designed using an external power supply. By using a shadow mask, the bias line structures are sheltered from the metal coating except the DC pads. Therefore, the bias lines have high resistance due to the high resistivity of the silicon layer. The bias line connected to the signal line is a meander structure to provide high inductance. The circuit size is 0.92 mm^2 (= 1.15 mm $\times$ 0.8 mm) in area, excluding the bias lines.

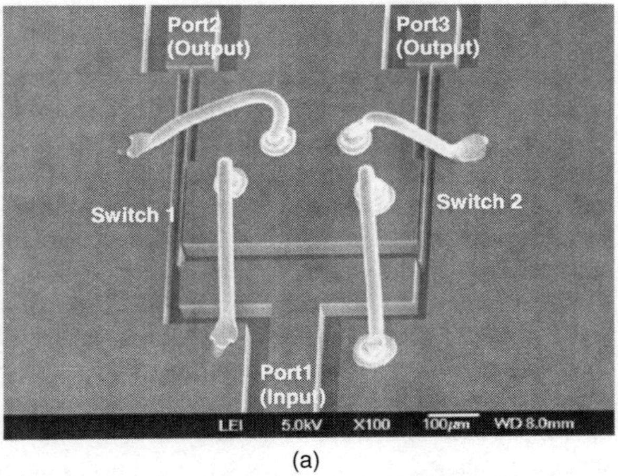

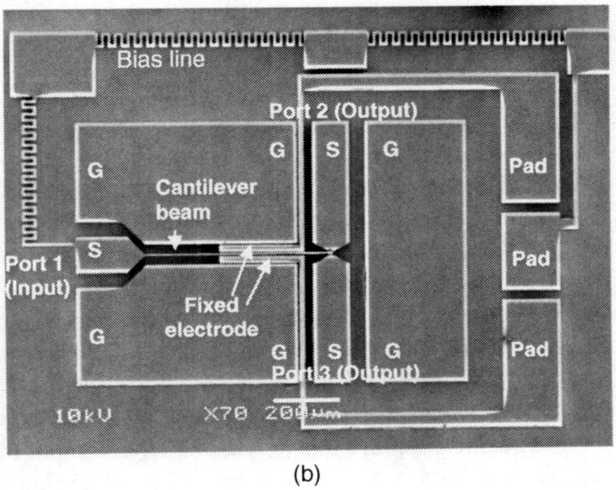

Fig. 5.11 SEM images of (**a**) the parallel SP2T switching circuit and (**b**) the single-beam SP2T switching circuit. Copyright/used with permission of/courtesy of Institute of Physics and IOP Publishing Limited

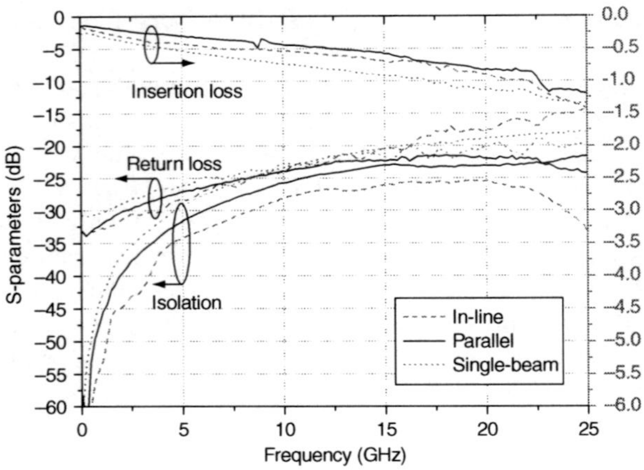

Fig. 5.12 Comparison of the measured *S*-parameters of the three SP2T switching circuits

A comparison of the measured *S*-parameters among the three different SP2T switching circuits is provided in Fig. 5.12. The insertion loss of the parallel SP2T switching circuit is less than 1 dB up to 22 GHz. Both the return loss and the isolation are above 22 dB up to 25 GHz. The parallel SP2T switching circuit is found to have the best insertion loss and return loss which implies that the parallel SP2T switching circuit provides the best impedance matching and conducts the RF power most effectively.

The insertion loss of the single-beam SP2T switching circuit is 0.52 dB at 5 GHz and 1 dB at 17 GHz. The return loss and isolation are above 17 and 20 dB up to 25 GHz, respectively. The insertion loss of the single-beam SP2T switching circuit is the least favorable when compared with other two designs. That is because the two fixed electrodes, which are located beside the cantilever beam, increase the parasitic shunt capacitance, C_g, of the switch. As a result, more RF power is guided to the ground plane.

5.4.3 SP3T Switching Circuit

The SEM image of the SP3T switching circuit is shown in Fig. 5.13a. It occupies 2.45 mm^2 (1.85 mm × 1.325 mm) in area. The measured *S*-parameters are shown in Fig. 5.13b, c. The measured insertion loss and the return loss of the SP3T switching circuit are 0.7 and 17 dB at 15 GHz, respectively. The isolation is higher than 20 dB up to 20 GHz. Depending on which port is turned on, the SP3T switching circuit exhibits either a "near port" (ports 2 and 4) or the "far port" (port 3) RF response. The insertion loss and return loss of the "far port" are better than the "near port." Since the "near port" is at an acute angle to the input port, there is some parasitic coupling that affects the input–output match. Hence, the "near port" exhibits a narrower return loss bandwidth than the "far port." The experimental results of the two "near ports" (ports 2 and 4) are slightly different, especially when

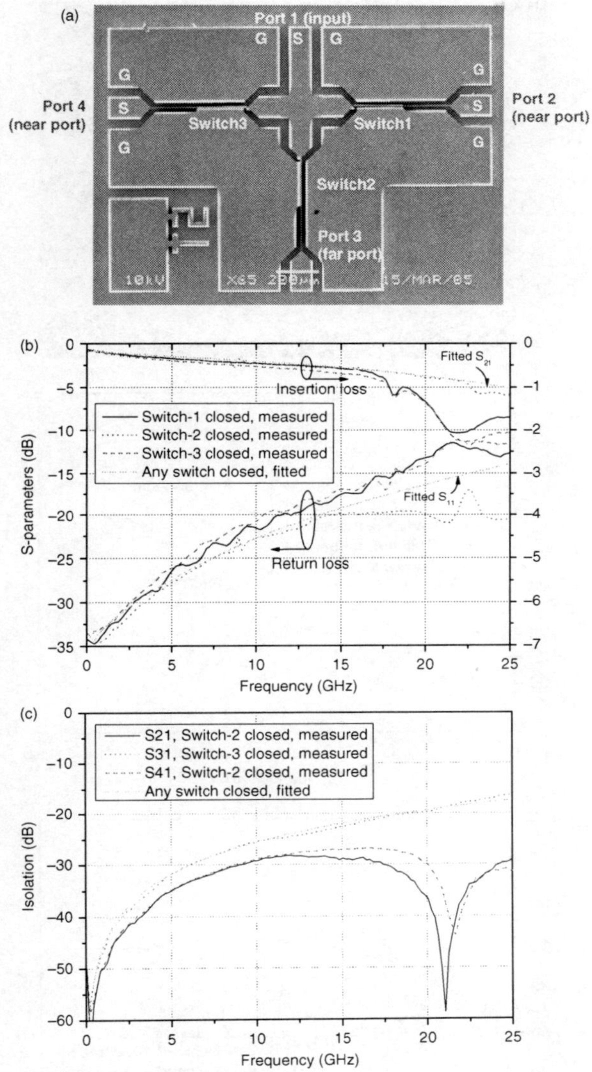

Fig. 5.13 Experimental and fitted results of an SP3T switching circuit: (**a**) SEM image; (**b**) insertion loss and return loss; and (**c**) isolation

the frequency is higher than 15 GHz. This is because two beams in two "near ports" are not exactly the same after fabrication. The circuit simulation results are compared with the measurement results in Fig. 5.13b, c. The circuit simulation results are in good agreement with the measurement results of the "far port," but differ with the "near port" results since circuit model cannot account for the coupling that exits at the acute angle. The measured insertion loss and return loss of the SP3T switching circuit are not as good when compared with the SP2T switching circuits. This is

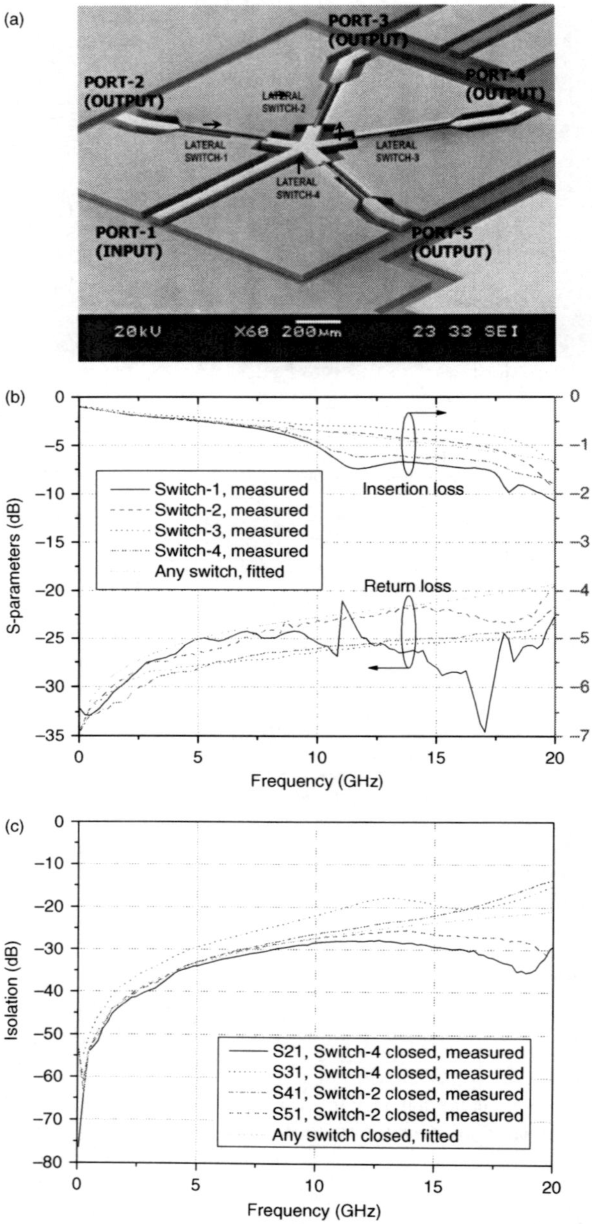

Fig. 5.14 Experimental and fitted results of an SP4T switching circuit: (**a**) SEM image; (**b**) insertion loss and return loss; and (**c**) isolation. Copyright/used with permission of/courtesy of IET

because the active circuit consists of two open stubs connecting the input port and the output port, which results in more capacitive coupling.

5.4.4 SP4T Switching Circuit

The SEM image of the SP4T switching circuit, which occupies 3.47 mm^2 (1.77 mm × 1.96 mm) in area, is shown in Fig. 5.14a. The comparison between the measured results and the simulated results is shown in Fig. 5.14b, c. The measured insertion loss is lower than 1 dB from 50 MHz to 10 GHz. Both the return loss and the isolation of the SP4T switching circuit are higher than 22 dB at 10 GHz. The measured *S*-parameters of the SP4T switching circuit show that the RF response depends on the location of the on-state switch. For the SP4T switching circuit, the "near ports" are ports 2 and 5 and the "far ports" are ports 3 and 4. The insertion losses of the "far ports" are 0.2 dB lower than that of "the near ports" at 10 GHz.

5.5 Summary

By directly applying the Si-core CPW and lateral switches, different types of SPMT switching circuits are designed, fabricated, and measured. The SP2T switching circuit with two cantilever beams is designed with the equivalent circuit analysis and the full-wave EM simulation. The insertion loss of the SP2T switching circuit with three bond wires is less than 1 dB up to 20 GHz. The parallel SP2T switching circuit with two cantilever beams demonstrates lowest insertion loss, which is less than 1 dB at 22 GHz. This is because the parallel SP2T switching circuit has better circuit matching.

The SP3T and SP4T switching circuits are also designed and fabricated. The SP3T switching circuit has an insertion loss of 0.7 dB and a return loss of 17 dB at 15 GHz. The isolation is higher than 20 dB up to 20 GHz. The SP4T switching circuit has an insertion loss of less than 1 dB up to 10 GHz. Both the return loss and the isolation are higher than 22 dB at 10 GHz.

Table 1.2 shows the RF performance comparison of various SPMT switching circuits. The lateral SPMT switching circuits are found to have more reliable mechanical performance and simpler fabrication process and the integration of the lateral switches and the Si-core CPW transmission lines provide a low-loss low-cost SPMT switching circuit.

References

1. M. Tang, A. Q. Liu, A. Agarwal, Z. S. Liu and C. Lu, "A single-pole double-throw (SPDT) circuit using lateral metal-contact micromachined switches," Sensor. Actuator. Phys. A Vol. 121, 2005, pp. 187–196.

2. M. Tang, A. Q. Liu and J. Oberhammer, "A silicon-on-glass single-pole double-throw (SPDT) switching circuit integrated with silicon-core metal-coated transmission line," J. Micromech. Microeng. Vol. 18, 2008, 095024 (9pp).
3. J. Oberhammer, M. Tang, A. Q. Liu and G. Stemme, "Mechanically Tri-stabel, true single-pole-double-throw (SPDT) switches," J. Micromech. Microeng. Vol. 16, 2006, pp. 2251–2258.
4. M. Tang, A. Q. Liu and A. Agarwal, "A low-loss single-pole double-throw (SPDT) switch circuit," Int. Solid-State Sensors, Actuators and Microsystems Conf., 2007 (Transducer'07 & Eurosensors XXI), Lyon, France, 10–14 June 2007, pp. 679–682.
5. J. Oberhammer, M. Tang, A. Q. Liu and G. Stemme, "Mechanically tri-stable in-line single-pole-double-throw all-metal switch," 19th IEEE International Conference on Micro-electro-mechanical Systems (MEMS 2006), Istanbul, Turkey, 22–26 January 2006, pp. 898–901.
6. A. Q. Liu, W. Palei, M. Tang and A. Alphones, "Single-pole-four-throw switch using high-aspect-ratio lateral switches, Electron. Lett. Vol. 40(18), 2004, pp. 1125–1126.

Chapter 6
Tunable Electromagnetic Band Gap Bandstop Filter

In this chapter, design and experiments of the coplanar waveguide (CPW)-based electromagnetic band gap (EBG) bandstop filter and the micromachined tunable EBG filters are discussed. These new CPW EBG structures include an unloaded and a loaded design. The frequency characteristics of the EBG are demonstrated using different circuit parameters. An equivalent circuit model for the unloaded and loaded unit structures is derived and a dispersion diagram is obtained for both structures using commercial software and Floquet's theorem. The electromagnetic wave behavior within a unit is analyzed. The bandstop filter is designed by cascading the EBG structures.

A micromachine-based tunable bandstop filter with low insertion loss is designed. An equivalent circuit model of the tunable bandstop filter is derived. For the first time, MEMS capacitive bridges are incorporated in the design of the tunable bandstop filter. A MEMS surface micromachining process is developed for the fabrication of tunable filter. Finally, the measurement results are discussed in terms of flat responses of the stopband, passband, and the tunable range of the filter.

6.1 Unloaded EBG Bandstop Filter

An unloaded lattice shape unit EBG structure is shown in Fig. 6.1. The substrate is high-resistivity silicon ($\rho = 4000\ \Omega \times \text{cm}$) with dielectric permittivity of $\varepsilon_r = 11.9$ and thickness of $H = 200\ \mu\text{m}$. The EBG structure is designed for $50\ \Omega$ in CPW configuration using a signal line with width of $W = 108\ \mu\text{m}$ and a gap width of $G = 60\ \mu\text{m}$.

In the unloaded unit EBG structure, a square slot is etched in the ground plane with a side length a. The square-etched slot is connected to the gap by a narrow transverse slot with length of w_s and width of d_s. The center cutoff frequency, which

Muhammad Faeyz Karim and Ai Qun Liu

A.Q. Liu, *RF MEMS Switches and Integrated Switching Circuits*,
MEMS Reference Shelf 5, DOI 10.1007/978-0-387-46262-2_6,

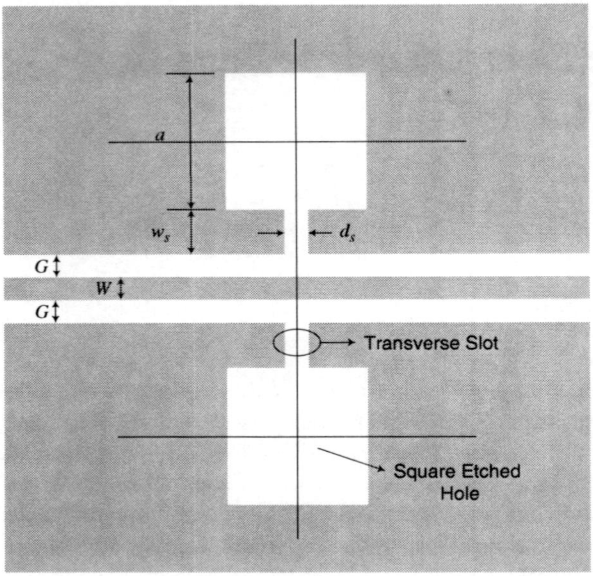

Fig. 6.1 Schematic of unloaded unit EBG structure [1, 2]. Copyright/used with permission of/courtesy of IET/ Institute of Physics and IOP Publishing Limited

is the resonant frequency, depends on the transverse slot and the square-etched hole in the ground plane.

To investigate the influences of these parameters on the frequency characteristics, two separate designs are constructed. The frequency characteristic for the proposed unit EBG structure is studied and analyzed using Agilent's momentum (an EM simulation tool based on method of moment) and the equivalent circuit model.

6.1.1 Effects of the Unloaded Square Aperture Size

Three different sets of dimensions are evaluated for the unloaded unit EBG structure. The influence of the square aperture size a on its frequency characteristic is first investigated. The dimensions $d_S = 60\,\mu m$ and $w_S = 200\,\mu m$ are constant for all the three different square aperture sizes $a = 350$, 500, and 600 μm.

From the simulation results shown in Fig. 6.2, the resonant frequency of the EBG unit decreases as the etched area of the square aperture increases. This frequency characteristic can also be explained using the parallel resonant circuit. As the size of the square aperture increases, the inductance also increases, whereas d_S and w_S are responsible for the capacitance. Since d_S and w_S are the same for all the three designs, the capacitance does not vary much compared to the inductance. Thus, as a

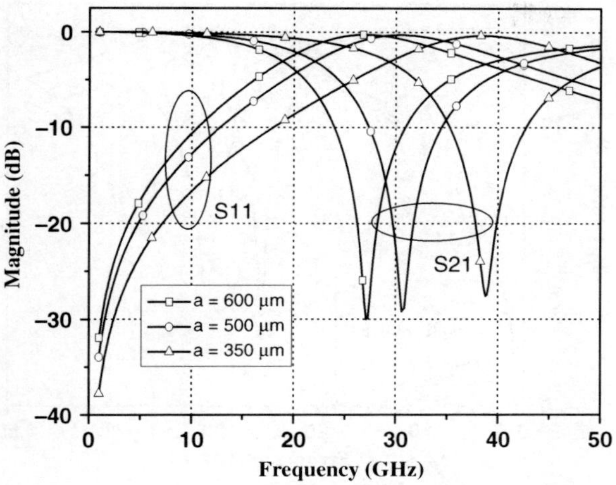

Fig. 6.2 Simulated *S*-parameters of the unloaded unit EBG structure versus variation of square aperture size [1]. Copyright/used with permission of/courtesy of IET

increases, the inductance increases and in turn reduces the resonance frequency of the equivalent parallel circuit and vice versa.

6.1.2 Influence of the Transverse Slot Width

The influence of the transverse slot width d_s is investigated in this section. The square aperture size, a, is kept constant at 500 μm × 500 μm, and the length of the transverse slot w_s is fixed at 200 μm. The width of the transverse slot is allowed to vary and set at 10, 60, and 200 μm, respectively. From Fig. 6.3, it can be observed that the resonant frequency shifts from 25.48 to 37.98 GHz when d_s changes from 10 to 200 μm. Since the size of the unloaded square aperture is fixed for all the three designs, the inductance variation is smaller. Thus, as d_s increases, the parallel capacitance decreases and the resonance frequency of the equivalent parallel circuit increases and vice versa.

6.1.3 Influence of the Transverse Slot Length

The effect of the transverse slot length w_s on the frequency characteristic is discussed. The parameters $d_s = 60$ μm and $w_s = 200$ μm are kept constant, but the transverse slot length w_s is allowed to change from 200, 400, and 600 μm. The simulation results show that the frequency shifts from 30 to 20 GHz, as shown in Fig. 6.4. By increasing the length of the transverse slot, the frequency response is shifted to a lower frequency.

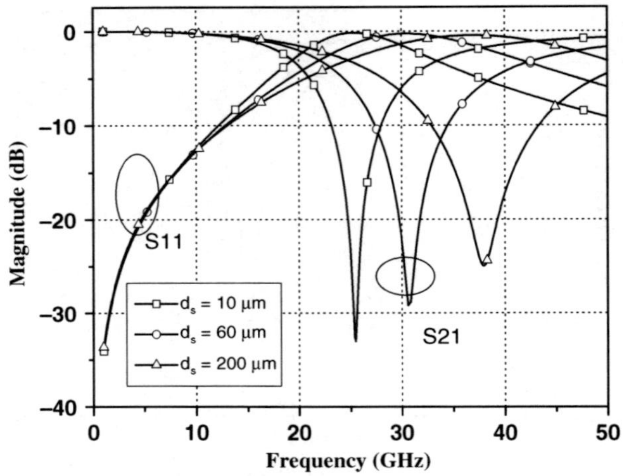

Fig. 6.3 Simulated *S*-parameter of the unloaded unit EBG structure versus variation of the transverse slot width d_S [1]. Copyright/used with permission of/courtesy of IET

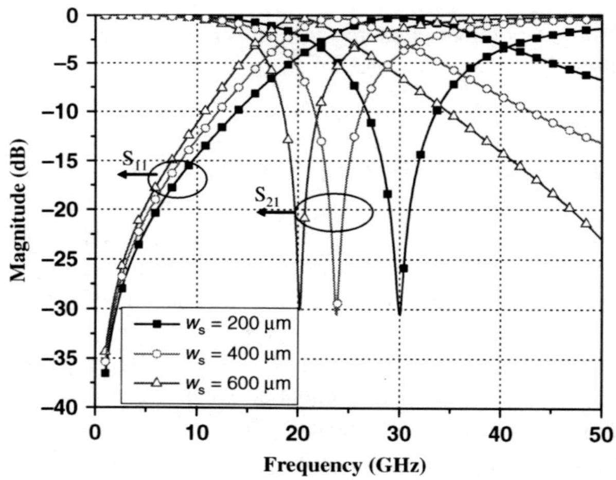

Fig. 6.4 Simulated *S*-parameter of the unloaded unit EBG structure versus variation of the transverse slot length w_S

6.1.4 Modeling of the Unloaded EBG

An equivalent parallel *LC* circuit is used to model the unloaded unit EBG structure as shown in Fig. 6.5. It consists of a series inductor and a parallel capacitor. From the practical point of view, the unloaded unit EBG structure can serve as a replacement for the parallel *LC* resonant circuits in many applications as it exhibits bandstop characteristics [3]. To apply the unloaded EBG unit to the practical circuit, it is

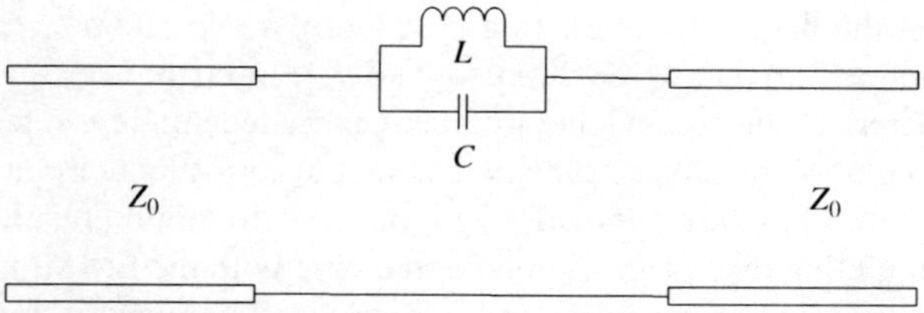

Fig. 6.5 Equivalent parallel resonant circuit for the unloaded unit EBG structure [1]. Copyright/used with permission of/courtesy of IET

necessary to extract the equivalent circuit parameters, which can be obtained from the simulation results of the unloaded unit EBG structure. The lumped capacitance C is mainly contributed by the transverse slot on the ground, while the inductance L is related to the magnetic flux passing through the apertures on the ground.

The equivalent impedance equation of the single resonant model can be expressed as

$$Z = \left(j\omega C + \frac{1}{j\omega L} \right)^{-1} \tag{6.1}$$

The resonant frequency of the parallel circuit is defined as

$$\omega_0 = \frac{1}{\sqrt{LC}} \tag{6.2}$$

The 3-dB cutoff angular frequency, ω_C, can be determined by

$$|S_{21}|\,|_{\omega=\omega_C} = \left| \frac{2Z_0}{2Z_0 + Z} \right| = \frac{1}{\sqrt{2}} \tag{6.3a}$$

Substituting Eq. (6.1) into Eq. (6.3a), it can be further expressed as

$$|S_{21}|\,|_{\omega=\omega_C} = \frac{2Z_0}{\sqrt{4Z_0^2 + \left(\frac{1}{\omega_c C - 1/\omega_c L} \right)^2}} = \frac{1}{\sqrt{2}} \tag{6.3b}$$

Substituting Eq. (6.2) into Eq. (6.3b), the capacitance can be obtained as

$$C = \frac{\omega_C}{2Z_0\left(\omega_0^2 - \omega_C^2\right)} \tag{6.4}$$

The inductance can be determined by

$$L = \frac{1}{\omega_0^2 C} \tag{6.5}$$

To illustrate the parameter extraction procedure, a, w_s, and d_s (see Fig. 6.1) are chosen so that $a = 350$ μm, $w_s = 200$ μm, and $d_s = 60$ μm.

The parameters of the equivalent circuit are extracted and presented in Table 6.1. The circuit simulation results are obtained using the equivalent circuit parameters as shown in Fig. 6.6. These are compared with the EM simulation results. The equivalent circuit simulation results are in good agreement with the EM simulation results. Thus, the derived equivalent circuit and the extracted parameters for the unloaded EBG structure can be accurately adapted to design the practical circuits as well as to analyze the dimensions of any proposed structures.

Table 6.1 Extracted equivalent circuit parameters for the unloaded unit EBG structure, with different square aperture size a

	$d_s = 60$ μm, $w_s = 200$ μm		
Square aperture size, a (μm)	350	500	600
Resonant frequency (GHz)	38.79	30.6	27.13
3-dB cutoff frequency (GHz)	29.12	21.92	18.78
Inductance (nH)	0.229	0.342	0.434
Capacitance (pF)	0.073	0.079	0.079

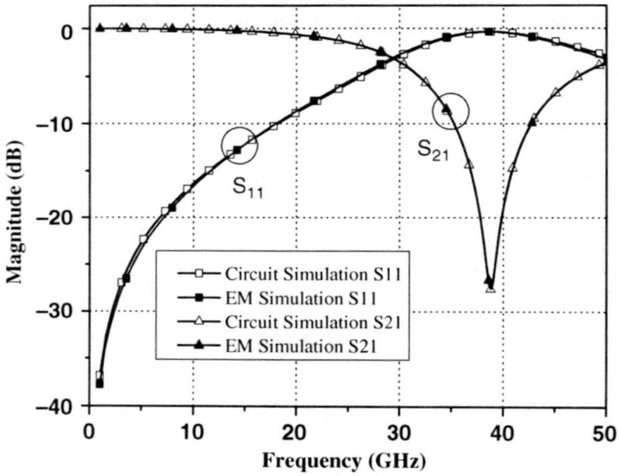

Fig. 6.6 Comparison on parameter extraction and EM simulation results of the unloaded unit EBG structure

The resonant frequency of the proposed unit EBG is affected by the variation of the unloaded square aperture dimensions. Based on the tabulated results (see Table 6.1) and the simulated results (see Fig. 6.2), it can be deduced that by varying from $a = 350$ to 600 μm, the change in capacitances is merely 15%. This is different from the case of inductance variation, where a variation as large as 178% is observed. The inductance variation rate is nearly 10 times greater than that of the capacitance. These observations prove that the dimension of the transverse slot is responsible for the parallel capacitor and the size of the square aperture relates directly to the series inductor. The series inductance is significantly affected by the size of the square aperture.

For the second design, only the width of the transverse slot d_s is also allowed to vary. Table 6.2 presents the extracted equivalent circuit parameters.

Table 6.2 Extracted equivalent circuit parameters of the unloaded EBG unit structure with variation of the transverse slot width d_s

	$w_s = 200\ \mu m$, $a = 500\ \mu m$		
Transverse slot width, d_s (μm)	10	60	200
Resonant frequency (GHz)	25.48	30.6	37.98
3-dB cutoff frequency (GHz)	19.38	21.92	24.56
Inductance (nH)	0.339	0.342	0.357
Capacitance (pF)	0.115	0.079	0.049

The simulation result is shown in Fig. 6.3. From the case of $d_s = 10\ \mu m$ to the case of $d_s = 200\ \mu m$, the inductance varies by 5%, whereas the capacitance changes by almost 57%. The capacitance variation rate is nearly 10 times greater than that of the inductance. This again verifies that the dimension of the transverse slot controls the capacitance and the size of the square aperture affects the inductance of the equivalent circuit. Since the width of the transverse slot d_s is varied, the parallel capacitance also changes accordingly.

6.1.5 Propagation Characteristics of the Unloaded Unit EBG

The unloaded EBG units are cascaded as a bandstop filter as shown in Fig. 6.7. The propagation characteristic on the unloaded EBG is analyzed using Floquet's theorem [4]. By simply considering the unit structure, the EM simulation can be used to obtain the dispersion diagram of the proposed unloaded EBG structure. The propagation factor in this case is $e^{-\gamma\Lambda}$, where $\Lambda = 2020\ \mu m$ is the period of the

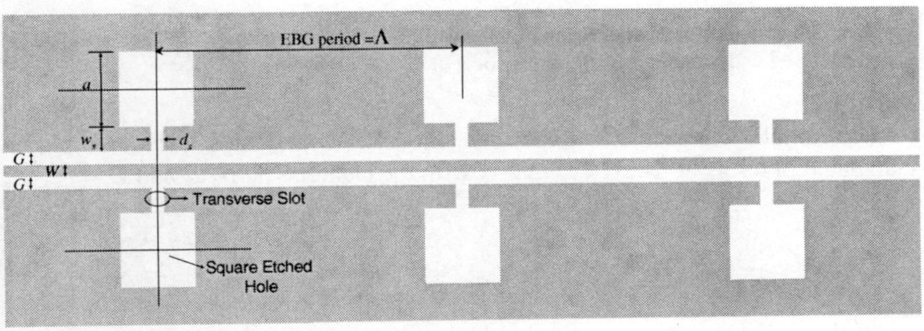

Fig. 6.7 Schematic of unloaded EBG bandstop filter

EBG structure, and $\gamma = \alpha + j\beta$ is the complex propagation constant in the direction of propagation that can be expressed as

$$\gamma = \frac{1}{\Lambda}\cosh^{-1}\left(\frac{\left[(1+S_{11})(1-S_{22})+S_{12}S_{21}+(Z_{01}/Z_{02})(1-S_{11})(1+S_{22})+S_{12}S_{21}\right]}{4S_{21}}\right) \quad (6.6)$$

where Z_{01} and Z_{02} are the impedance of ports 1 and 2 which is 50 Ω in the design.

The scattering matrix of the *m*th mode propagating in the *z*-direction is calculated using full-wave simulator Zeland IE-3D [4]. The normalized phase constant β/k_0 and the attenuation constant α/k_0 of a single EBG unit are illustrated in Fig. 6.8. When β/k_0 is close to the Bragg condition ($\beta\,\Lambda = \pi$), the signal is strongly attenuated. The stopband is very wide and the propagation of any signal with frequency inside it is prohibited. It can also be seen from Fig. 6.8 that considerable rejection is expected in the stopband, when the number of units is increased. There is a significant rise in the loss factor in the stopband due to radiation. Simulation results of a higher frequency range (i.e., larger than 43 GHz) are unfavorable compared to those of a lower frequency range (i.e., less than 18 GHz).

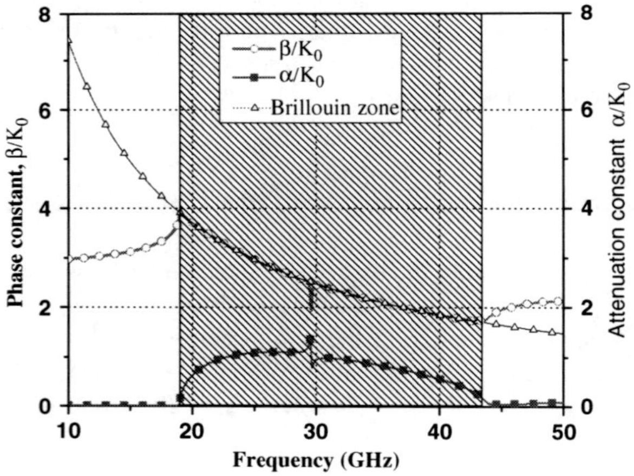

Fig. 6.8 Simulated dispersion diagram of unloaded EBG structure with $d_s = 60$ and $a = 500$ [1]. Copyright/used with permission of/courtesy of IET

The complex characteristic impedance, $Z_c = \mathrm{Re}(Z_c) + j\,\mathrm{Im}(Z_c)$, can be explicitly calculated in terms of the *ABCD* matrix with the elements *A*, *B*, *C*, and *D*:

$$Z_c = \sqrt{\frac{B}{C}} \quad (6.7)$$

when the *ABCD* matrix is converted to *S*-parameters, the complex characteristic impedance can be expressed as

$$Z_c = Z_0 \sqrt{\frac{(1+S_{11})(1+S_{22}) - S_{12}S_{21}}{(1-S_{11})(1-S_{22}) - S_{12}S_{21}}} \tag{6.8}$$

where Z_0 is the characteristic impedance which is 50 Ω. At the low-frequency range, Re (Z_c) starts to rise slowly from 72 Ω in an exponential way, while Im (Z_c) is kept to zero. When the frequency increases, the bandstop occurs, Re (Z_c) and Im (Z_c) become infinite as shown in Fig. 6.9. For the uniform CPW characteristic, the impedance is 48 Ω throughout the frequency range.

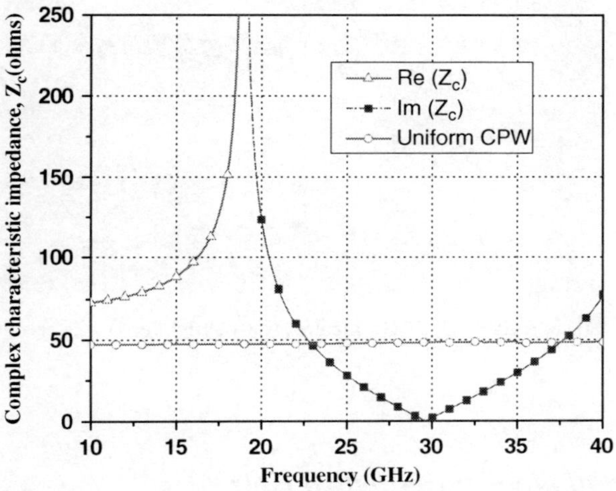

Fig. 6.9 Simulation results of complex characteristic impedance of the unloaded unit EBG structure [1]. Copyright/used with permission of/courtesy of Institute of Physics and IOP Publishing Limited

6.2 Loaded EBG Bandstop Filter

In this section, the loaded unit EBG structure is shown in Fig. 6.10. A ring slot is etched in the ground plane with side length a and ring width r_s. The ring slot is connected to the gap by a narrow transverse slot with length w_s and width d_s. In addition, a square-etched hole with side length b is etched in the center of the ring slot. Compared to the EBG structure without a center-etched hole (i.e., unloaded), the proposed EBG structure includes a ring slot and a square-etched hole and hence is named as loaded EBG structure.

When the side length of the square-etched hole and the ring slot side length are equal (i.e., at $b = a$), the square-etched hole and the ring slot merge to form a square aperture and then the unit cell becomes the unloaded EBG structure. Next, the effect of the dimensions of the square-etched hole (i.e., side length b) on the frequency characteristic, the equivalent circuit model, and the dispersion diagram of the loaded EBG is discussed.

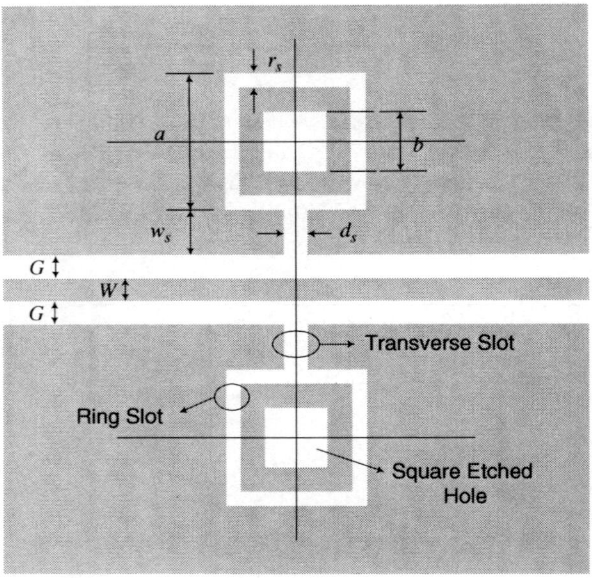

Fig. 6.10 Schematic of loaded unit EBG structure [1]. Copyright/used with permission of/courtesy of IET

6.2.1 Effect of the Square-Etched Hole

In this section, four different sets of dimensions are designed. The square aperture size a is 500 μm × 500 μm and the dimension of transverse slot is maintained at $d_s = 60$ μm and $w = 220$ μm for all the four different designs. A center-etched square hole is added in the area of the unloaded square aperture (see Fig. 6.10). While the width of the ring slot r_s is chosen at 30 μm for all the four designs, the length of the square-etched hole, b, is set at 0, 200, 380, and 440 μm, respectively. Based on the simulation results as shown in Fig. 6.11, all the three designs with $b = 0$, 200, and 440 μm exhibit similar behavior except for the design of $b = 380$ μm. The resonant frequencies for all the four designs are almost identical, but the frequency response begins to drift after the resonant frequency. The loaded structures in the square aperture area are the main contributor to such a variation. It is widely known that low insertion loss and high return loss are expected in the passband of the bandstop filter. An optimum width of the square-etched hole (i.e., at $b = 380$ μm) can clearly be deduced from Fig. 6.11 where the return loss is the highest and the insertion loss is the lowest among all the four different designs.

The existence of the loaded structure in the square aperture perturbs the whole circuit, which implies that the transverse slot is no longer the only factor contributing to the equivalent capacitor. Likewise, the square aperture can no longer act as the

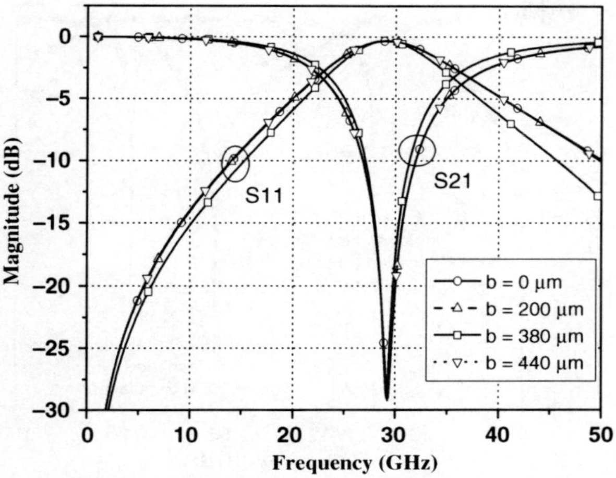

Fig. 6.11 Simulated *S*-parameters of the loaded unit EBG structure versus different center slots within the ring *b* [1]. Copyright/used with permission of/courtesy of IET

main determining factor to the equivalent inductor. Therefore, the additional structure in the square aperture area can significantly improve the passband insertion loss as well as the return loss. Single resonant equivalent circuit is no longer effective to model such a loaded structure.

6.2.2 Modeling of the Loaded EBG Structure

In the previous section, the single resonant parallel *LC* circuit model is employed to model the unloaded unit EBG structure. In the case of the loaded EBG unit, such a model would not be accurate. When the single resonant circuit model is used as shown in Fig. 6.12 the simulation results for the insertion and return losses cannot match with the EM simulation results, especially when the frequency becomes greater than the resonant frequency.

As a result, the loaded EBG unit has to be modeled using a different circuit. By cascading two parallel resonant circuits, the loaded unit EBG structure can be represented in Fig. 6.13.

The equivalent impedance of the cascaded parallel resonant circuits can be expressed as

$$Z = \left(j\omega C_1 + \frac{1}{j\omega L_1}\right)^{-1} + \left(j\omega C_2 + \frac{1}{j\omega L_2}\right)^{-1} \tag{6.9}$$

where subscripts "1" and "2" denote the two different resonant circuits. The reflection coefficient, S_{11}, is defined as

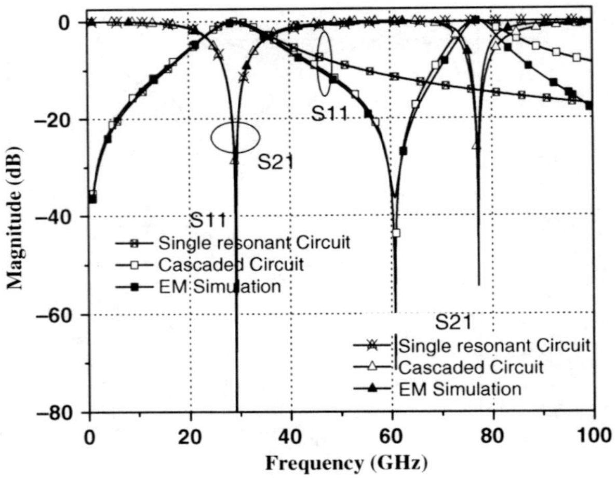

Fig. 6.12 Comparison on parameter extraction and EM simulation results of the loaded unit EBG structure [1]. Copyright/used with permission of/courtesy of IET

$$|S_{11}| = \left|\frac{Z}{2Z_0 + Z}\right| = \frac{\left|\frac{1}{\omega C_1 - 1/\omega L_1} + \frac{1}{\omega C_2 - 1/\omega L_2}\right|}{\sqrt{4Z_0^2 + \left(\frac{1}{\omega C_1 - 1/\omega L_1} + \frac{1}{\omega C_2 - 1/\omega L_2}\right)^2}} \tag{6.10}$$

The maximum value of $|S_{11}|$ can be obtained by differentiating it with respect to the angular frequency, ω, and the two roots obtained are

$$\omega_1 = (L_1 C_1)^{-\frac{1}{2}}, \qquad \omega_2 = (L_2 C_2)^{-\frac{1}{2}} \tag{6.11}$$

The insertion loss is given by

$$|S_{21}| = \left|\frac{2Z_0}{2Z_0 + Z}\right| = \frac{2Z_0}{\sqrt{4Z_0^2 + \left(\frac{1}{\omega C_1 - 1/\omega L_1} + \frac{1}{\omega C_2 - 1/\omega L_2}\right)^2}} \tag{6.12}$$

By differentiating S_{21}, the first notch angular frequency, ω_n, can be obtained as

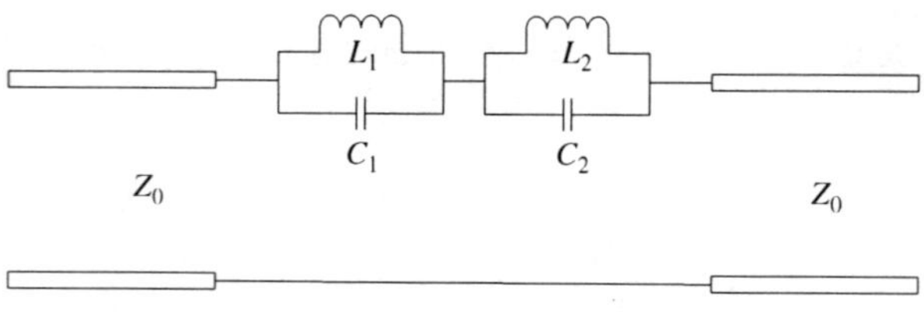

Fig. 6.13 Equivalent circuit for a unit cell of the cascaded parallel resonant EBG structure [1]. Copyright/used with permission of/courtesy of IET

$$\omega_n = \frac{\sqrt{L_1 + L_2}}{\sqrt{L_1/\omega_2^2 + L_2/\omega_1^2}} \tag{6.13}$$

The 3-dB cutoff angular frequency, ω_c, can be determined as

$$|S_{21}||_{\omega=\omega_c} = \frac{2Z_0}{\sqrt{4Z_0^2 + \left(\frac{1}{\omega_c C_1 - 1/\omega_c L_1} + \frac{1}{\omega_c C_2 - 1/\omega_c L_2}\right)^2}} = \frac{1}{\sqrt{2}} \tag{6.14}$$

From Eqs. (6.11), (6.13), and (6.14), the equivalent circuit parameters can be extracted, whereby

$$L_1 = \frac{2Z_0\left(\omega_n^2 - \omega_1^2\right)\left(\omega_1^2 - \omega_c^2\right)\left(\omega_2^2 - \omega_c^2\right)}{\omega_c\omega_1^2\left(\omega_2^2 - \omega_1^2\right)\left(\omega_n^2 - \omega_c^2\right)} \tag{6.15}$$

and

$$L_2 = \frac{2Z_0\left(\omega_2^2 - \omega_n^2\right)\left(\omega_1^2 - \omega_c^2\right)\left(\omega_2^2 - \omega_c^2\right)}{\omega_c\omega_2^2\left(\omega_2^2 - \omega_1^2\right)\left(\omega_n^2 - \omega_c^2\right)} \tag{6.16}$$

The capacitance values can be obtained as

$$C_1 = 1/\omega_1^2 L_1, \quad C_2 = 1/\omega_2^2 L_2 \tag{6.17}$$

All the four parameters L_1, L_2, C_1, and C_2 can be calculated using Eqs. (6.15), (6.16), and (6.17), respectively.

To illustrate the parameter extraction procedure, the values of a, r_s, w_s, d_s, and b are selected as 500, 30, 220, 60, and 380 μm, respectively. The parameters for both the single resonant circuit model and the cascaded circuit model are extracted and summarized in Table 6.3.

Table 6.3 Extracted equivalent circuit parameters of the loaded EBG unit structure

		d_s= 60 μm, w_s= 220 μm, a = 500 μm, r_s= 30 μm			
Square-etched hole, b (μm)		0	200	380	440
First resonant frequency (GHZ)		29.24	29.24	29.24	29.19
First 3-dB cutoff frequency (GHz)		22.42	22.42	23.20	22.34
Second resonant frequency (GHz)		81.77	81.42	77.26	77.77
First notch frequency (GHz)		66.67	66.67	60.76	65.13
Single resonant model	L(nH)	0.285	0.285	0.243	0.286
	C(pF)	0.104	0.104	0.123	0.103
Cascaded model	L_1(nH)	0.282	0.282	0.243	0.285
	C_1(pF)	0.105	0.105	0.122	0.104
	L_2(nH)	0.023	0.022	0.028	0.021
	C_2(pF)	0.168	0.172	0.152	0.195

In order to determine the suitability of the two models for the loaded EBG structure, simulation results of the single resonant and cascaded models are shown in Fig. 6.12. The cascaded circuit simulation results are in good agreement with the EM simulation results at frequency which is below the notch frequency (i.e., at 60.76 GHz). However, the simulation results for the single resonant circuit model drifts always from the EM simulation results after 40 GHz. Therefore, the cascaded circuit model is more appropriate than the single resonant circuit model for the loaded EBG structure and is valid up to the first notch frequency. The extracted equivalent circuit parameters are simulated in Fig. 6.12. The parameters of the single and the cascaded resonant model are listed in Table 6.3. For frequencies below the first notch frequency, the resonant circuit of Fig. 6.13 with the parameters L_1 and C_1 are dominated and compared with the parameters L_2 and C_2.

For the design of parameters L_1 and C_1, the resonant frequency of the proposed circuit does not vary with the change in size of the loaded square-etched hole varies. The extracted parameters for the three designs of b variations, i.e., $b = 0$, 200, and 440 μm, are almost the same. However, a significant difference is noted for the case when $b = 380$ μm. The capacitance C_1 is increased by 18% while the inductance L_1 is decreased by 15% with the same resonant frequency. The increase in capacitance and the decrease in inductance imply that the greater the amount of signals pass through the circuit for both higher and lower frequencies, the lesser the amount of signals get reflected.

This characteristic is desirable in the passband design for the bandstop filters. From both the EM simulation results and the equivalent circuit extraction parameters, the optimum value of the square-etched hole width is that value among all the valid width values for the unit EBG at which the passband performance is the optimum and, for this case, this optimum value is $b = 380$ μm.

6.2.3 Propagation Characteristics of the Loaded Unit EBG

The loaded bandstop filter is designed by cascading the loaded EBG units as shown in Fig. 6.14. The dispersion diagram for the loaded EBG periodic structures, which

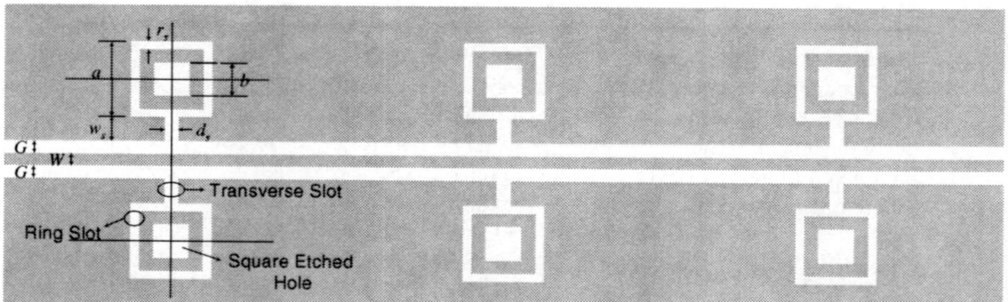

Fig. 6.14 Schematic of loaded EBG bandstop filter

is also known as Brillouin diagram, is useful in analyzing the guided and leaky wave characteristics of the structures. The period of the loaded structure is the same as that of the unloaded EBG. The shaded region from 18 to 42 GHz shows that the electromagnetic wave is prohibited in the periodic structures as shown in Fig. 6.15. There exists a spurious peak in the phase constant curve obtained through the EM simulation and can be overcome by increasing the mesh resolution. The band gap region of the loaded EBG is smaller than the unloaded EBG. The complex characteristic impedance of the loaded structure is shown in Fig. 6.16.

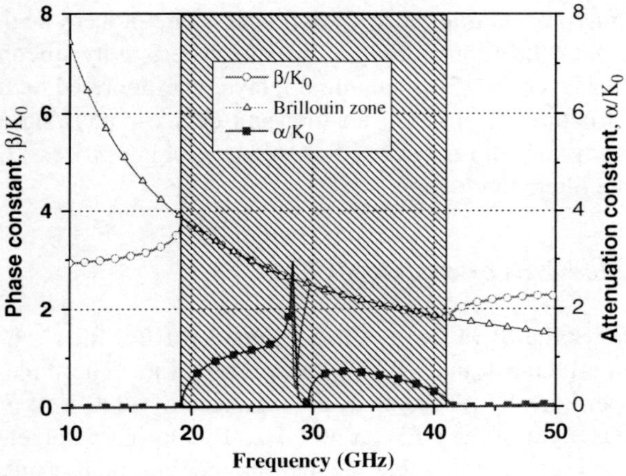

Fig. 6.15 Simulated dispersion diagram of the loaded unit EBG structure [1]. Copyright/used with permission of/courtesy of IET

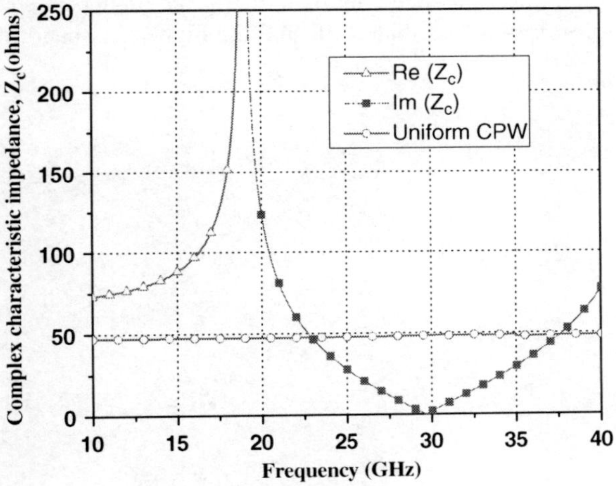

Fig. 6.16 Complex characteristic impedance of the loaded unit EBG structure

The response of the Re (Z_c) and Im (Z_c) impedance is almost identical with that of the unloaded structure, because it is also a periodic structure but with some geometrical differences.

6.3 Experimental Results and Discussions

Based on the equivalent circuit and the extracted parameters for the unit EBG structures (both unloaded and loaded), bandstop filters are designed and fabricated using the proposed EBG unit structure [1, 5, 6]. The bandstop filters are fabricated using surface micromachining fabrication process. The process begins by growing a 0.5-μm-thick SiO_2 layer on a 200-μm-thick high-resistivity silicon substrate that serves as a buffer layer. A 1.5-μm aluminum layer is evaporated on the buffer layer to define the structure. Photoresist coating and device patterning are performed. Finally, aluminum thin film is removed using wet etching process. Only one mask is needed for the filters fabrication.

6.3.1 Unloaded EBG Bandstop Filter

Figure 6.17 shows the SEM image of the unloaded EBG filter which consists of three EBG unit structures and a CPW transmission line. The three cascaded unit structures are periodically placed apart at a distance of 2020 μm, which is half of the guided wavelength of the CPW at 30 GHz. The design parameters of the filter a, w_s, and d_s are presented in Table 6.2 and the measurement results are shown in Fig. 6.18

The measurement result shows that the 20-dB stopband range is from 24.4 to 40.6 GHz. The stopband return loss is less than 2 dB. The return loss is greater than 10 dB for the lower passband and is 6 dB for the higher passband. The lower passband insertion loss is less than 2 dB and the higher passband insertion loss is below 5 dB.

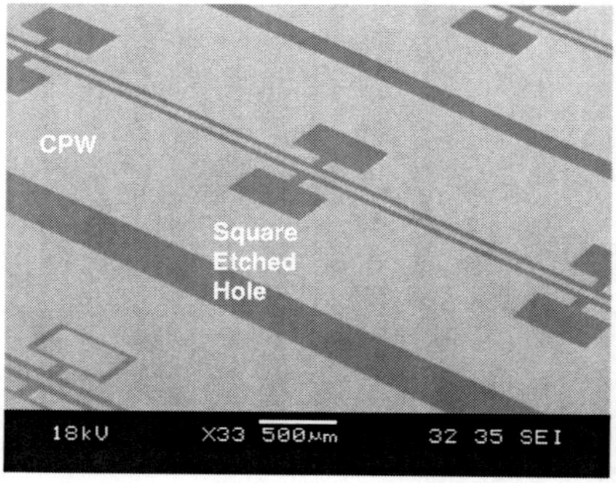

Fig. 6.17 SEM image of the unloaded EBG bandstop filter

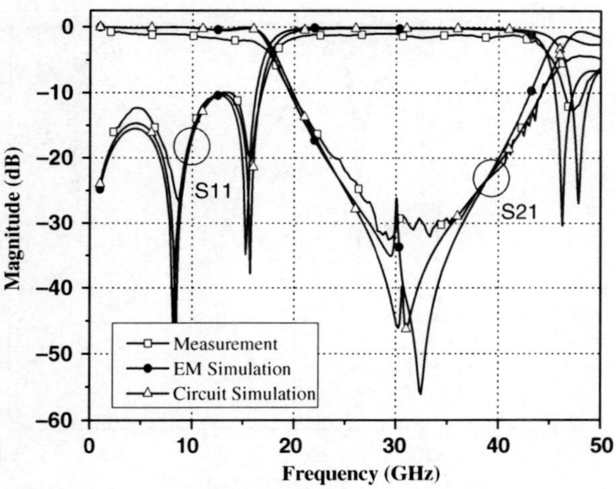

Fig. 6.18 Comparison of measurement and the simulation results (EM and circuit simulations) of the unloaded EBG bandstop filter [1]. Copyright/used with permission of/courtesy of IET

6.3.2 Loaded EBG Bandstop Filter

Figure 6.19 shows the SEM image of the loaded EBG filter which consists of three loaded EBG unit structures and a CPW transmission line. The design parameters of the loaded EBG a, r_S, w_S, d_S, and b are presented in Table 6.3. The measurement and simulation results of the EBG bandstop filter are compared as shown in Fig. 6.20. The simulation results and the equivalent circuit are in good agreement with the measured results, which show the validity of the proposed equivalent circuit.

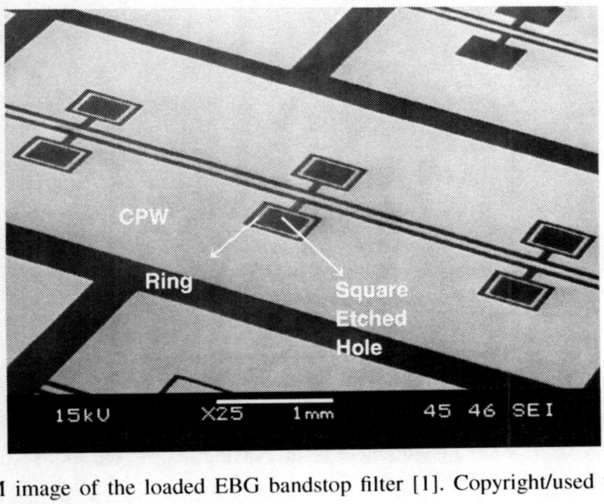

Fig. 6.19 SEM image of the loaded EBG bandstop filter [1]. Copyright/used with permission of/courtesy of IET

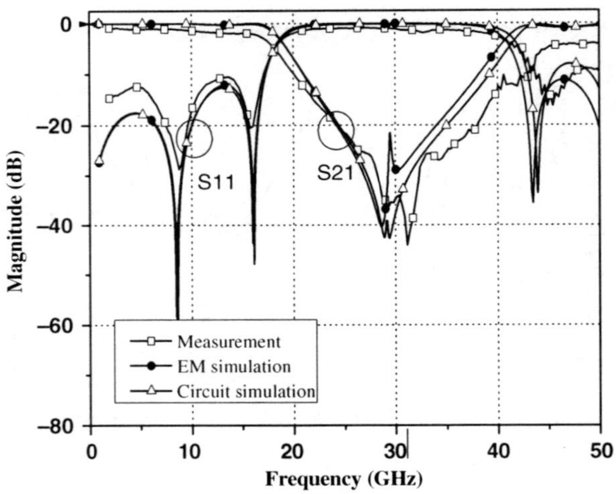

Fig. 6.20 Comparison of the measurement and simulation results of the loaded EBG band-stop filter [1]. Copyright/used with permission of/courtesy of IET

The measured results show a relatively wide 20-dB stopband from 24.8 to 38 GHz with the return loss of less than 2 dB. For the passband, the return loss is higher than 10 dB except for those at the higher frequencies. The lower passband insertion loss is less than 2 dB and the higher passband insertion loss is less than 4.5 dB. The measurement results of the loaded and the unloaded EBG bandstop filters are compared as shown in Fig. 6.21. Except for the 20-dB stopband width, it is

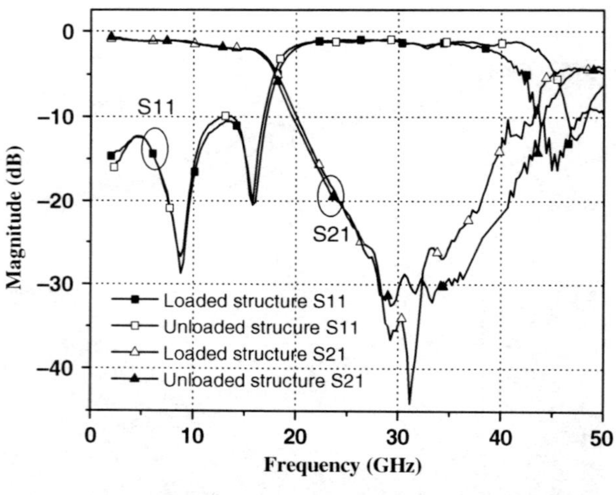

Fig. 6.21 Comparison of the measurement results of the loaded and unloaded EBG bandstop filters

noted that the performances of the unloaded EBG bandstop filter are more inferior compared to the loaded EBG bandstop filter in all aspects. In particular, the higher passband performances of the loaded EBG filter are better than the unloaded EBG filter.

The reason for this improvement lies in the equivalent inductance of the loaded unit EBG structure, which is much lower than the unloaded unit EBG structure. As for the equivalent capacitance, the loaded unit yields a higher capacitance than the unloaded unit. It is noted that both the passband insertion loss and the stopband return loss characteristics are very flat and contain less than 0.2 ripples for the two different filter designs.

6.4 Design of Tunable Bandstop Filter

The bandstop filter consists of the cascaded unit EBG. The number of the cascaded EBG units depends on three different factors such as insertion loss, rejection level, and bandwidth. Figure 6.22a, b shows that the rejection level and the bandwidth increase with the number of the unit EBG cells, which also leads to an increase of the insertion loss because of higher conductor loss. Based on the optimum design, the bandstop filter with three unit EBG cells has a high rejection level of 40 dB with insertion loss of 0.45 dB.

In the tunable EBG bandstop filter, the CPW transmission line has a signal strip with a width of 70 μm and a gap width of 115 μm as shown in Fig. 6.23. The transmission line has high impedance of 65 Ω. This is because the dimension of the CPW is designed by taking into account of the periodical loading effects of MEMS capacitive bridges. There are a total of eight micromachined bridges in each transmission line between the two EBG structures. The width and length of the bridges are 50 and 300 μm, respectively [7–9]. The periodic spacing between the two bridges is s =200 μm.

Each micromachined bridge acts as a tuning element because of the changing bridge height, i.e., varying the capacitance, when the DC bias voltage is applied between the signal line and the ground plane. Hence, the frequency of the band rejection can be tuned with different bias levels. An equivalent *LC* circuit is used to describe the tunable EBG structure as shown in Fig. 6.24. The lumped capacitance C_1 is basically due to the transverse slot on the ground, while the inductance L_1 is related to the magnetic flux passing through the aperture on the ground, i.e., a. Since the micromachined bridges are shunt capacitive, they can be modeled as an equivalent parallel capacitor C_2 and in series with a inductor L_2 in each resonator section, as shown in Fig. 6.24.

The high impedance transmission line Z_h compensates for the decrease in impedance due to the increase in shunt capacitance, when the bridges are slowly moved down by the driven voltage. The equivalent circuit parameters are summarized in Table 6.4. These parameters are numerically extracted from the circuit simulation results of the EBG structure using Agilents ADS software.

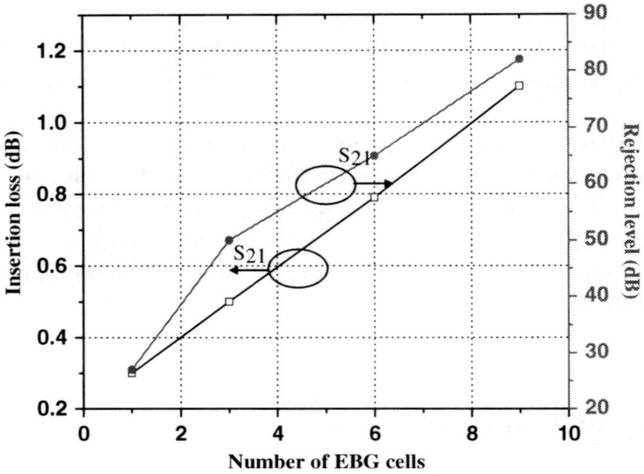

(a)

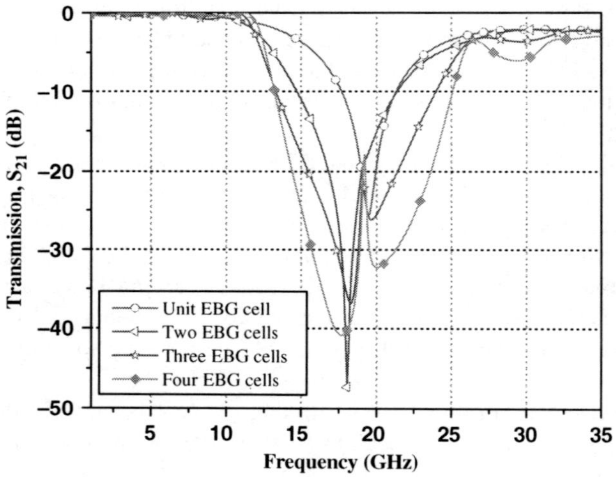

(b)

Fig. 6.22 Simulation results of the number of unit EBG cells: (**a**) insertion loss and rejection level and (**b**) bandwidth variation [2]. Copyright/used with permission of/courtesy of Institute of Physics and IOP Publishing Limited

Current distribution for this structure is obtained using IE-3D software as shown in Fig. 6.25. It is noted that the transmission power cannot pass through the circuit and maximally attenuated at the bandstop frequency of 19 GHz. The current distribution on the micromachined bridges also shows the same phenomena of negligible current flowing on the micromachined bridges.

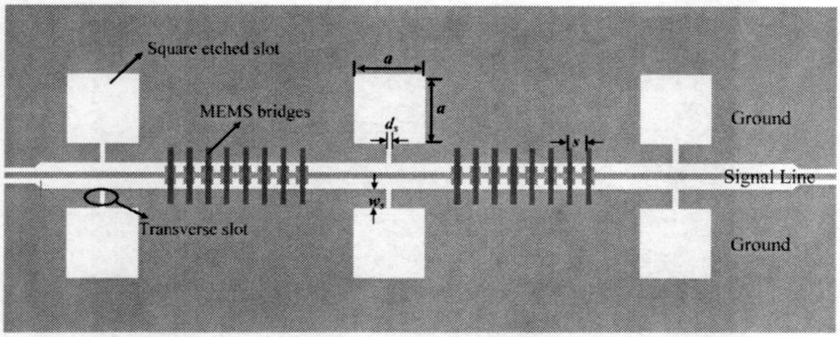

Fig. 6.23 Schematic of the tunable bandstop filter using MEMS bridges [2]. Copyright/used with permission of/courtesy of Institute of Physics and IOP Publishing Limited

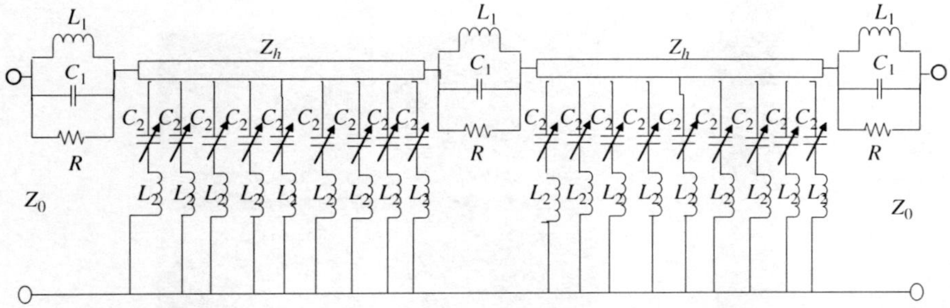

Fig. 6.24 Equivalent parallel resonant circuit for the tunable bandstop filter [2]. Copyright/used with permission of/courtesy of Institute of Physics and IOP Publishing Limited

Table 6.4 Extracted equivalent circuit parameters of the tunable EBG structure

Variable capacitance C_2 (pF)	0.015	0.017	0.022	0.025
Resonant frequency (GHz)	19.10	18.70	17.6	17.4
Inductance L_1(nH)	0.49	0.49	0.49	0.49
Fixed capacitance C_1 (pF)	0.13	0.13	0.13	0.13
Resistance (kΩ)	1.80	1.82	1.85	1.86

An important phenomenon, the propagation characteristic of the micromachined bridges, is analyzed for the first time. Micromachined bridges are periodically placed so the slow wave factor that varies with the number of bridges and the distance between the bridges are investigated. A high impedance CPW transmission line without any perturbation is loaded with the micromachined bridges. Figure 6.26a shows the dispersion diagram of the periodically loaded micromachined bridges. The distance between the bridges is fixed at 200 μm while the number of bridges is varied. The band gap for the eight bridges starts around 15 GHz while for the six and four bridges at 16 and 16.5 GHz, respectively. Although the

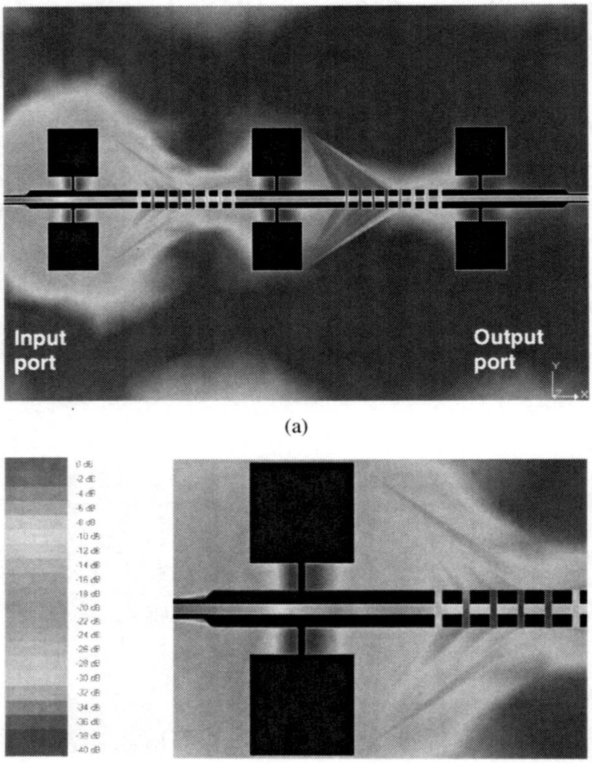

(b)

Fig. 6.25 Simulation result of the current distribution in the micromachined tunable filter: (**a**) overview and (**b**) zoom view [2]. Copyright/used with permission of/courtesy of Institute of Physics and IOP Publishing Limited

bandstop behavior of the eight bridges starts early, there is very little variation in the dispersion at higher frequencies. The results for the bridges with a periodic distance of 400 and 600 μm are also shown in Fig. 6.26b, c, respectively. The bridges with a periodic distance of 200 μm has a phase constant of 100 compared to 50 and 25 for the 400 and 600-μm distance bridges, respectively. Therefore, eight bridges with a periodic distance of 200 μm are used on each section of the tunable EBG bandstop filter.

6.4.1 Experimental Results and Discussions

The SEM image of the MEMS tunable bandstop is shown in Fig. 6.27. Figure 6.28 shows the measurement results of the *S*-parameters when the bias voltage between

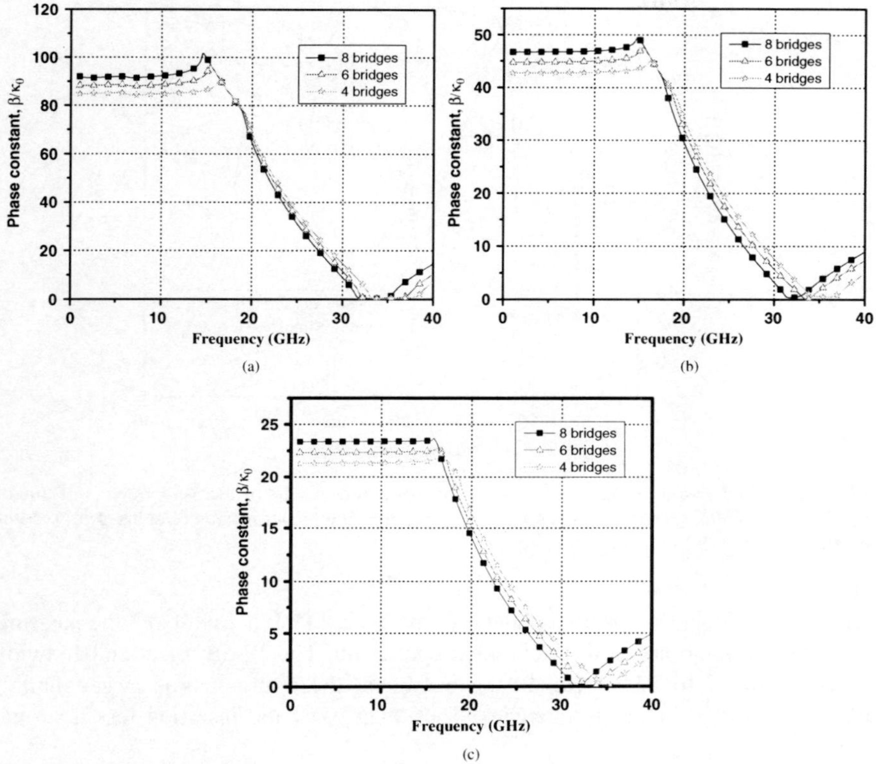

Fig. 6.26 Dispersion diagram of the micromachined bridges loaded on CPW transmission line (**a**) at a period of 200 μm, (**b**) at a period of 400 μm, and (**c**) at a period of 600 μm [2]. Copyright/used with permission of/courtesy of Institute of Physics and IOP Publishing Limited

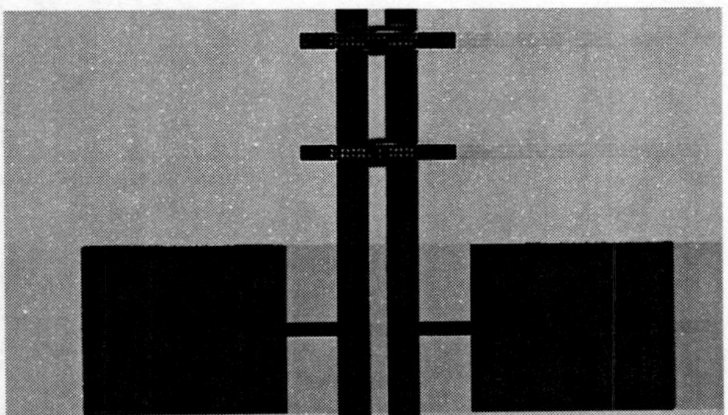

Fig. 6.27 SEM image of the fabricated MEMS tunable EBG bandstop [2]. Copyright/used with permission of/courtesy of Institute of Physics and IOP Publishing Limited

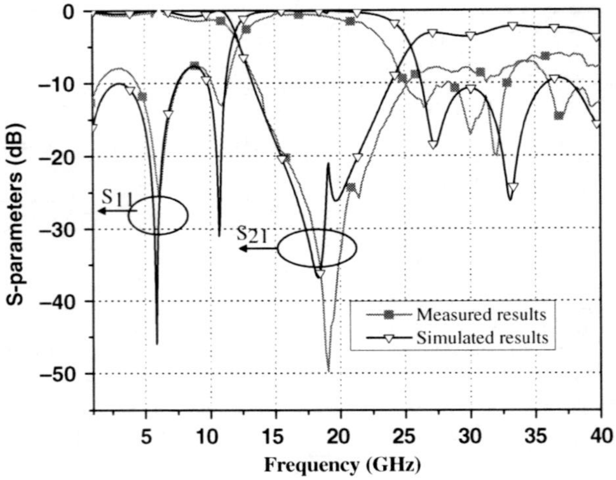

Fig. 6.28 Comparison of the simulation and measurement results of the *S*-parameters of tunable EBG bandstop filter (zero bias voltage) [2]. Copyright/used with permission of/courtesy of Institute of Physics and IOP Publishing Limited

the signal strip and the ground plane is zero [2, 10, 11]. It is found that the measurement and simulation results are in good agreement. The 20-dB rejection bandwidth varies from 17 to 22.5 GHz. In the passband, the return loss is larger than 10 dB except in the higher frequencies. The lower passband insertion loss is around 0.7–1.1 dB.

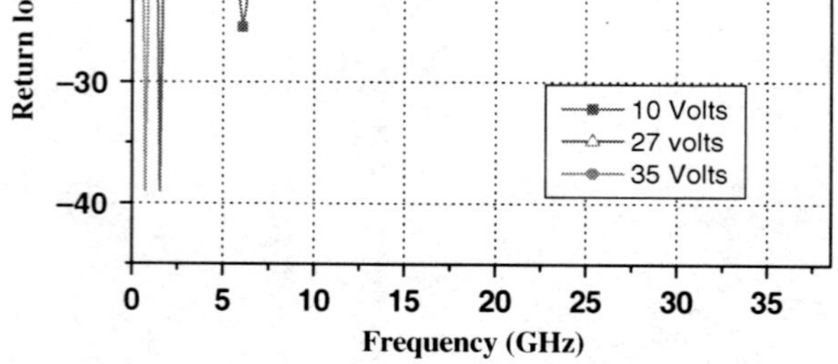

(a)

Fig. 6.29 Measurement results of the tunable bandstop filter: (**a**) return loss, S_{11}, and (**b**) insertion loss, S_{21} [2]. Copyright/used with permission of/courtesy of Institute of Physics and IOP Publishing Limited

Fig. 6.29 (continued)

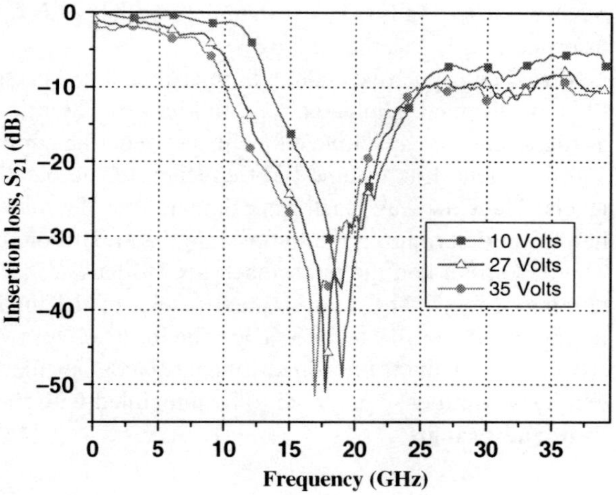

(b)

The measurement results of the fabricated tunable EBG bandstop filter are shown in Fig. 6.29. There are eight bridges in each section. The initial gap height of the bridges is 2 μm. The 20-dB optimum bandwidth of the EBG bandstop filter is 5.5 GHz. When a bias voltage of 0–35 V is applied to the bridges and the frequency is tuned from 19 to 17.3 GHz, this is basically an analog tuning in which the resonant frequency is varied continuously by changing the capacitance of the MEMS bridges. Overlaps between the tuning frequencies are observed because of continuous tuning, as shown in Fig. 6.29. Most often, continuous tuning is desirable in MEMS applications so that desired frequency bands can be rejected or selected. The lower passband insertion loss varies from 0.7 to 2.2 dB due to the change in the gap height of the bridges.

6.5 Summary

In this chapter, the bandstop filter using unloaded and loaded structures is investigated. The frequency characteristics of the EBG unit were studied using the equivalent circuit model. The extracted parameters show that the band gap effect of the EBG units can be accurately interpreted based on the circuit analysis theory. The complex characteristic impedance is estimated which is helpful for analyzing the effect of impedance change in the EBG structure. The bandstop filter is also designed by cascading the unloaded and loaded EBG units. The measured results of the loaded structures show enhanced passband performance compared to the unloaded structure. Due to higher equivalent capacitance and lower equivalent inductance, it helps in decreasing the insertion loss and increasing the return loss at higher frequencies. The 20-dB stopband is as wide as 13.2 GHz. The lower passband insertion loss is less than 2 dB with 0.2 dB ripple, and the higher passband insertion loss is less than 4.5 dB.

Based on capacitive change of micromachined bridges, a tunable EBG bandstop filter is designed, simulated, and fabricated. The micromachined bridges are used as high-contrast capacitive elements between the coplanar waveguide ground plane and the signal line to tune the frequency of the band rejection with different bias levels. The slow wave factor is calculated for the micromachined bridges, which is helpful in determining the tuning range and periodic behavior of the tunable filter. The EBG filter and the micromachined bridges are fabricated by surface micromachining process. The measurement results of the tunable bandstop filter show the tuning range from 19 to 17.3 GHz. The insertion loss varies from 0.7 to 2.2 dB. The advantages of micromachined tunable bandstop filters are low cost, compactness and has an immense potential to be integrated with the CMOS devices for wireless front-end systems.

References

1. M. F. Karim, A. Q. Liu, A. B. Yu, A. Alphones and X. J. Zhang, "CPW bandstop filter using unloaded and loaded EBG structures," IEE Proc. Microwave Antenna Propag. Vol. 152(6), 2005, pp. 434–440.
2. M. F. Karim, A. Q. Liu, A. Alphones and A. B. Yu, "A tunable bandstop filter via the capacitance change of micromachined switches," J. Micromech. Microeng. Vol. 16(4), 2006, pp. 851–861.
3. L. Garde, M. J. Yabar and C. D. Rio, "Simple modeling of DGS to design 1D PBG low pass filter," Microwave Optical Tech. Lett. Vol. 37(3), 2003, pp. 228–232.
4. S. G. Mao and M. Y. Chen, "Propagation characteristics of finite width conductor backed coplanar waveguides with periodic electromagnetic bandgap cells," IEEE Trans. Microwave Theory Tech. Vol. 50, 2002, pp. 2624–2628.
5. X. J. Zhang, M. F. Karim, A. Q. Liu, A. B. Yu and C. L. Law, "RF MEMS band-stop filter based on photonic bandgap (PBG) structure," Asia-Pacific Conference of Transducers and Micro-Nano Technology, Sapporo, Japan, 25–27 July 2004.
6. X. J. Zhang, A. Q. Liu, M. F. Karim, A. B. Yu and Z. X. Shen, "MEMS based photonic bandgap (PBG) bandstop filter," IEEE Microwave Theory and Techniques Society Digest (MTT-S), Boston, MA, USA, 8–12 June 2004, pp. 1463–1466.
7. A. B. Yu, A. Q. Liu, X. Q. Zhang, L. Zhu, A. Alphones and S. A. Peter, "Improvement of isolation for RF MEMS capacitive shunt switch via membrane planarization," Sensor. Actuator. Phys. A Vol. 19, 2005, pp. 206–213.
8. A. B. Yu, A. Q. Liu, Q. X. Zhang and H. M. Hosseini, "Effects of surface roughness on electromagnetic characteristics of capacitive switches," J. Micromech. Microeng. Vol. 16, 2006, pp. 2157–2166.
9. A. B. Yu, A. Q. Liu, J. Oberhammer, Q. X. Zhang and H. M. Hosseini, "Characterization and optimization of dry releasing for the fabrication of RF MEMS capacitive switches," J. Micromech. Microeng. Vol. 17, 2007, pp. 2024–2030.
10. M. F. Karim, A. Q. Liu, A. Alphones and A. B. Yu, "Micromachined tunable bandstop filters for wireless sensor networks," IEEE Sensors, USA, 30 October–3 November, 2005, pp. 151–154.
11. A. Q. Liu, A. B. Yu, M. F. Karim and M. Tang, "RF MEMS switches and integrated switching circuit," J. Semicond. Technol. Sci. Vol. 7(3), 2007, pp. 166–176.

Chapter 7
Tunable Electromagnetic Bandpass Filter and Reconfigurable Circuit

In this chapter, a tunable Fabry–Perot (F–P) bandpass filter and a reconfigurable micromachined switching filter are studied. The F–P bandpass filter is designed by cascading the unit EBG structures. The theoretical aspects of the F–P cavity and the EBG structures are discussed for the filter design. A defect or F–P cavity is created in the periodic structures to generate bandpass phenomena. Parametric analysis on the defect length and slot dimensions is performed and the design of the F–P filter is optimized. Micromachined switches are employed in the defect cavity as tuning elements for bandpass frequency tuning.

The reconfigurable filter based on DC-contact micromachined switches is designed and fabricated. An equivalent circuit model is derived for the reconfigurable unit structure. Extracted parameters show the characteristics of both bandpass and bandstop filters which can be accurately analyzed using circuit analysis. The bandpass filter is designed by cascading the unit structures. This bandpass filter can be reconfigured as a bandstop filter using the DC-contact micromachined switches. Dispersion characteristics are obtained to analyze the electromagnetic wave behavior within the unit structure for the bandpass and bandstop filters based on Floquet's theorem. The rest of the chapter presents the experimental results and discussions.

7.1 Theory of the Fabry–Perot of Tunable Filter

The Fabry–Perot interferometer is traditionally associated with optical resonators or etalons which consists of two parallel and partially reflecting plates separated by a dielectric of thickness, known as a cavity. When collimated light is coupled through the plates, resonance occurs whenever they are separated by an integral number of the half wavelength of the light [1, 2]. This concept transfers virtually intact into the microwave region; the difference in implementation arises from the fact that the dimension of microwave instrumentation is typically comparable with the wavelength of the electromagnetic radiation [3]. For microwave transmission line

Muhammad Faeyz Karim and Ai Qun Liu

A.Q. Liu, *RF MEMS Switches and Integrated Switching Circuits*,
MEMS Reference Shelf 5, DOI 10.1007/978-0-387-46262-2_7,

filters, Q decreases by $1/\sqrt{f}$ due to high conductor losses. Therefore, the physical size of a cavity operating in low-order mode is too small to be practical at high frequencies. Moreover, when a high-order mode is used, the resonance of nearby modes is very close in frequency and there is no separation between them, making such resonators unusable.

A conceptual way of avoiding these difficulties is to remove the sidewalls of the cavity resonator, which has the effect of reducing conductor losses and the number of possible resonant modes, as shown in Fig. 7.1. This is also known as Fabry–Perot cavity. For the resonator, the plates must be parallel and the extent must be large enough so that no significant radiation leaves the region between the plates. These constraints can be reduced by using spherical- or parabolic-shaped reflecting mirrors to focus and confine the energy to a stable mode pattern. For example, a quasi-optical resonator is useful at millimeter wave and sub-millimeter wave frequencies applicable at infrared and visible wavelengths.

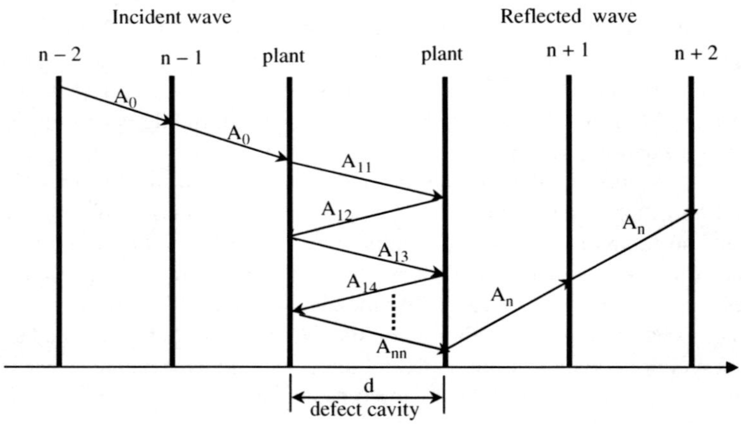

Fig. 7.1 Schematic of a Fabry–Perot filter with multilayer parallel plates

Figure 7.1 shows the schematic of a Fabry–Perot filter with multilayer parallel plates; the transverse electromagnetic (TEM) standing wave field between the two plates of the defect cavity is expressed as

$$E_x = E_0 \sin k_0 z \tag{7.1}$$

$$H_y = j\frac{E_0}{\eta_0} \cos k_0 z \tag{7.2}$$

where E_0 is an arbitrary amplitude constant, $\eta_0 = 377\,\Omega$ is the intrinsic impedance of free space, and k_0 is the wave number. These fields satisfy the boundary condition $E_x = 0$ at $z = 0$ and the boundary conditions of $E_x = 0$ at $z = d$. The resonant frequency can be expressed as

$$k_0 d = l\pi, \qquad l = 1,\ 2,\ 3, \ldots, n \tag{7.3}$$

$$f_0 = \frac{ck_0}{2\pi} = \frac{cl}{2d}, \quad l = 1, 2, 3, \ldots, n \tag{7.4}$$

where l is only a single index for these modes, as opposed to three indices for rectangular or cylindrical cavities, which is a consequence without the conducting walls.

The stored electric energy per square meter of cross section is

$$W_e = \frac{\varepsilon_0}{4} \int_{z=0}^{d} |E_x|^2 dz = \frac{\varepsilon_0 |E_0|^2 d}{8} \tag{7.5}$$

and the stored magnetic energy per square meter of cross section is

$$W_m = \frac{\mu_0}{4} \int_{z=0}^{d} |H_y|^2 dz = \frac{\varepsilon_0 |E_0|^2 d}{8} \tag{7.6}$$

which is equal to the stored electric energy. The power lost per square meter in both conducting plates is

$$P_c = 2\left(\frac{R_s}{2}\right) |H_y(z=0)|^2 = \frac{R_s |E_0|^2}{\eta_0^2} \tag{7.7}$$

Therefore, the conductor loss quality factor Q_c can be determined as

$$Q_c = \frac{\omega (W_e + W_m)}{P_c} = \frac{\pi l \eta_0}{4R_s} \tag{7.8}$$

Equation (7.8) shows that Q_c increases in proportion to the mode number, l; which is often several thousands or more for such resonators. If the region between the two plates is filled with a dielectric material with a loss factor, $\tan\delta$, the quality factor for the dielectric loss Q_d can be expressed as

$$Q_d = \frac{1}{\tan\delta} \tag{7.9}$$

This equation is useful for finding the quality factor for the EBG F–P filter.

7.2 Design of the EBG Circuit

Periodic structures or EBG structures are basically an infinite transmission line or waveguide periodically loaded with reactive elements. The loading elements are formed as discontinuities in the transmission line; the model of lumped reactance across the transmission line is shown in Fig. 7.2. The periodic structures support

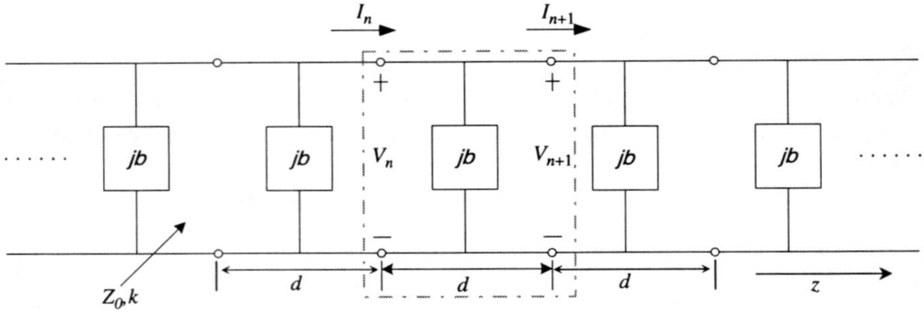

Fig. 7.2 Equivalent circuit of a periodically loaded transmission line

slow wave propagation (slower than the phase velocity of the unloaded line), which can provide the function of passband and stopband behavior that can be used to design the bandpass and bandstop filters.

7.2.1 Capacitively Loaded Transmission Line Circuit Analysis

The wave propagation characteristic of the infinite loaded line is shown in Fig. 7.2. Each unit structure of this line consists of a length d of transmission line and a shunt susceptance across the midpoint of the line; the shunt susceptance b is normalized to the characteristic impedance Z_0. The capacitively loaded transmission line can be analyzed to conceive the idea on periodic structures. The infinite line can be considered as a cascade of identical two-port networks. The voltage and current on either side of the nth unit structure matrix can be expressed as

$$\begin{bmatrix} V_n \\ I_n \end{bmatrix} = \begin{bmatrix} A & B \\ C & D \end{bmatrix} \begin{bmatrix} V_{n+1} \\ I_{n+1} \end{bmatrix} \quad n = 1, 2, 3, \ldots, N \tag{7.10}$$

where A, B, C, and D are the matrix parameters, V_n is the voltage at the nth unit structure, and I_n is the current at the nth unit structure.

As shown in dotted lines of Fig. 7.2, the $ABCD$ matrix calculation for the shunt susceptance b sandwitched between two transmission lines of length $d/2$ can be expressed as [4]

$$\begin{bmatrix} A & B \\ C & D \end{bmatrix} = \begin{bmatrix} \cos\frac{\theta}{2} & j\sin\frac{\theta}{2} \\ j\sin\frac{\theta}{2} & \cos\frac{\theta}{2} \end{bmatrix} \begin{bmatrix} 1 & 0 \\ jb & 1 \end{bmatrix} \begin{bmatrix} \cos\frac{\theta}{2} & j\sin\frac{\theta}{2} \\ j\sin\frac{\theta}{2} & \cos\frac{\theta}{2} \end{bmatrix} \tag{7.11}$$

and then Eq. (7.11) can be simplified as

$$\begin{bmatrix} A & B \\ C & D \end{bmatrix} = \begin{bmatrix} \left(\cos\theta - \frac{b}{2}\sin\theta\right) & j\left(\sin\theta + \frac{b}{2}\cos\theta - \frac{b}{2}\right) \\ j\left(\sin\theta + \frac{b}{2}\cos\theta + \frac{b}{2}\right) & \left(\cos\theta - \frac{b}{2}\sin\theta\right) \end{bmatrix} \tag{7.12}$$

where $\theta = kd$ and k is the propagation constant of the unloaded line. For any wave propagating in the $+z$ direction, the voltage and current can be expressed as

$$V(z) = V(0)e^{-\gamma z} \tag{7.13a}$$

$$I(z) = I(0)e^{-\gamma z} \tag{7.13b}$$

for a phase reference at $z = 0$. Since the structure has an infinite length, the voltage and current at the nth terminals can differ from the voltage and current at the $n + 1$ terminals only by the propagation factor, $e^{-\gamma d}$. Thus, it can be expressed as

$$V_{n+1} = V_n e^{-\gamma d} \tag{7.14a}$$

Substituting Eq. (7.14) into Eq. (7.10), it can be expressed as

$$\begin{bmatrix} V_n \\ I_n \end{bmatrix} = \begin{bmatrix} A & B \\ C & D \end{bmatrix} \begin{bmatrix} V_{n+1} \\ I_{n+1} \end{bmatrix} = \begin{bmatrix} V_{n+1}e^{\gamma d} \\ I_{n\_+1}e^{\gamma d} \end{bmatrix} \tag{7.15a}$$

or

$$\begin{bmatrix} A - e^{-\gamma d} & B \\ C & D - e^{-\gamma d} \end{bmatrix} \begin{bmatrix} V_{n+1} \\ I_{n+1} \end{bmatrix} = 0 \tag{7.15b}$$

For a nontrivial solution, the determinant of the above matrix can be expressed as

$$AD + e^{2\gamma d} - (A + D)e^{\gamma d} - BC = 0 \tag{7.16}$$

Using the property of $ABCD$ matrix for the reciprocal network, $AD - BC = 1$, then Eq. (7.16) can be simplified as

$$1 + e^{2\gamma d} - (A + D)e^{\gamma d} = 0 \tag{7.17}$$

$$e^{-\gamma d} + e^{\gamma d} = A + D \tag{7.18}$$

Therefore,

$$\cosh \gamma d = \frac{A + D}{2} = \cos\theta - \frac{b}{2}\sin\theta \tag{7.19}$$

If $\gamma = \alpha + j\beta$, Eq. (7.19) can be expressed as

$$\cosh \gamma d = \cosh \alpha d \cos \beta d + j \sinh \alpha d \sin \beta d = \cos \theta - \frac{b}{2} \sin \theta \tag{7.20}$$

Since the right-hand side of Eq. (7.20) is real, then either $\alpha = 0$ or $\beta = 0$.

When $\alpha = 0$ and $\beta \neq 0$, the structure has a non-attenuating behavior and the waves propagate along the periodic structure and it generates a passband phenomena. Hence, Eq. (7.20) can be expressed as

$$\cos \beta d = \cos \theta - \frac{b}{2} \sin \theta \tag{7.21}$$

When the magnitude of the right-hand side is less than or equal to unity, there are an infinite number of values of β that satisfy Eq. (7.21).

When $\alpha \neq 0$ and $\beta = 0$, there is no wave propagation, i.e., the waves are reflected back to the input of the transmission line. This defines the stopband phenomena. Hence, Eq. (7.21) is simplified to

$$\cos \alpha d = \left| \cos \theta - \frac{b}{2} \sin \theta \right| \geq 1 \tag{7.22}$$

Equations (7.21) and (7.22) are very important for the understanding of the propagation of EM waves through periodic structures. Attenuated waves generate the appearance of stopbands while the propagation of unattenuated waves exhibits passband regions. Therefore, these derivations can help to explain the stopband and passband characteristics of EBG.

The EBG bandpass filter is designed based on the F–P mathematical model.

7.3 Design of Fabry–Perot Tunable Bandpass Filter

A uniplanar 1D EBG structure is realized on a CPW structure. These unit EBG structures are placed at a period which is equal to half of the center frequency wavelength. A defect cavity is realized when the periodicity is changed. This is shown in Fig. 7.3. The center resonant line (defect cavity) with periodic EBG reflectors on both sides is analogous to Fabry–Perot (F–P) resonators. The prohibition of wave propagation in the forbidden gap makes the periodic structures to act as reflectors. The waves are bounded back and forth within the defect cavity. The defect cavity L_R is defined as

$$L_R = 1.5\, d \tag{7.23}$$

where d is the period, i.e., the distance between each unit EBG structures.

The arrangement of the unit EBG structures satisfies the Bragg conditions. The propagation constant of the stopband generated is given by

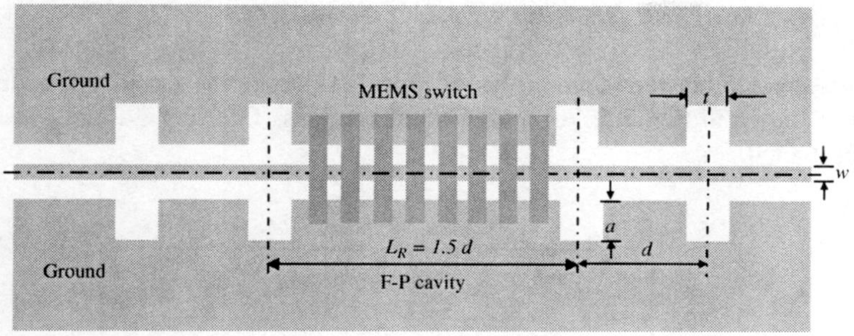

Fig. 7.3 Schematic of EBG Fabry–Perot filter

$$\beta = \frac{2\pi}{\lambda_g} \tag{7.24}$$

where λ_g is the guided wavelength and

$$d = \frac{\lambda_g}{2} \tag{7.25}$$

An equivalent circuit model of the EBG Fabry–Perot filter is shown in Fig. 7.4. Each unit EBG structure is modeled with a parallel LC resonant circuit. L_1 and C_1 are the inductance and capacitance of the etched EBG on the left- and right-hand sides of the defect cavity, respectively. The defect cavity, L_R, is loaded with MEMS capacitive switches and modeled as C_2 and L_2. The extracted parameters of the F–P EBG filter are listed in Table 7.1.

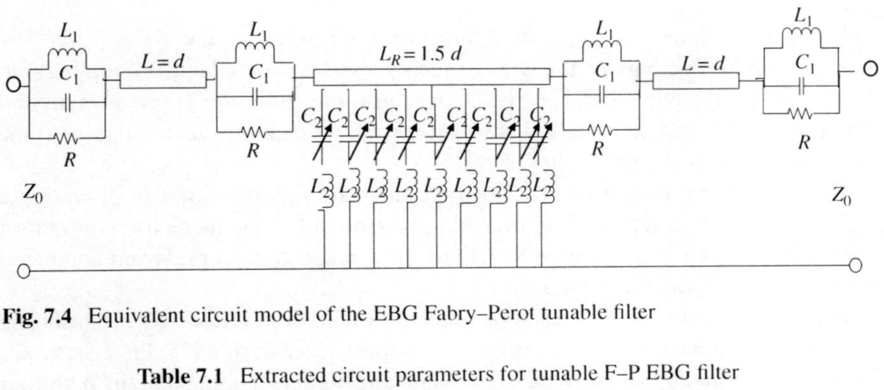

Fig. 7.4 Equivalent circuit model of the EBG Fabry–Perot tunable filter

Table 7.1 Extracted circuit parameters for tunable F–P EBG filter

Variable capacitance C_2 (pF)	0.015	0.017	0.022	0.025
Resonant frequency (GHz)	11.10	10.70	9.6	9.4
Inductance L_1 (nH)	0.8	0.8	0.8	0.8
Fixed capacitance C_1 (pF)	0.13	0.13	0.13	0.13

7.3.1 Effects of the Number of EBG Structures

The relationship between the number of unit EBG structures associated with the Q factor is shown in Fig. 7.5. For an optimum bandpass filter design, the Q-factor is greater than 40.

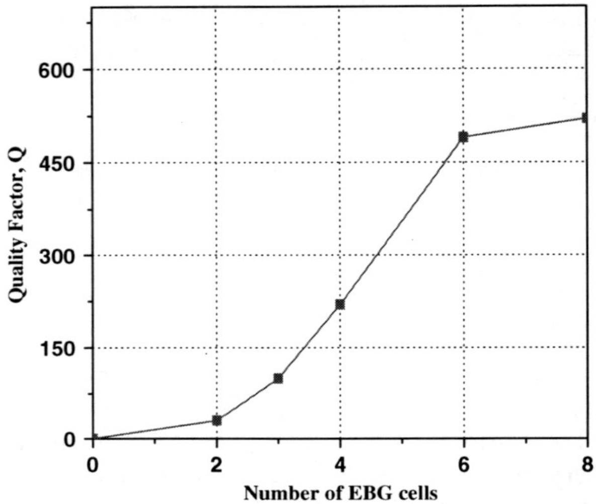

Fig. 7.5 Simulation results for the relationship between the Q factor versus the number of EBG structures

When the number of EBG structures on both sides of the defect cavity is increased, the reflectivity of the Bragg mirrors also increases, but the coupling into the cavity is reduced. When four EBG structures are used, the Q factor is around 180. It increases approximately to 480, after which the energy accumulation in the cavity saturates as shown in Fig. 7.5.

The dielectric constant of the substrate plays an important role in providing a high Q factor. Specifically, Q is inversely proportional to the dielectric constant of the substrate. As a result, a high-resistivity silicon substrate is preferred compared to low-resistivity silicon substrate.

Figure 7.6 shows the influence of numbers of EBG structures on the insertion loss and the attenuation. For an efficient bandpass filter design [5], the criteria for insertion loss should be less as possible while attenuation should be high. With four EBG structures, the insertion loss and attenuation are 0.9 and 23 dB, respectively, while for eight EBG structures, the insertion loss and attenuation are 10.1 and 40 dB, respectively. It is observed that the attenuation increases at the expense of the insertion loss. The reason behind this is that the conductor loss increases with the size of

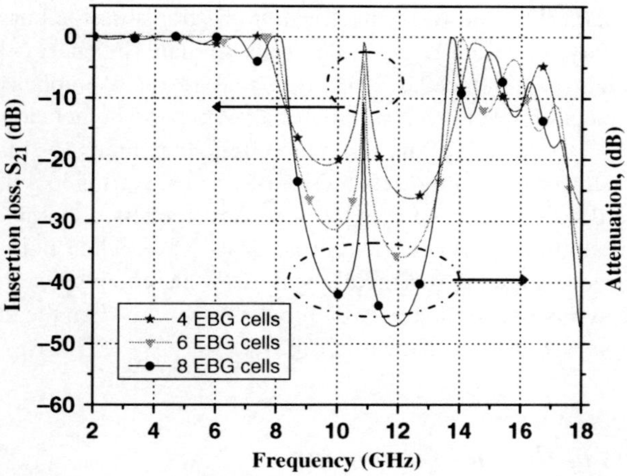

Fig. 7.6 Simulation for insertion loss and attenuation versus the number of EBG structures

EBG filter. The insertion loss and the attenuation are optimum when the number of EBG structures is 4.

7.3.2 Effect of the Defect Length

In this section, the defect length, L_R, is changed and its effect on the filter characteristics is analyzed. When there are four units of EBG structures, the defect length L_R changes from $L_R = d$ to 2 d, as shown in Fig. 7.7a. At $L_R = d = \lambda_g/2$, corresponds

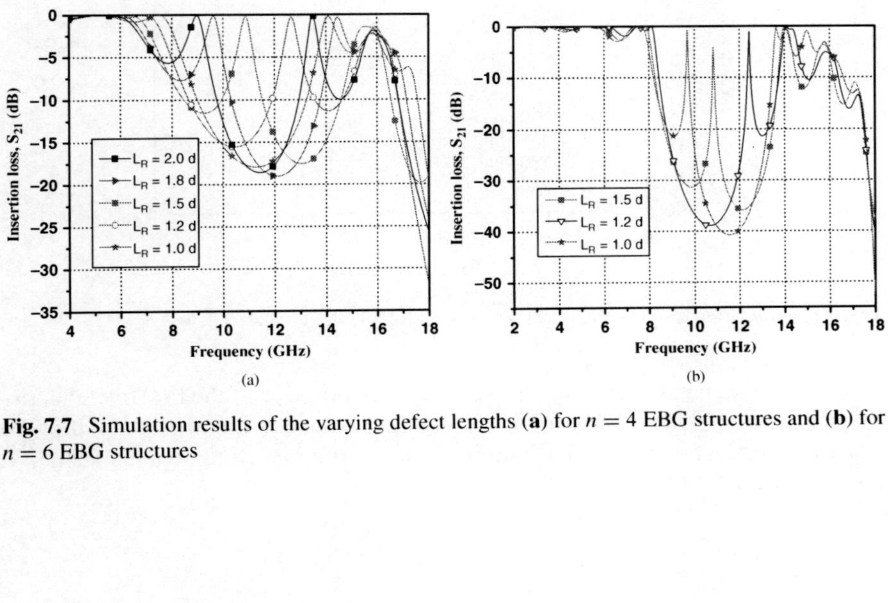

Fig. 7.7 Simulation results of the varying defect lengths (**a**) for $n = 4$ EBG structures and (**b**) for $n = 6$ EBG structures

to Eqs. (7.24) and (7.25), no resonance peak in the transmission curve is observed, as no defect cavity is present to generate or accumulate the energy. However, with only a slight change in defect $L_R = 1.2\ d$, a peak resonance appears at 12.6 GHz but the upper rejection level is not symmetrical. When the defect cavity is increased further from $L_R = 1.8\ d$ to $2\ d$, the resonance frequency tends to shift to lower frequency, and the lower rejection level deteriorates. The insertion loss does not change by varying the defect length but the F–P resonant frequency is tuned from 9.71 to 8.98 GHz. The optimum resonant length is $L_R = 1.5\ d$, when both the upper and lower rejection levels are approximately symmetrical. The results are quite similar in terms of the symmetry of the upper and lower rejection when the number of EBG structures, i.e., $n = 4$ and 6 as shown in Fig. 7.7a, b.

7.3.3 Effects of the Slot Dimension

The EBG structures designed on the ground plane of the CPW transmission line are square, with length and width equal to $\lambda_g/4$. In this section, the width is changed to analyze its influence on the transmission curve. When the slot is perfectly square, $t = 1220\ \mu\text{m}$, the insertion loss is 2 dB, and attenuation is greater than 25 dB. When the slot $t = 610\ \mu\text{m}$ and a remains as $\lambda_g/4$, the insertion loss improves but with attenuation less than 20 dB as shown in Fig. 7.8.

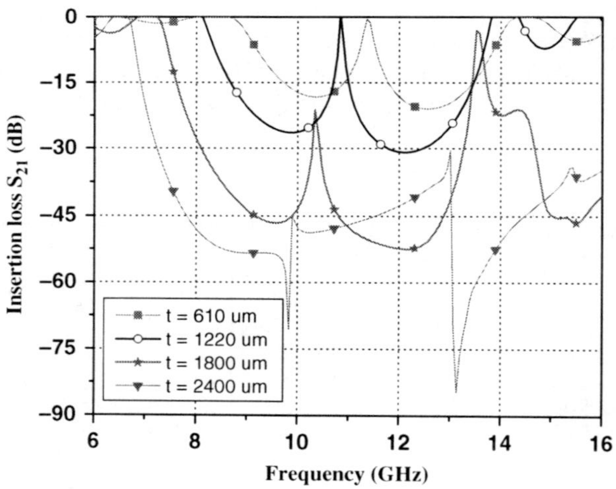

Fig. 7.8 Simulation result of the effect of varying slot length t on insertion loss S_{21}

On the other hand, when $t = 2400\ \mu\text{m}$, the performance of the F–P filter deteriorates in terms of insertion loss of 45 dB but the attenuation improves to 50 dB. The optimum design for F–P filter with minimum insertion loss and high attenuation is $t = 1220\ \mu\text{m}$.

7.3.4 Effect of Propagation Characteristics

The propagation characteristic of the F–P EBG is analyzed using the Floquet's theorem. The propagation factor is $e^{-\gamma \Lambda}$, where $\Lambda = 4878\ \mu\text{m}$ is the period of the EBG structure and $\gamma = \alpha + j\beta$ is the complex propagation constant in the direction of transmission.

The normalized phase constant β/k_0 and the attenuation constant α/k_0 are shown in Fig. 7.9. When β/k_0 is close to the Bragg condition ($\beta\,\Lambda = \pi$), the signal is highly attenuated. The stopband is very wide and the propagation is prohibited. The attenuation is almost zero which indicates the presence of the bandpass filter

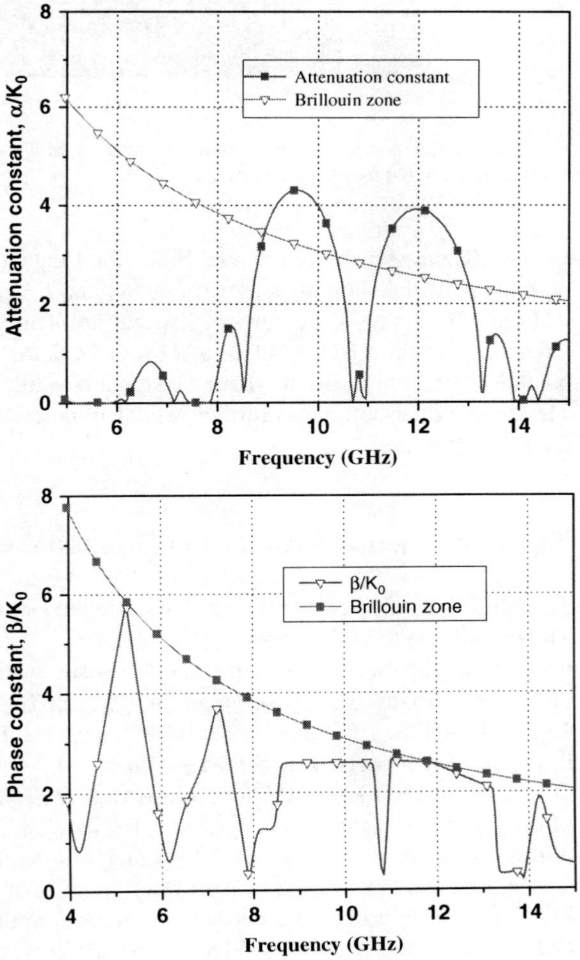

Fig. 7.9 Dispersion diagram of EBG F–P filter: (**a**) attenuation and (**b**) propagation

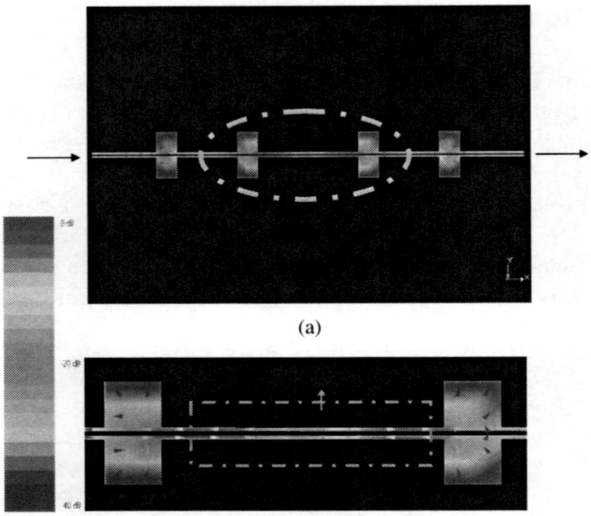

(a)

(b)

Fig. 7.10 Simulation results of current distribution of the tunable F–P filter at 11.4 GHz: (**a**) overview and (**b**) magnified image

at 11.4 GHz as shown in Fig. 7.9a. When the frequency is higher than 17 GHz, the periodic structure supports fast-wave mode, i.e., $\beta/k_0 < 1$, with small attenuation.

Figure 7.10 shows the current distribution of the F–P EBG filter. There is no attenuation in the EBG structure at 11.4 GHz as the signal can pass through. The band gap phenomena occur where the signal is totally rejected at other frequencies. The current distribution has further validated the existence of the bandpass filter in the EBG structures.

7.4 Experimental Results and Discussions

The EBG F–P tunable filter is fabricated using the surface micromachining process. The circuit length is dependent on the number of EBG structures, the central frequency, and the dielectric constant. High-resistivity silicon substrate is used with a dielectric constant of 11.9 and height of 725 μm. The stopband center frequency is 12.5 GHz and the EBG unit period is 4878 μm. The EBG structure is connected to input and output ports using 50 Ω feed lines.

The fabricated EBG F–P filter consists of four EBG structures. The *S*-parameters of the F–P filter is shown in Fig. 7.11. The insertion loss is 4 dB at 11.4 GHz. The return loss is greater than 12 dB. The *Q* factor of the filter is 70.

The tunable F–P EBG bandpass filter consists of MEMS bridges as shown in Fig. 7.12. These bridges are capacitively loaded on the defect cavity. The width *w* and length *l* of the bridges are 50 and 300 μm, respectively. There are eight bridges

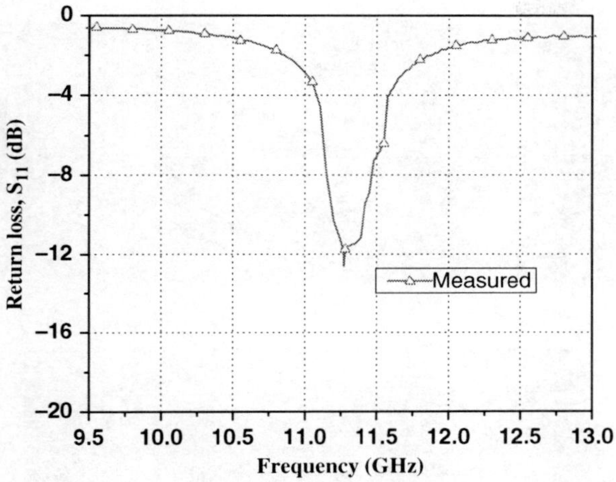

(a)

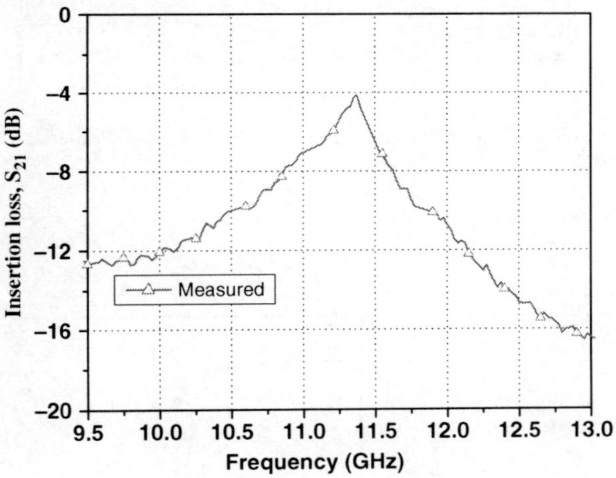

(b)

Fig. 7.11 Comparison between the simulation and measurement results: (**a**) return loss, S_{11}, and (**b**) insertion loss, S_{21}

and the periodic spacing between the bridges, *s*, is 500 μm. The initial height of the bridges is 2 μm. The filter is simulated using commercial software IE-3D.

The measured results are shown in Fig. 7.13. When the drive voltage is 5 V, the central frequency is 11.39 GHz, the insertion loss is 4.1 dB while *Q* factor is 70. When the voltage increases from 20 to 24 V, the frequency is shifted from 11.39 to 11.04 GHz with insertion loss ranging from 4.5 to 5.5 dB.

When the applied voltage increases from 24 to 34 V, the insertion loss increases significantly. The reason is that some of the bridges are pulled down completely.

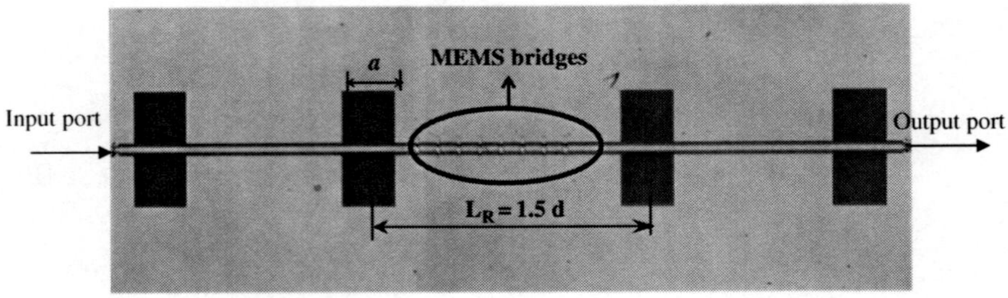

(a)

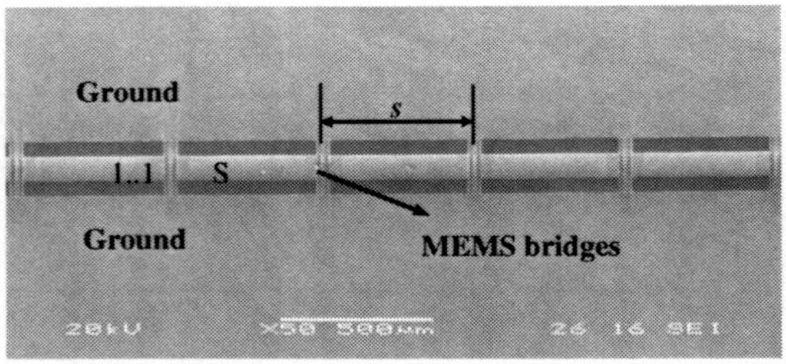

(b)

Fig. 7.12 Tunable Fabry–Perot EBG filter: (**a**) photograph of the F–P filter and (**b**) enlarged view of MEMS bridges

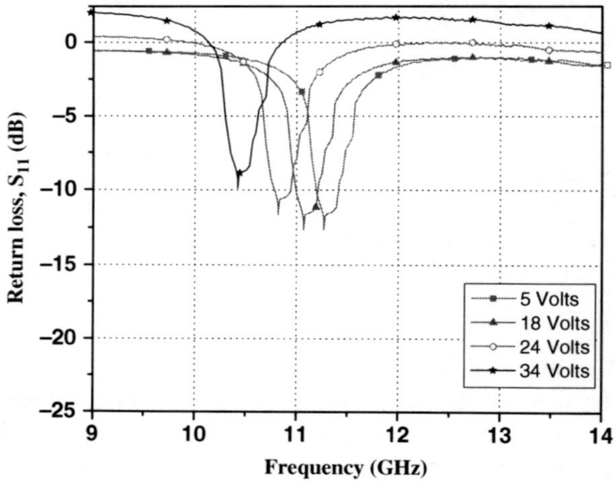

(a)

Fig. 7.13 Measurement result of the tunable F–P EBG filter: (**a**) return loss, S_{11}, and (**b**) insertion loss, S_{21}

Fig. 7.13 (continued)

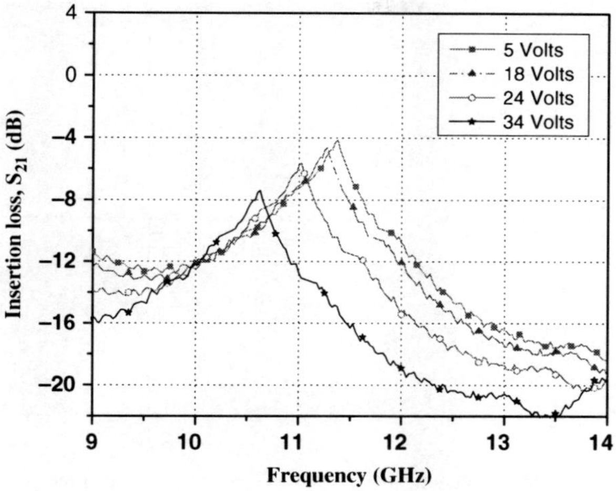

(b)

Table 7.2 Measurement results of the tunable F–P EBG filter

Bias voltage (V)	5	18	24	34
Frequency (GHz)	11.39	11.04	10.85	10.4
Insertion loss (dB)	4.1	4.6	5.6	7.35
Return loss	12.2	11.8	10.9	9.5

The tuning range is around 7.5% in the X-band as shown in Fig. 7.13. The measurement results are summarized in Table 7.2. The central frequency, insertion loss, and return loss are listed for different values of the bias voltage, *V*. The difference between the simulation and the measurement results is due to the fabrication tolerance and the substrate loss.

7.5 Reconfigurable Switching Filter

The reconfigurable filter structure consists of three unit EBG structures and DC-contact MEMS switches, as shown in Fig. 7.14. These unit EBG structures are placed periodically at a distance equal to half of the wavelength of the central frequency. Before analyzing the reconfigurable filter, the unit EBG structure is investigated in detail. A lattice-shaped unit EBG structure is shown in Fig. 7.15. In the unit cell structure, a small square slot is etched in the ground plane with a side length *a*. The smaller square-etched slot is connected to the gap by a narrow transverse slot with length of w_s and width of d_s as shown in Fig. 7.15 with dotted lines.

Another bigger square slot of size *b* and the transverse slot are etched on top of the smaller square slot. The resonance frequency depends on the size of the

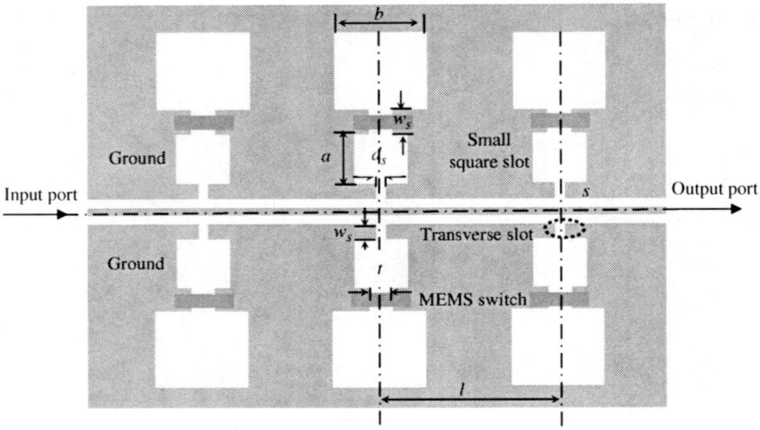

Fig. 7.14 Schematic of the MEMS reconfigurable filter [6]. Copyright/used with permission of/courtesy of IEEE

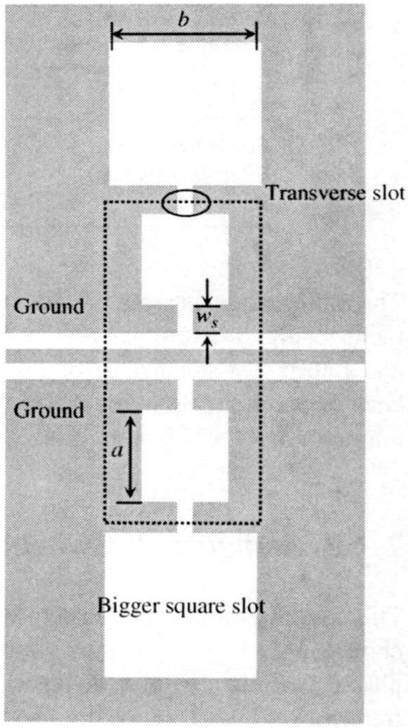

Fig. 7.15 Schematic of the unit reconfigurable cell structure [6]. Copyright/used with permission of/courtesy of IEEE

transverse slot and the square-etched hole in the ground plane. The EM and circuit simulation are performed with the help of commercial software, advanced design system (ADS) from Agilent. The circuit modeling shows a bandpass appearing in the characteristics when a bigger square slot of size b is used together with the transverse slot, otherwise the structure behaves like a bandstop filter.

7.5.1 Unit Reconfigurable Structure and Its Modeling

An equivalent parallel LC circuit is shown in Fig. 7.16 to model the EBG structure. It consists of a series inductor, parallel capacitor, and a resistor. From a practical point of view, the unit structure can serve as a replacement for the parallel LC resonant circuit in many applications. To apply the EBG structure to a practical circuit, it is necessary to extract the equivalent circuit parameters, which can be obtained from the simulation result of the unit cell structure. The lumped capacitance C_1 is mainly contributed by the transverse slot on the ground, while the inductance L_1 is related to the magnetic flux passing through the small square aperture of size a on the ground. The equivalent circuit parameters are extracted and presented in Table 7.3. The circuit simulation result is obtained using the equivalent circuit parameters as shown in Fig. 7.17. The circuit simulation and the EM simulation show good agreement. The response exhibits a bandstop response at 19.6 GHz. Thus, the derived equivalent circuit for the EBG structure can be used accurately to design practical circuits.

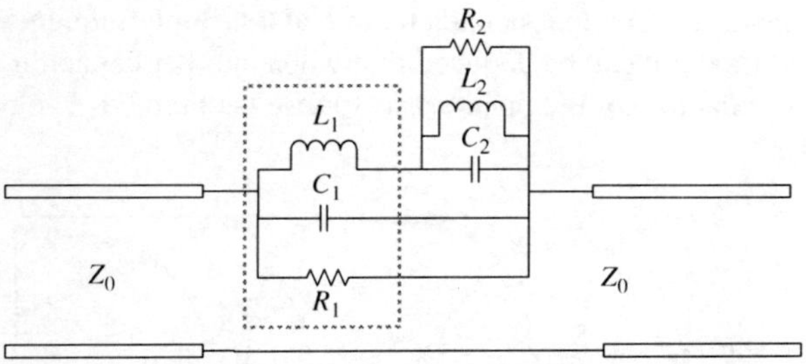

Fig. 7.16 Equivalent circuit model of the unit reconfigurable cell structure [6]. Copyright/used with permission of/courtesy of IEEE

Table 7.3 Extracted circuit parameters of the reconfigurable filter

Parameters	$d_s = 60\,\mu m$, $w_s = 200\,\mu m$
Smaller square size, a (μm)	762.00
Bigger square size, b (μm)	1200.00
Resonant frequency (GHz)	17.34
Inductance, L_1 (nH)	0.54
Inductance, L_2 (nH)	1.00
Capacitance, C_1 (pF)	0.13
Capacitance, C_2 (pF)	0.21
Resistance, R_1 (kΩ)	1.8
Resistance, R_2 (kΩ)	2.0

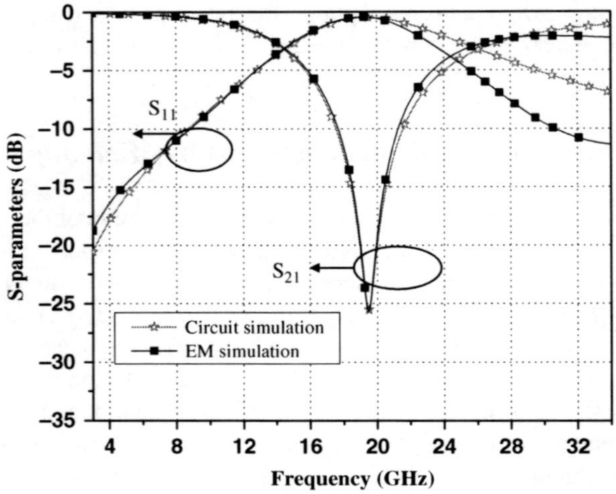

Fig. 7.17 Comparison of parameter extraction and EM simulation results of the unit reconfigurable structure [6]. Copyright/used with permission of/courtesy of IEEE

The whole unit cell structure which includes both the smaller and the bigger square slots is discussed. Both the circuit simulation and EM simulation shown in Fig. 7.18 demonstrate a bandpass characteristic at the same frequency of 19.6 GHz. From the circuit model it can be deduced that when another capacitor C_2 is added in parallel with an inductor L_2, the bandpass response is obtained. The performance of

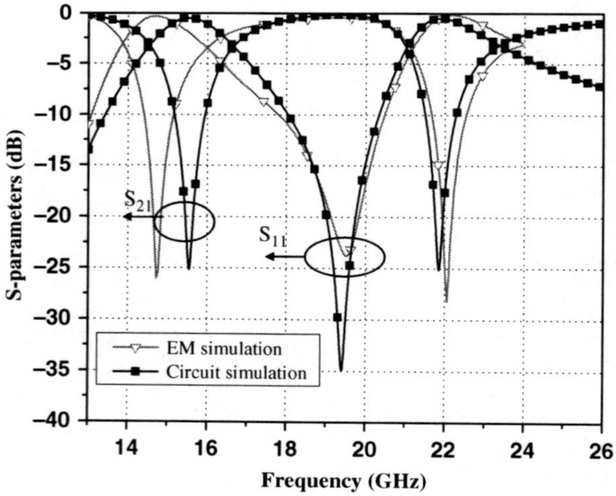

Fig. 7.18 Comparison of circuit and EM simulation results of the unit reconfigurable structure [6]. Copyright/used with permission of/courtesy of IEEE

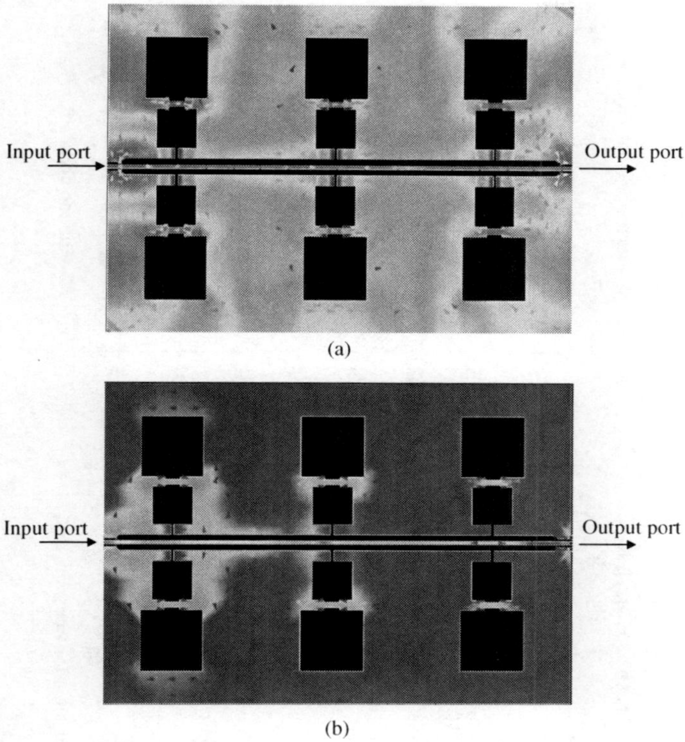

Fig. 7.19 Simulation results of the current distribution for reconfigurable filter: (**a**) bandpass response and (**b**) bandstop response [6]. Copyright/used with permission of/courtesy of IEEE

the bandpass response improves when the relative size of the bigger square slot with respect to that of the smaller square slot is increased, in other words L_2 increased and the transverse slot C_2 changes slightly. Therefore, under these circumstances two transmission zeros occur at 15.2 and 22 GHz. Figure 7.19 shows the current distribution of the reconfigurable filter. For the bandpass state, all the signals are passing through the transmission line as shown in Fig 7.19a. For the bandstop state, the energy is totally attenuated at the resonant frequency as shown in Fig. 7.19b.

7.5.2 Propagation Characteristics

Figure 7.20a, b shows the dispersion diagram of the EBG with the smaller slot and the inclusion of the bigger slot, respectively. Figure 7.20a shows the typical bandstop propagation characteristics, whereas Fig. 7.20b validates that the periodic structure has shown the bandpass behavior at the same frequency of the band gap. The curve

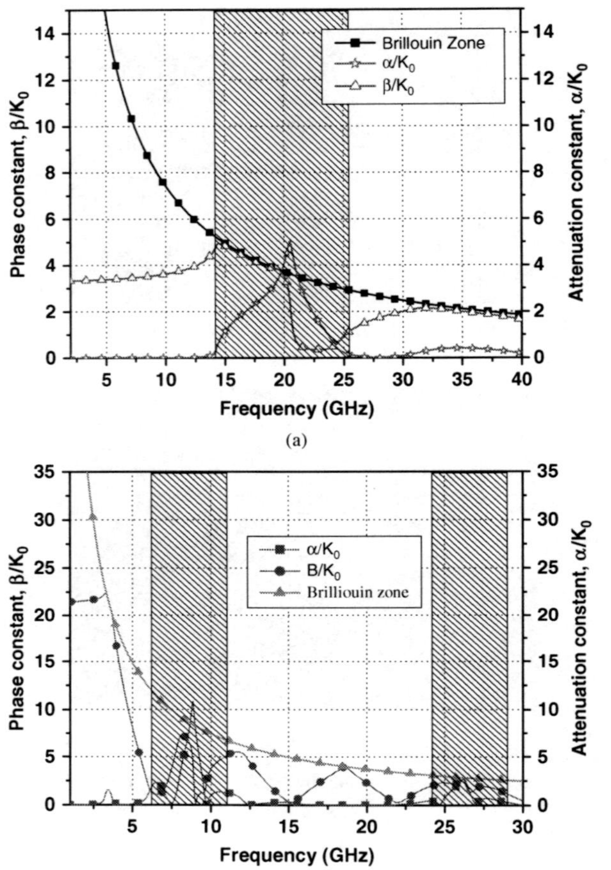

(b)

Fig. 7.20 Simulated dispersion characteristics of unit cell structure: (**a**) bandstop filter phenomena and (**b**) bandpass filter phenomena [6]. Copyright/used with permission of/courtesy of IEEE

starts to follow a Brillouin zone boundary at 3 GHz. From the simulation results of the unit structure, the band gap occurs at 8.5 GHz. The attenuation curve indicates a normalized absolute value of 12. After that the curves start to fall when the bandpass phenomenon is approaching. The attenuation value of the bandpass filter is zero at the central frequency of 18 GHz. The attenuation curve starts to rise again when it approaches the second pole. However, it has an absolute value of 4 due to the low rejection level in the second pole. The phase response rises fast due to the band gap at 4 GHz, and then it fluctuates within the band gap and the bandpass regions. Figure 7.21 shows the phase response of the reconfigurable filter, which is linear in the bandpass and bandstop regions, respectively.

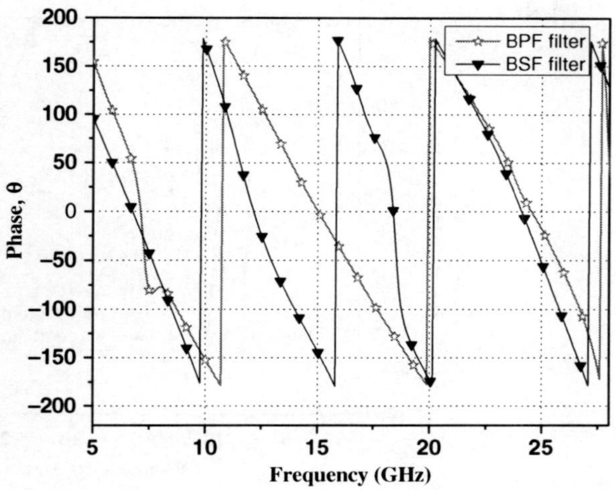

Fig. 7.21 Simulated phase response of the reconfigurable filter [6]. Copyright/used with permission of/courtesy of IEEE

7.5.3 *Effects of the Bigger Slot Width*

In this section, the dimension of the bigger slot is studied. From the equivalent circuit, the change in the dimensions affects the inductance L_2 value. The change in the inductance will shift the resonant frequency. The inductance L_2 changes when the dimension b is varied from 762, 1200, and 1500 μm. The simulation results are shown in Fig. 7.22a, b. The change in the inductance affects the return loss and

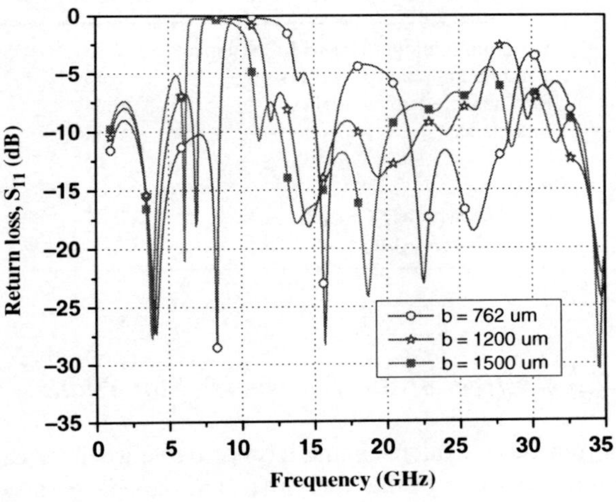

Fig. 7.22 Simulation results showing the influence of the change in the bigger slot width b: (**a**) return loss S_{11} and (**b**) insertion loss S_{21} [6]. Copyright/used with permission of/courtesy of IEEE

Fig. 7.22 (continued)

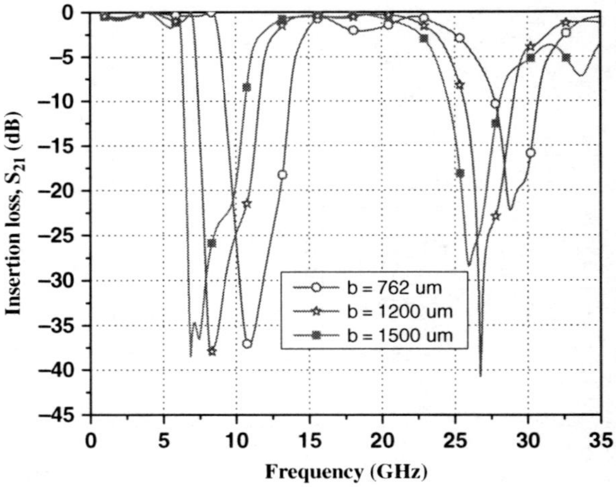

(b)

the transmission curve. When the slot width is 762 μm, the resonant frequency is 20.33 GHz. Ripples are observed and the rejection level in the second zero is less than 21 dB. The simulated inductance value using the equivalent circuit is 0.45 nH and the other parameters remain constant. The bigger slot value is increased from 1200 to 1500 μm. The results indicate that when the value of the inductance L_2 is 1 nH, the filter resonates at the center frequency of 17.43 GHz.

The insertion loss curve, S_{21}, is smooth in the passband region and shows rejection greater than 35 dB on the upper and lower sides of the passband. The extracted parameters of the filter are listed in Table 7.4.

Table 7.4 Extracted parameters of the variation in the bigger slot width b

Variation in bigger slot width b (μm)	762	1200	1500
Variable inductance L_2 (nH)	0.45	1	1.9
Resonant frequency (GHz)	20.33	17.43	16.12
Inductance L_1 (nH)	0.54	0.54	0.54
Fixed capacitance, C_1 (pF)	0.13	0.13	0.13
Resistance, R_1 (kΩ)	1.85	1.80	1.82
Fixed capacitance C_2 (nH)	0.21	0.21	0.21
Resistance, R_2 (kΩ)	2	2.1	2.1

7.5.4 *Effect of the Transverse Slot Width*

From the equivalent circuit, it is noted that when the capacitance of the filter changes with the width of the transverse slot, the change in width affects the resonant frequency. The shift in resonance frequency is less compared to the case when the inductance is changed. The width is changed from 120 to 300 μm. The effective change in capacitance is 0.18–0.27 pF. The simulation results in Fig. 7.23a show that for a width of 120 μm the return loss S_{11} is less than –10 dB at the resonant frequency and ripples are observed in the passband.

The optimum result occurs at the width of 180 μm when the rejection on both sides of the passband is greater than 35 dB and the insertion loss in passband is at

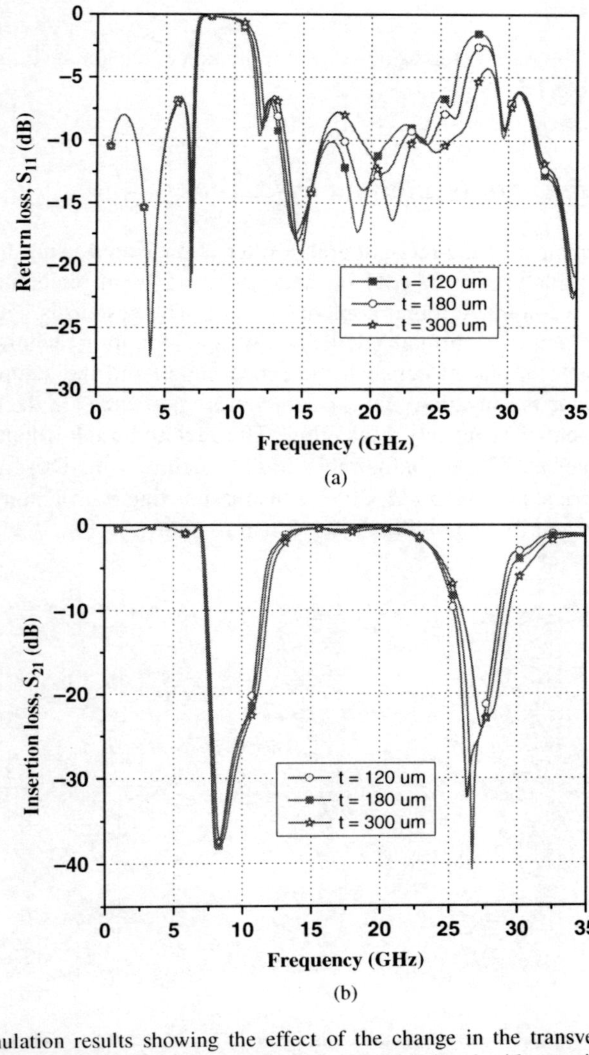

Fig. 7.23 Simulation results showing the effect of the change in the transverse slot width *t*: (**a**) return loss S_{11} and (**b**) insertion loss S_{21} [6]. Copyright/used with permission of/courtesy of IEEE

Table 7.5 Extracted parameters of the variation of the transverse slot *t*

Variation of transverse slot *t* (μm)	120	180	300
Variable capacitance C_2 (pF)	0.16	0.21	0.26
Resonant frequency (GHz)	17.2	17.43	17.62
Inductance L_1 (nH)	0.54	0.54	0.54
Fixed capacitance C_1 (pF)	0.13	0.13	0.13
Resistance R_1 (kΩ)	1.85	1.80	1.82
Fixed Inductance L_2 (nH)	1.00	1.00	1.00
Resistance R_2 (pF)	2.00	2.1	2.10

the minimum. The parameter values of the inductance, capacitance, and resistance are listed in Table 7.5.

7.5.5 Experimental Results and Discussions

The RF performance of the reconfigurable filter is measured using the HP 8510C vector network analyzer with gold tip and with a 150-μm pitch for the cascade microtech ground–signal–ground coplanar probes. The system is calibrated using standard short-open-load-through (SOLT) on wafer calibration technique. A 5-mm-thick plastic plate is placed between the probe chuck and the sample to remove higher order mode propagation. All experiments are performed in the room temperature environment without any packaging. The reconfigurable filter is fabricated using gold material. The reconfigurable filter structure with DC-contact MEMS switches is fabricated on a 50-Ω CPW transmission line with a signal line width of $W = 70$ μm and the gap width $G = 115$ μm as shown in Fig. 7.24a. The values

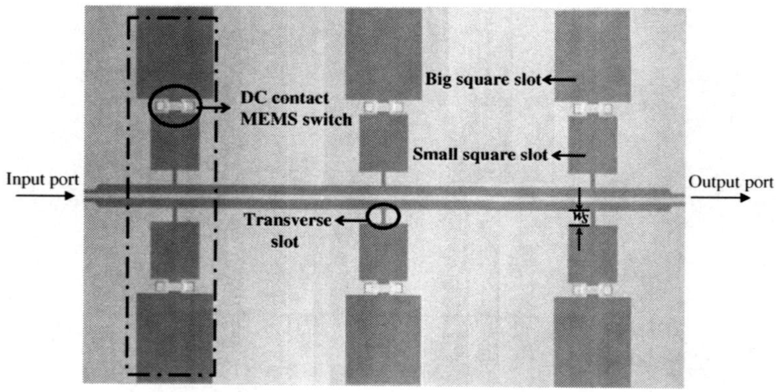

(a)

Fig. 7.24 SEM image of the reconfigurable filter with DC-contact MEMS switches: (**a**) overview of the reconfigurable filter and (**b**) zoomed view of rectangular unit structure [6]. Copyright/used with permission of/courtesy of IEEE

Fig. 7.24 (continued)

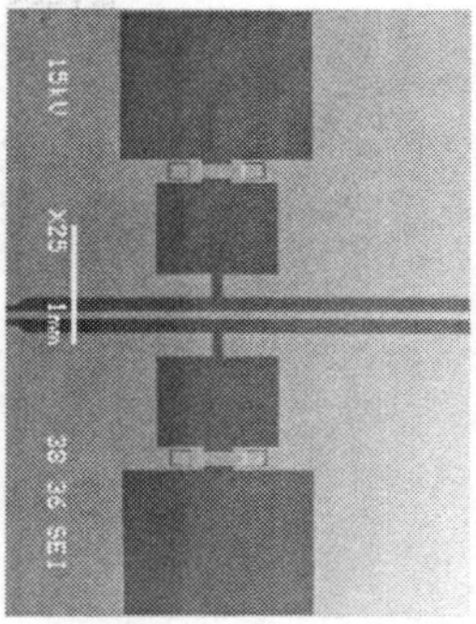

(b)

of a, b, w_s, and d_s of the filter are presented in Table 7.3. The zoom view of the part of the switch of the reconfigurable filter is shown in Fig 7.24b. In order to show the validity of the equivalent circuit and the extracted parameters for the proposed unit structure, the reconfigurable filter is formed by cascading three unit structures with the DC-contact MEMS switches C_s placed at the transverse slot as shown in Fig. 7.25.

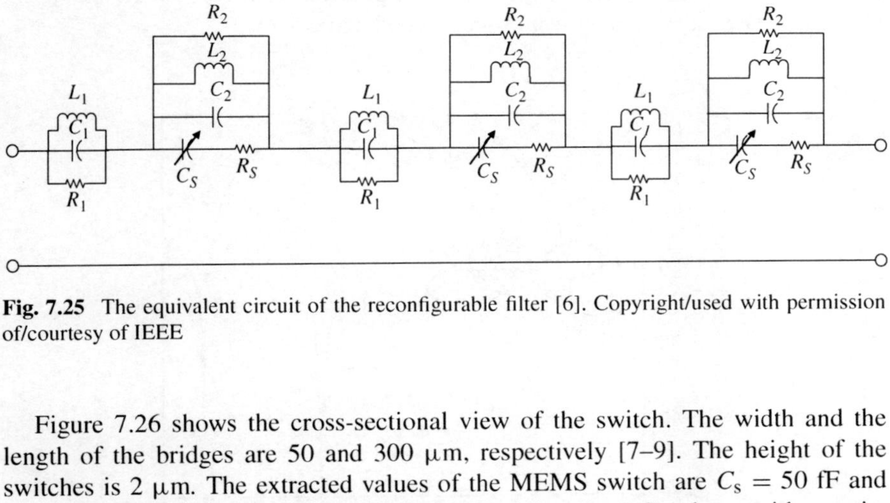

Fig. 7.25 The equivalent circuit of the reconfigurable filter [6]. Copyright/used with permission of/courtesy of IEEE

Figure 7.26 shows the cross-sectional view of the switch. The width and the length of the bridges are 50 and 300 μm, respectively [7–9]. The height of the switches is 2 μm. The extracted values of the MEMS switch are C_s = 50 fF and R_s = 1 Ω. The bias lines are designed for providing the DC voltage with a resistance of 1000 Ω. When the DC-contact MEMS switches are in the off-state, the filter behaves like a bandpass filter. When the DC-contact MEMS switches are in the on-state, the filter behaves like a bandstop filter. The simulation results of the reconfigurable filter are shown in Fig. 7.27. It has an insertion loss of 0.9 dB at 19.8 GHz and the 3-dB bandwidth is 9.6 GHz. The bandstop filter has an insertion loss of 0.6 dB. The measurement results of the reconfigurable filter are shown

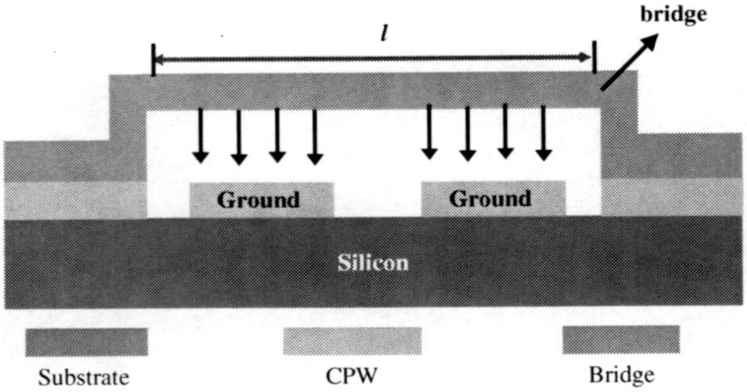

Fig. 7.26 Cross section of the MEMS DC-contact switch [6]. Copyright/used with permission of/courtesy of IEEE

in Fig. 7.28. When the DC-contact MEMS switches are at the off-state, the filter provides the function of the bandpass filter with the insertion loss of 1.55 dB at 20.1 GHz. The 3-dB bandwidth is 9.2 GHz and rejection is greater than 20 dB. When the DC-contact MEMS switches are at the on-state with applied DC bias voltage of 38 V, the top transverse slot and bigger square slot, *b*, are shortened to ground, and it provides the function of the bandstop filter. The measurement results show that the resonant frequency of the bandstop is 19.8 GHz and the insertion loss is 1.2 dB. The 20-dB rejection bandwidth changes from 17 to 22.5 GHz [6, 10]. The simulation and measurement results are in good agreement.

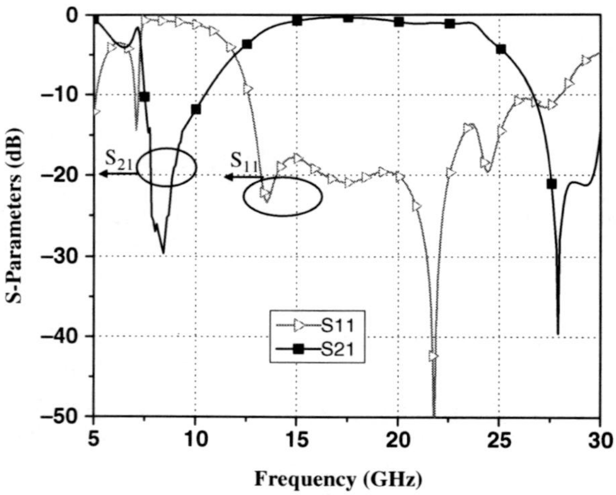

(a)

Fig. 7.27 Simulation results of the *S*-parameters of the reconfigurable filter: (**a**) off-state of switch, reconfigurable filter behaves as a bandpass filter and (**b**) on-state of switch, reconfigurable filter behaves as a bandstop filter [6]. Copyright/used with permission of/courtesy of IEEE

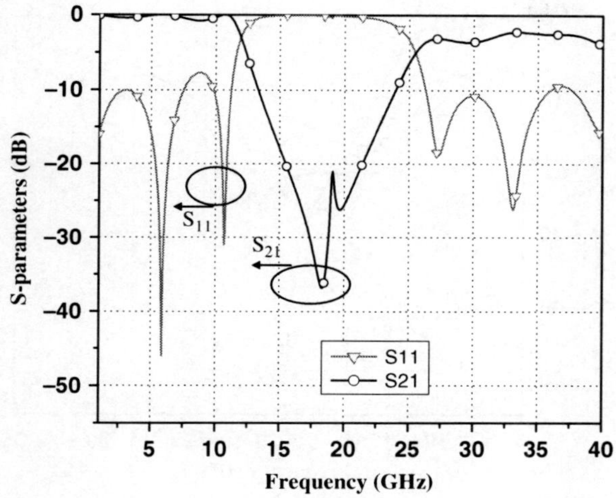

(b)

Fig. 7.27 (continued)

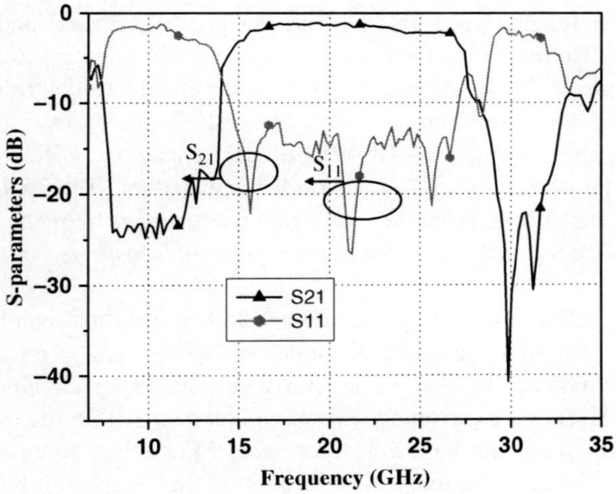

(a)

Fig. 7.28 Measurement results of *S*-parameters of the reconfigurable filter: (**a**) off-state of switch, reconfigurable filter behaves as a bandpass filter and (**b**) on-state of switch, reconfigurable filter behaves as a bandstop filter [6]. Copyright/used with permission of/courtesy of IEEE

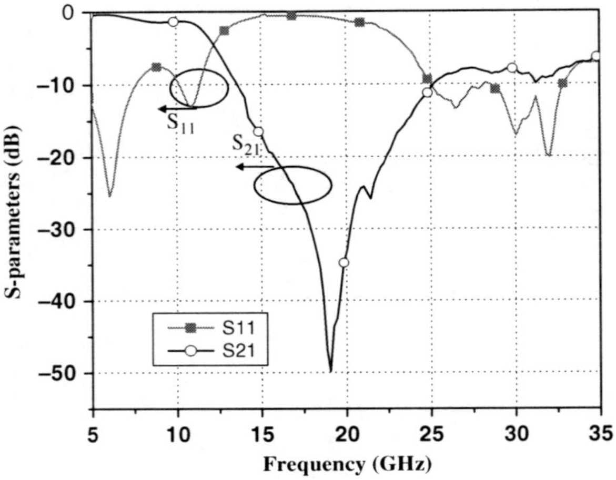

(b)

Fig. 7.28 (continued)

7.6 Summary

The Fabry–Perot tunable bandpass filters are designed, fabricated, and experimented in this chapter. The F–P filters are realized by creating a defect cavity in the EBG periodic structures. A mathematical model is developed to determine the Q factor of the filter. The measurement result of the F–P filter shows a quality factor Q of 70 with an insertion loss of 4.0 dB. The capacitive MEMS switches are fabricated at the defect cavity to develop a tunable filter. The switch heights are varied with a bias voltage from 0 to 34 V and the resonant frequency is tuned. The result shows a tuning range of 7.5% with an insertion loss that varies from 4.1 to 7.35 dB.

The micromachined switching reconfigurable bandpass and bandstop filters are also presented. An equivalent circuit model is derived. The extracted parameters show both bandpass and bandstop characteristics. The dispersion characteristics are obtained to analyze the electromagnetic wave behavior within the unit structure of the bandpass filter and the bandstop filter using Floquet's theorem. The measurement results of the reconfigurable filter show that the insertion loss is 1.55 dB for the bandpass filter whereas the band rejection level is >20 dB and the insertion loss in the passband is 1.2 dB for the bandstop filter.

References

1. R. Clarke and C. B. Rosenberg, "Fabry-Perot and open resonators at microwave and millimeter wave frequencies, 2-300 GHz," J. Phys. Sci. Instrum. Vol. 15, 1982, pp. 9–23.
2. J. Correia, Optical Microsystem in Silicon Based on a Fabry-Perot Resonance Cavity, Delft University Press, Delft, 1999.

3. H. Cam, S. Toutain and G. Landrac, "Study of a Fabry-Perot cavity in the microwave frequency range by boundary element method," IEEE Trans. Microwave Theory Tech. Vol. 40, 1992, pp. 298–304.
4. D. M. Pozar, Microwave Engineering, Wiley, New York, NY, 1998.
5. G. L. Matthaei, L. Yong and E. M. T. Jones, Microwave Filters, Impedance-Matching Networks, and Coupling Structures, McGraw-Hill, New York, NY, 1964.
6. M. F. Karim, A. Q. Liu, A. Alphones and A. B. Yu, "A reconfigurable micromachined switching filter using periodic structures," IEEE Trans. Microwave Theory Tech. Vol. 55(6), 2007, pp. 1154–1162.
7. A. B. Yu, A. Q. Liu, X. Q. Zhang, L. Zhu, A. Alphones and S. A. Peter, "Improvement of isolation for RF MEMS capacitive shunt switch via membrane planarization," Sensor. Actuator. Phys. A Vol. 19, 2005, pp. 206–213.
8. A. B. Yu, A. Q. Liu, Q. X. Zhang and H. M. Hosseini, "Effects of surface roughness on electromagnetic characteristics of capacitive switches," J. Micromech. Microeng. Vol. 16, 2006, pp. 2157–2166.
9. A. B. Yu, A. Q. Liu, J. Oberhammer, Q. X. Zhang and H. M. Hosseini, "Characterization and optimization of dry releasing for the fabrication of RF MEMS capacitive switches," J. Micromech. Microeng. Vol. 17, 2007, pp. 2024–2030.
10. A. Q. Liu, A. B. Yu, M. F. Karim and M. Tang, "RF MEMS switches & integrated switching circuit," J. Semicond. Technol. Sci. (invited review paper) Vol. 7(3), 2007, pp. 166–176.

Chapter 8
Capacitive Switching Bandpass Filters

The objective of this chapter is to design and characterize tunable bandpass filters that integrate the capacitive switch (or the DC-contact switch) into the CPW transmission line structure. A capacitance tuning bandpass filter is designed. The filter consists of two series coupled resonators and a series of coupling capacitors and shunt capacitors in each resonator. Then, an inductance tuning bandpass filter is proposed. This two-pole bandpass filter consists of two shunt resonators. The two bandpass filters are discussed in this chapter.

8.1 Capacitance Tuning Bandpass Filter

Capacitive tuning is one of the most commonly used approaches for the tuning of the bandpass filter. Normally, the bandpass filter consists of shunt resonators and series coupling capacitors. For the tuning of the filter, all the inductors are kept constant while all the capacitors are tuned. Therefore, the design of tunable capacitor is critical in this tuning approach. Both capacitance value and tuning ratio of the tunable capacitor are important to obtain a large tuning range of the tunable bandpass filter [1].

In this design, DOG capacitive switches are used as tuning elements to change the capacitance of the coupling capacitors and the shunt resonators. The DOG switch does not suffer from the down-state capacitance degradation [2]. Different down-state capacitances can also be obtained precisely and easily. Therefore, the whole circuit design of the capacitance tuning bandpass filter is simplified.

8.1.1 Design of the Capacitance Tuning Bandpass Filter

Figure 8.1 shows the topology of a second-order bandpass filter which consists of two resonators that are coupled by an inductor (L_M). The input and output ports

---

Aibin Yu and Ai Qun Liu

A.Q. Liu, *RF MEMS Switches and Integrated Switching Circuits*,
MEMS Reference Shelf 5, DOI 10.1007/978-0-387-46262-2_8,

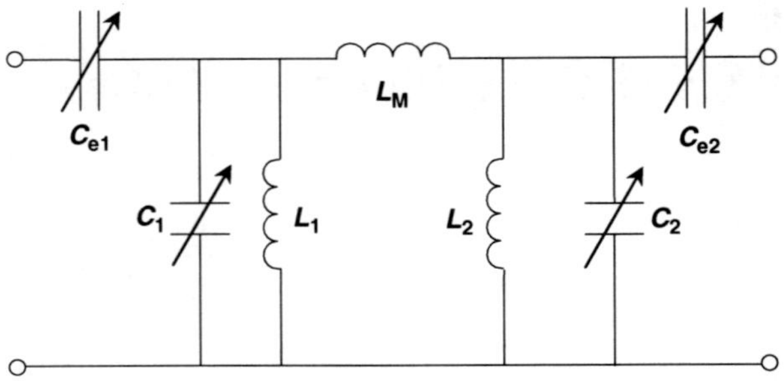

Fig. 8.1 Topology of two-pole tunable bandpass filter

of the two resonators are coupled by tunable switch-type capacitors C_{e1} and C_{e2}, respectively.

The specifications of the designed tunable filter are as follows: bandwidth $w = 40\%$; pole number $N = 2$; passband ripple $= 0.1$ dB; and initial center frequency $\omega_0 = 10$ GHz.

When the two shunt inductance L_i ($i = 1$ and 2) values are selected, according to the center frequency ω_0, the shunt capacitance can be given as [3]

$$C_i = \frac{1}{\omega_0^2 L_i}, \quad i = 1 \text{ and } 2 \tag{8.1}$$

Then the series coupling capacitance can be expressed as

$$\frac{C_e}{C_1} \cong \frac{2\Delta\omega}{\omega_0} \tag{8.2}$$

where $\Delta\omega$ is the 3-dB bandwidth.

In Fig. 8.1, L_M is realized by a quarter-wavelength series stub at initial center frequency of 10 GHz [3]. Based on Eq. (8.1), when the shunt inductance value is kept constant, different shunt capacitances can be calculated at different desired frequencies. After obtaining the shunt capacitor, the series capacitors C_e can be calculated from Eq. (8.2). In this design, the two resonators are inductively coupled by the inductor, L_M. When tuning the filter, the inductor, L_M, need not be tuned. Table 8.1 lists all the element values for the three tuning stages.

Figure 8.2 shows the SEM image of the fabricated tunable bandpass filter. The CPW material is Ti/Cu with the thickness of 250 Å/2 μm; two thin layers of Cr/Au (250 Å/0.5 μm) are deposited above the Ti/Cu to protect the Cu thin film from oxidizing. The metal bridge material is Al with the thickness of 0.6 μm. The sacrificial layer is photoresist with the thickness of 2 μm, which defines the initial gap between the metal bridge and the transmission line. The sacrificial layer is removed using oxygen plasma. A Ti thin film with thickness of 0.1 μm is deposited and patterned as pull-down electrodes.

Table 8.1 Design parameters of capacitance and inductance for different tuning stages

f_0 (GHz)	5	7.5	10
Series coupling inductor, L_M (nH)	2.0	2.0	2.0
Shunt inductor, L_1 (nH)	0.36	0.36	0.36
Shunt inductor, L_2 (nH)	0.36	0.36	0.36
Series coupling capacitor, C_{e1} (pF)	1.7	0.8	0.36
Series coupling capacitor, C_{e2} (pF)	1.7	0.8	0.36
Shunt capacitor, C_1 (pF)	2.8	1.25	0.65
Shunt capacitor, C_2 (pF)	2.8	1.25	0.65

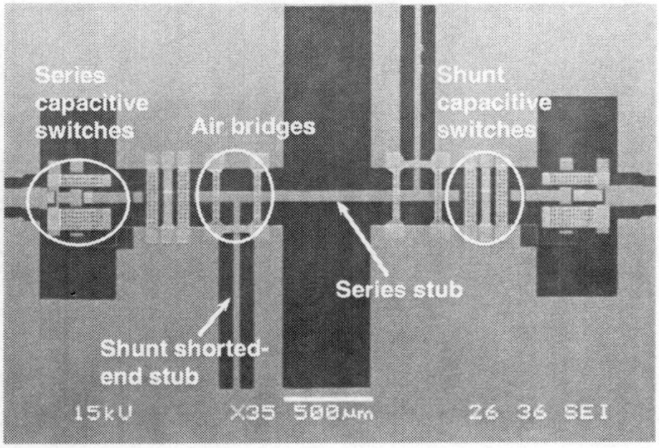

Fig. 8.2 SEM image of the capacitance tuning bandpass filter

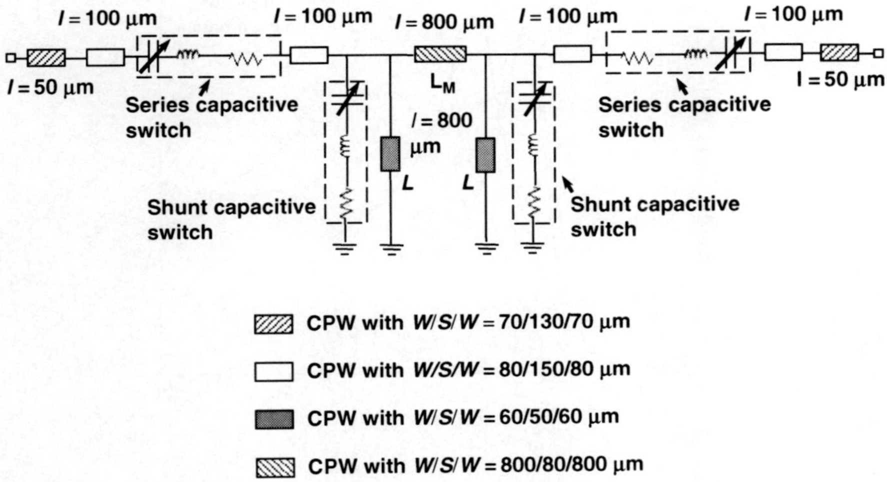

Fig. 8.3 Equivalent circuit of the tunable bandpass filter with capacitive tuning

Figure 8.3 shows the equivalent circuit of the tunable bandpass filter with capacitive tuning. Both the shunt and series capacitive switches are presented by an *RLC* network with the capacitance having up-state value (C_u) or down-state value (C_d), which have been discussed in Chapter 3. The series coupling inductance is determined by the length of the CPW line with $W/S/W = 800/80/800$ μm. The simulation results are shown in Fig. 8.4.

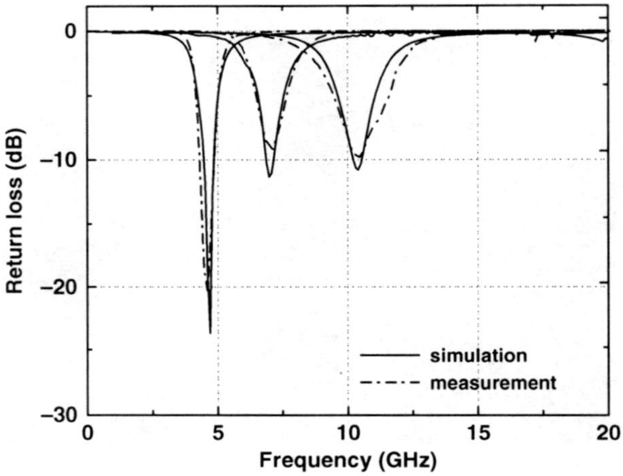

(a)

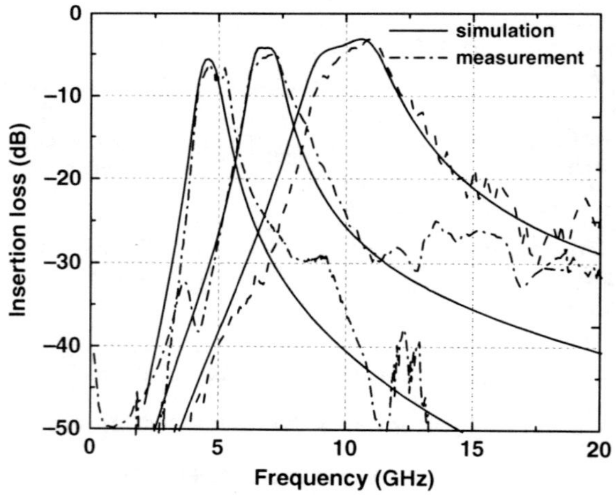

(b)

Fig. 8.4 Measured and simulated frequency response of the tunable bandpass filter: (**a**) return loss and (**b**) insertion loss

8.1.2 Experimental Results and Discussions

The experimental results and the circuit simulation of the tuning of the filter are shown in Fig. 8.4. When the series and the shunt capacitive switches are driven down, the series and shunt capacitance values are changed accordingly. As a result, the central frequency is shifted from 10.3 to 4.9 GHz with the insertion loss varying from 5 to 6 dB. The measured 3-dB bandwidth is 40% and the return loss varies from 25 to 10 dB. Figure 8.4 shows that the simulation results agree with the measurement results.

8.2 Inductance Tuning Bandpass Filter

For the capacitance tuning bandpass filter, it is more suitable for the tuning of a bandpass filter with middle bandwidth. This is because for the design of the bandpass filter with narrow bandwidth, the capacitance values of the coupling capacitor between the two resonators or between the resonators and outside network are smaller than 50 fF. This is a challenge for MEMS technology to build a tunable capacitor with such small capacitance value.

In this section, an inductance tuning bandpass filter is innovated. The inductance change is achieved by changing the length of the shunt shorted-end stubs with the DC-contact switch. Large inductance tuning ratio can then be achieved.

8.2.1 Design of the Inductance Tuning Filter

Figure 8.5 gives a topology of a two-pole capacitively coupled Chebyshev bandpass filter [4, 5]. The specifications of the tunable filter are as follows: bandwidth $w = 10\%$; pole number $N = 2$; passband ripple $= 0.1$ dB; $G_A = G_B = 1/50 = 0.02$. The detailed design procedures are summarized [3, 6].

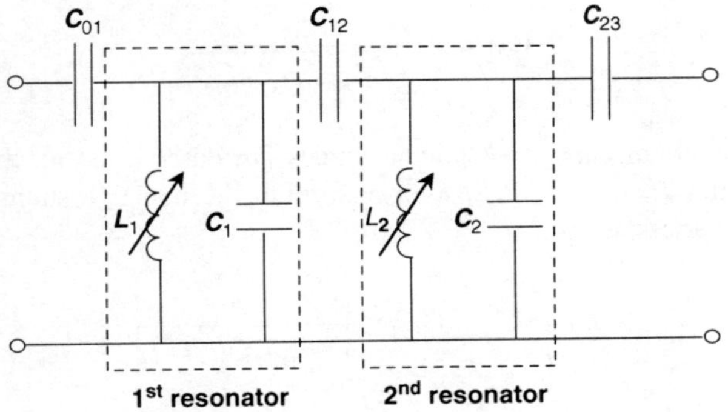

Fig. 8.5 Topology of the inductance tuning bandpass filter

First, select the normalized g element values for the filter g_0–g_3 such that

$$g_0 = 1, \quad g_1 = 0.843, \quad g_2 = 0.622, \quad g_3 = 1.354 \tag{8.3}$$

The normalized initial value for the shunt inductor, L_i, is selected. This value is arbitrary but can later be scaled to achieve the desired inductor and capacitor values. In this design, the initial value of the inductor is set as 1.8 nH and the center frequency is set as 8 GHz. Then the initial values for the shunt capacitor are expressed as

$$C_i^0 = \frac{1}{\omega_0^2 L_i} \tag{8.4}$$

Based on the normalized element values, the impedance inverter values are given as follows:

$$J_{01} = \sqrt{\frac{G_A \omega_0 C_1^0 w}{g_0 g_1 \omega_1'}} \tag{8.5a}$$

$$J_{12} = \sqrt{\frac{C_1^0 C_2^0}{g_1 g_2} \frac{w \omega_0}{\omega_1'}} \tag{8.5b}$$

$$J_{23} = \sqrt{\frac{G_B \omega_0 C_2^0 w}{g_2 g_3 \omega_1'}} \tag{8.5c}$$

where w is the bandwidth and ω_1' is equal to 1 rad/s as normalized frequency.

Then, the series coupling capacitance is given as follows:

$$C_{01} = \frac{J_{01}}{\omega_0 \sqrt{1 - (J_{01}/G_A)^2}} \tag{8.6a}$$

$$C_{12} = \frac{J_{12}}{\omega_0} \tag{8.6b}$$

$$C_{23} = \frac{J_{23}}{\omega_0 \sqrt{1 - (J_{23}/G_B)^2}} \tag{8.6c}$$

Finally, the shunt capacitor element values are determined by the normalized capacitor values C_i ($i = 1$ or 2) after absorbing the negative shunt effect of the impedance inverters as

$$C_1 = C_1^0 - C_{12} - \frac{C_{01}}{1 + (\omega_0 C_{01}/G_A)^2} \tag{8.7a}$$

$$C_2 = C_2^0 - C_{12} - \frac{C_{23}}{1 + (\omega_0 C_{23}/G_A)^2} \tag{8.7b}$$

Table 8.2 Values for capacitance and inductance for different tuning frequencies

f_0 (GHz)	8	9	10	11
Shunt inductor, L_1 and L_2 (nH)	1.6	1.3	1.1	0.9
Series coupling capacitor, C_{01} (fF)	115	115	115	115
Series coupling capacitor, C_{12} (fF)	34	34	34	34
Series coupling capacitor, C_{23} (fF)	115	115	115	115
Shunt capacitor, C_1 (fF)	105	105	105	105
Shunt capacitor, C_2 (fF)	105	105	105	105

From Eq. (8.6), when the inductance values of each resonator are changed and all the capacitance values are kept as constant, the center frequency is shifted. Table 8.2 lists all the capacitance and inductance values based on the calculation for different tuning stages.

Figure 8.6 shows the SEM image of the fabricated two-pole tunable bandpass filter. The three series coupling capacitors are realized by three metal–insulator–metal (MIM) capacitors with 1-μm-thick SiO_2 layer as the insulator. The dielectric constant of SiO_2 is 4.6. The shunt capacitors are realized by implementing the shunt metal bridges 2 μm above the center conductor of the CPW transmission line. The shunt inductors, however, are realized by the shunt shorted-end stubs. The dimensions of the CPW transmission line are $W/S/W = 80/150/80$ μm. The areas of the three series capacitors, C_{01}, C_{12}, C_{23}, are 55×55 μm^2, 30×30 μm^2, and 55×55 μm^2, respectively. The other design dimensions are listed in Table 8.3. Three DC-contact switches are implemented above each shunt shorted-end stub. When the DC-contact switch is driven down one by one, the metal–metal contact between the switches and the shunt shorted-end stub leads to the change in the length of the stub,

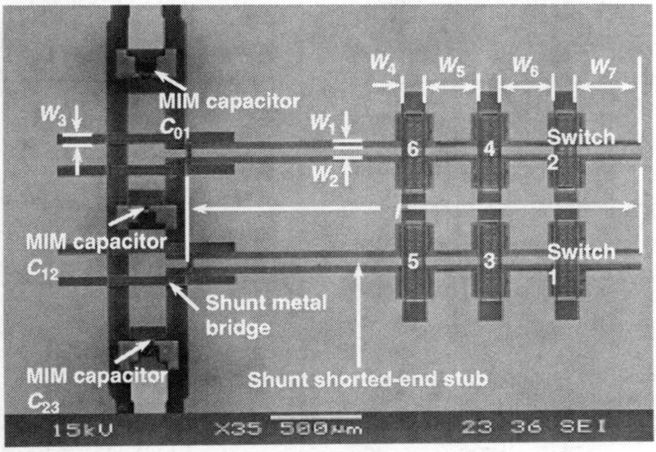

Fig. 8.6 SEM image of two-pole tunable bandpass filter

Table 8.3 Dimensions of the tunable bandpass filter

Parameters	Value
Width of slot of the shunt shorted-end stub, W_1 (μm)	30
Width of the shunt shorted-end stub, W_2 (μm)	40
Width of the shunt metal bridge, W_3 (μm)	60
Width of the DC-contact switch, W_4 (μm)	80
Space between 4th and 6th DC-contact switches, W_5 (μm)	200
Space between 2nd and 4th DC-contact switches, W_6 (μm)	250
Length from the end of the stub and the 2nd DC-contact switch, W_7 (μm)	300
Length of the shunt shorted-end stub, l (μm)	2100

which further results in the change of the inductance of each resonator. As a result, the center frequency of the bandpass filter can be tuned by the shunt switch.

Figure 8.7 gives the equivalent circuit of the tunable bandpass filter. The DC-contact switch can be replaced by an $R_uL_uC_u$ circuit when the metal bridge is at the up-state position, whereas it is replaced by an R_dL_d circuit when the metal bridge is at the down-state position where R_u, L_u, C_u, R_d, L_d are the up-state resistance,

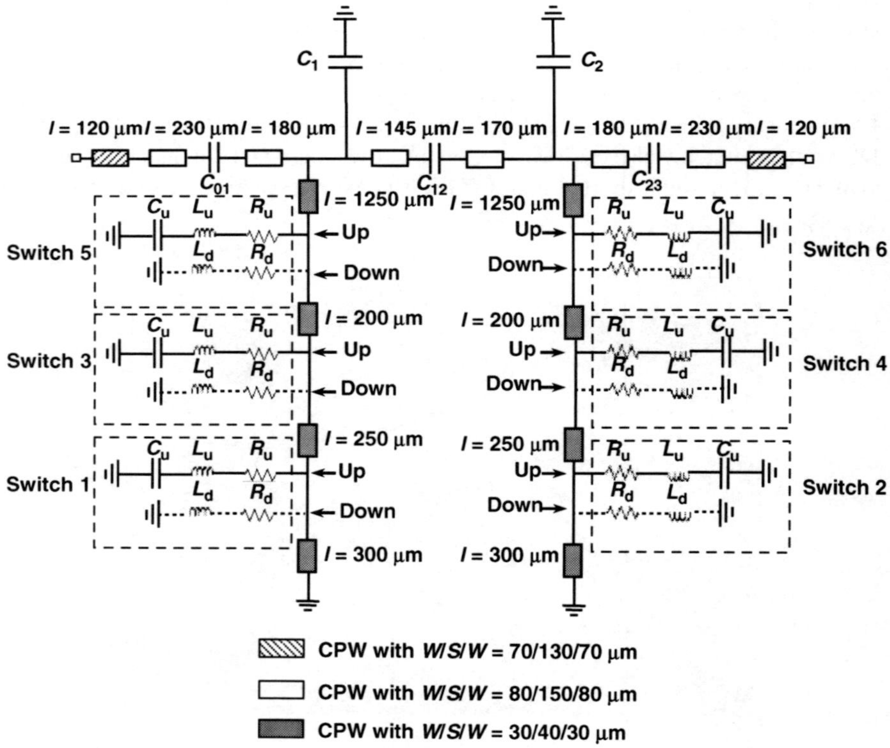

Fig. 8.7 Equivalent circuit of the two-pole tunable bandpass filter

inductance, capacitance and the down-state resistance and inductance, respectively. Those *RLC* values can be extracted from the frequency response of a DC-contact shunt switch at the up-state and down-state positions, respectively. The capacitance values of the two shunt capacitors, C_1 and C_2, can be extracted from the simulation or the measurement results of the shunt switch at the up-state with 2-μm air gap.

From Fig. 8.7, the RF performance of the designed bandpass filter changes with the *RLC* values can be modeled using commercial software of Agilent ADS [7]. When the metal bridge is at the up-state position, Fig. 8.8a gives the relationship between the frequency response and R_u of the switch. It is evident that R_u has little effect on the insertion loss in the passband; Fig. 8.8b shows that L_u has no effect on the insertion loss in the passband. Figure 8.8c indicates the insertion loss increases from 6 to 7.5 dB and the second resonant frequency lowers down from 30 to 17.5 GHz when C_u increases from 20 to 200 fF. When C_u is smaller than 50 fF, it has little effect on the characteristics of the tunable BPF.

Figure 8.9a shows the simulation results of R_d effect on the RF performance of the filter. It is assumed that the first and the second switches are driven down which

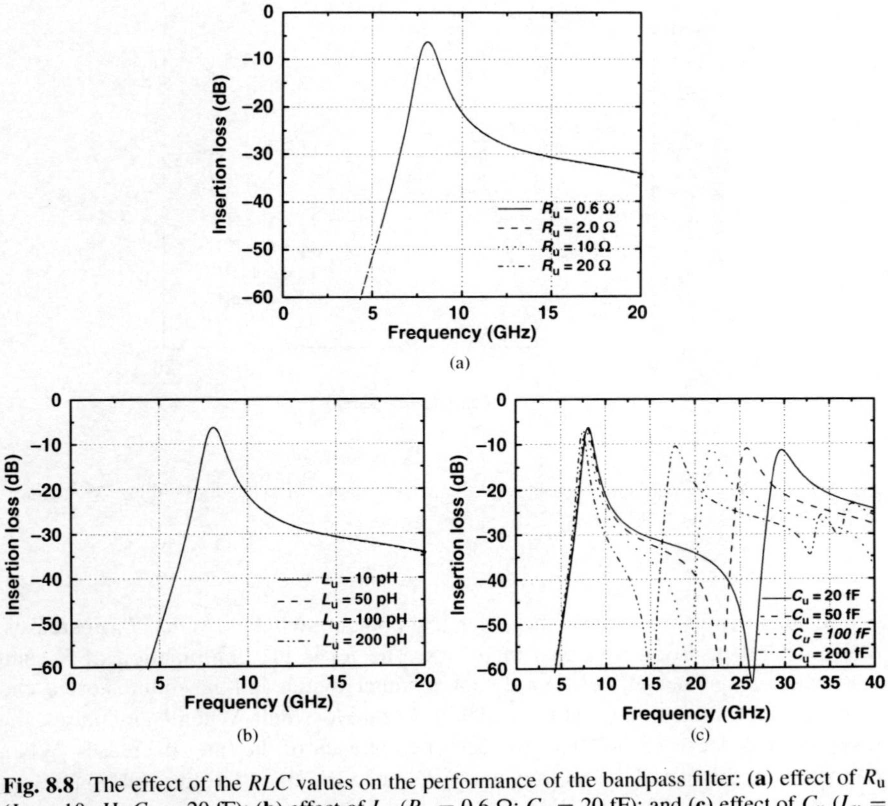

Fig. 8.8 The effect of the *RLC* values on the performance of the bandpass filter: (**a**) effect of R_u ($L_u = 10$ pH; $C_u = 20$ fF); (**b**) effect of L_u ($R_u = 0.6$ Ω; $C_u = 20$ fF); and (**c**) effect of C_u ($L_u = 10$ pH; $R_u = 0.6$ Ω)

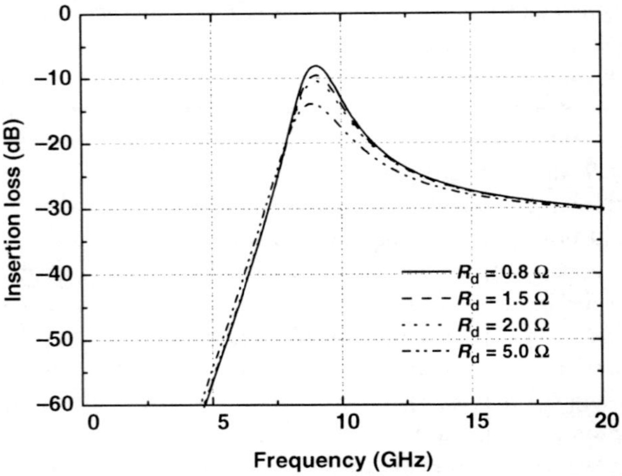

(a)

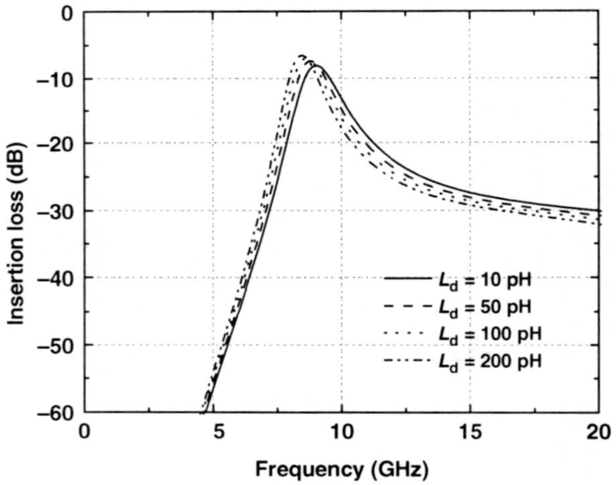

(b)

Fig. 8.9 Effects of the R_d and L_d values on the performance of the bandpass filter: (**a**) R_d (L_d = 10 pH) and (**b**) L_d ($R_d = 0.8\ \Omega$)

is equivalent to driving down the other DC-contact switches. When R_d increases, the passband insertion loss also increases. The R_d is the combination of R_u and the contact resistance R_c between the DC-contact switch and the shunt shorted-end stub. Therefore, it is important to keep both R_u and R_c small. When L_d increases, the insertion loss decreases whereas the center frequency of the filter decreases. When L_d increases, the overall shunt inductance L increases, since L is the combination of L_d and the inductance of the shunt shorted-end stub as shown in Fig. 8.9b.

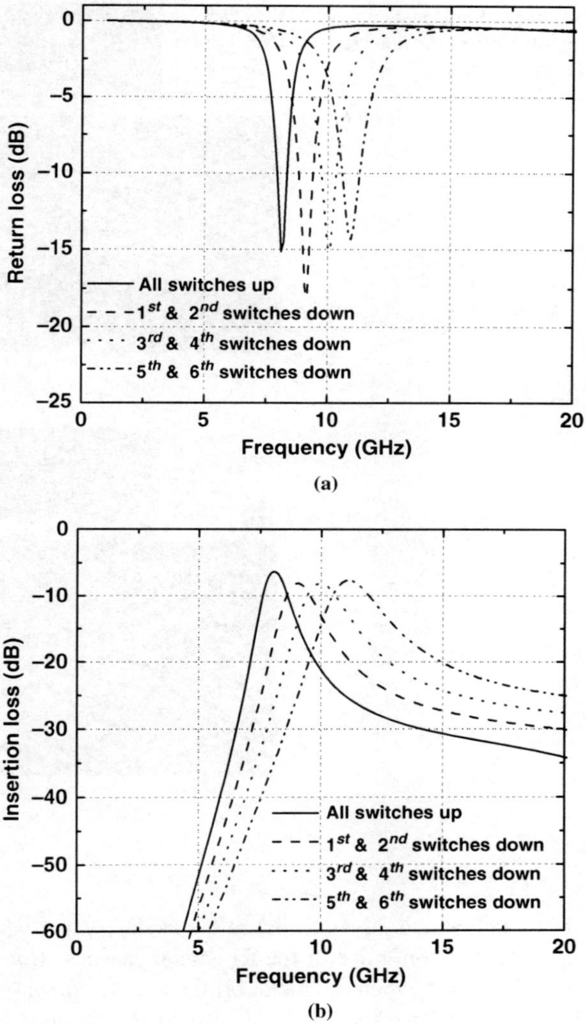

Fig. 8.10 Simulated frequency response of the tunable bandpass filter: (**a**) return loss and (**b**) insertion loss

The simulation results of the frequency tuning are shown in Fig. 8.10. The center frequency is shifted from 8 to 11 GHz with the passband insertion loss varying from 5.5 to 7.5 dB. The return loss in the entire tuning range is higher than 15 dB. The bandwidth can be kept at 10%, but the band rejection at the higher frequency side is not as sharp as that of the lower frequency side.

Figure 8.11a shows the simulated current distribution of the filter when all the switches are at the up-state position. Figure 8.11b shows the simulated current distribution of the filter when the first and second switches are driven down. When the switch is driven down, the current is distributed at the front side of the part of

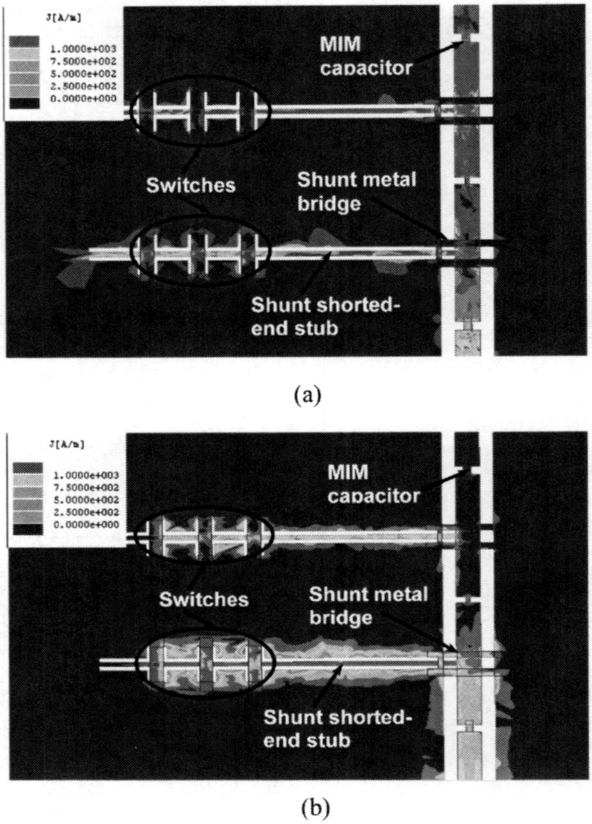

Fig. 8.11 Simulated current distribution for the tunable bandpass filter

the switch above the slot. Figure 8.12a shows that when the metal bridge is at the up-state position, most of the RF signal can pass from the input port to the output port. However, when the metal bridge is at the down-state position, most of the RF signal will reflect back, and the current in the metal bridge is concentrated on the front edge as shown in Fig. 8.12b.

8.2.2 Experimental Results and Discussions

Figure 8.13 shows the SEM images of the DC-contact shunt switch and the design parameters of the switch are shown in Table 8.4. The measured *S*-parameters for the DC-contact switch are shown in Fig. 8.14. The measured pull-down voltage is 26.8 V. From Fig. 6.16, when the metal bridge is at the up-state position, the insertion loss is 0.2 dB at 10 GHz. When the metal bridge is driven down, the isolation is –28 dB at 10 GHz. Based on the measurement results of the *S*-parameters, the

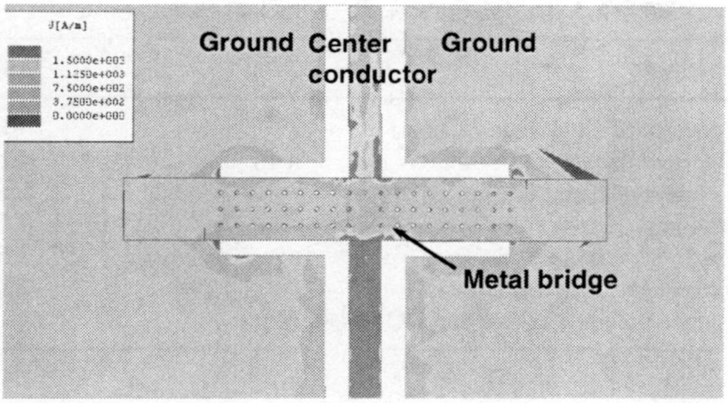

(a)

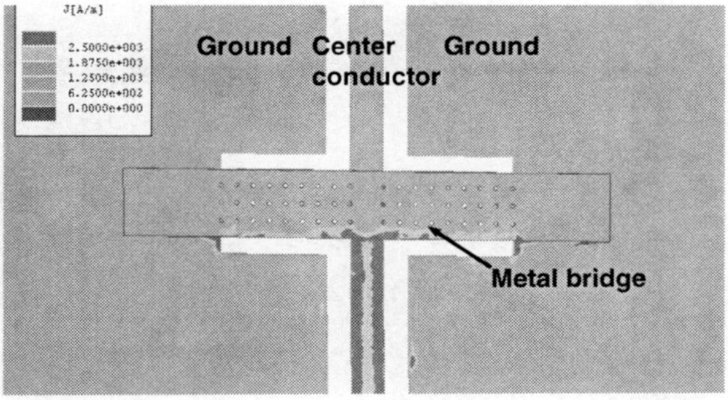

(b)

Fig. 8.12 Simulated current distribution of the DC-contact shunt switch

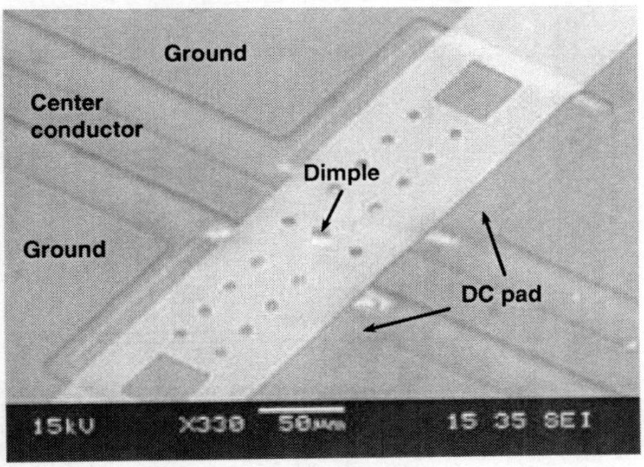

Fig. 8.13 SEM image of the DC-contact switch

Table 8.4 Design parameters of the DC-contact switch

Parameter	Value
Slot width of CPW line, W (μm)	30
Center conductor width of CPW line, S (μm)	40
Thickness of metal bridge, t (μm)	0.6
Initial gap, g_0 (μm)	2
Length of metal bridge, l (μm)	390
Width of DC pad, d (μm)	130
Distance between bridge and DC pad, l_1 (μm)	10

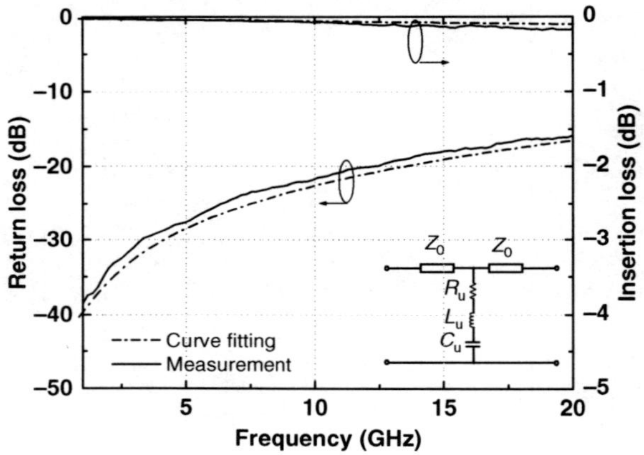

(a)

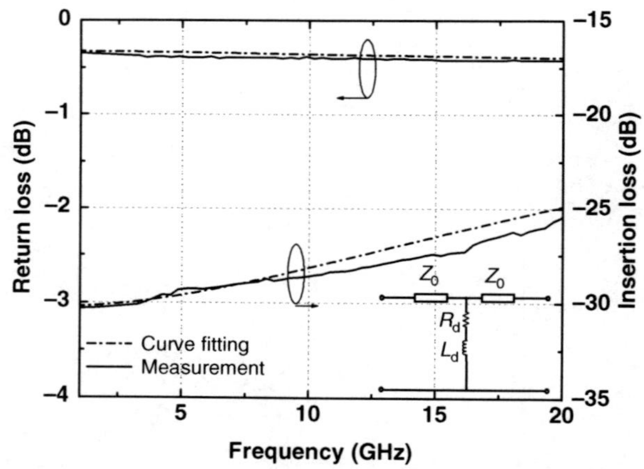

(b)

Fig. 8.14 Comparison of curve-fitting and measurement results of DC switch curve-fitted *RLC* values for the up-state are $R_u = 0.6\ \Omega$, $L_u = 10$ pH, and $C_u = 20$ fF; whereas the curve-fitted *RL* values for the down-state are $R_d = 1.1\ \Omega$ and $L_d = 10$ pH. The problem of the stiction and contact of the DC-contact switch can be solved by properly designing the stiffness of the metal bridge and the contact material [8].

Figure 8.15 is the SEM image of a MIM capacitor. The measurement results of S_{21} of MIM capacitor with different capacitance values are shown in Fig. 8.16. The capacitance can be extracted for different dimensions of MIM capacitors [9]. These extracted capacitance values can match well with the calculated values.

Figure 8.17 shows the measurement results of the insertion loss changes with the frequency for different driven voltages applied on the DC-contact switch (for all

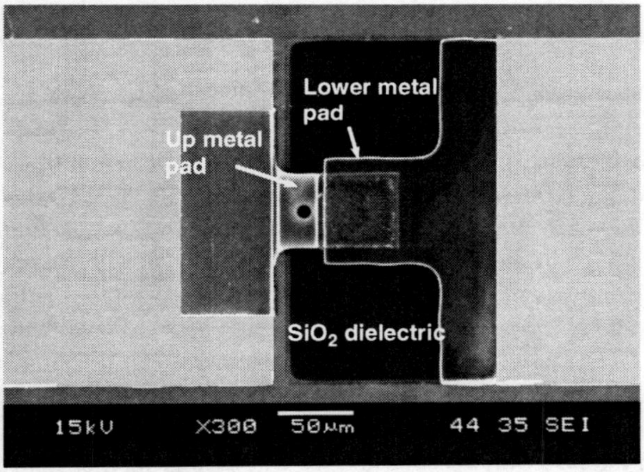

Fig. 8.15 SEM image of the capacitor

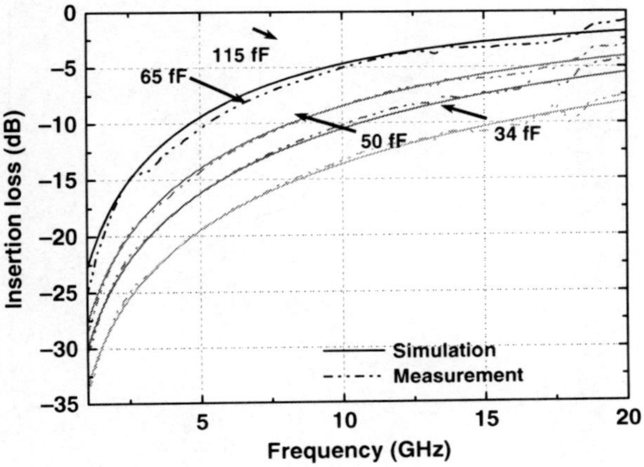

Fig. 8.16 Measurement results of different MIM capacitors

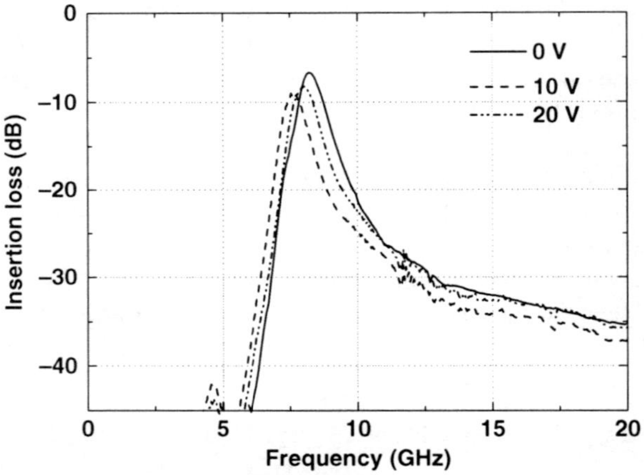

Fig. 8.17 Measured frequency response of filter with different aplied voltages

switches). When the driven voltage increases, the insertion loss at center frequency increases from 5.5 to 6.5 dB. When the driven voltage increases but smaller than the pull-down voltage (26.8 V), the metal bridge is driven down to the lower DC pad and the capacitance of the DC switch increases. The capacitance can be extracted from the insertion loss of the DC switch with different applied voltages, as shown in Fig. 8.18. When the applied driven voltage increases to 20 V, the insertion loss is 0.25 dB at 20 GHz. Based on Eq. (3.6), the capacitance can be calculated to increase from 20 to 35 fF. According to the simulation results in Fig. 8.8c, the insertion loss of the filter increases too. As mentioned in this section, the up-state capacitance has little effect on the insertion loss of the filter when C_u is smaller than 50 fF.

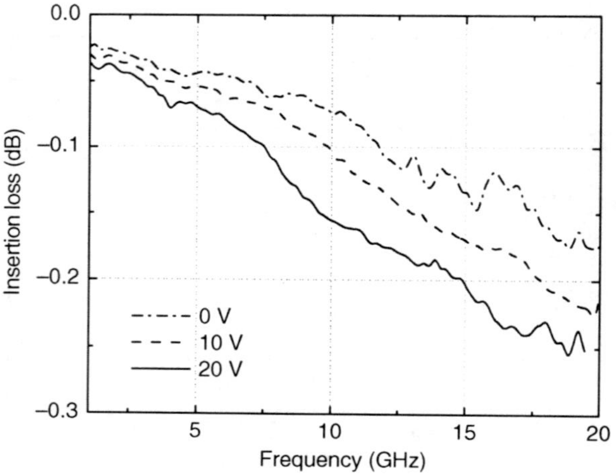

Fig. 8.18 Measured insertion loss of the DC-contact switch with different voltages

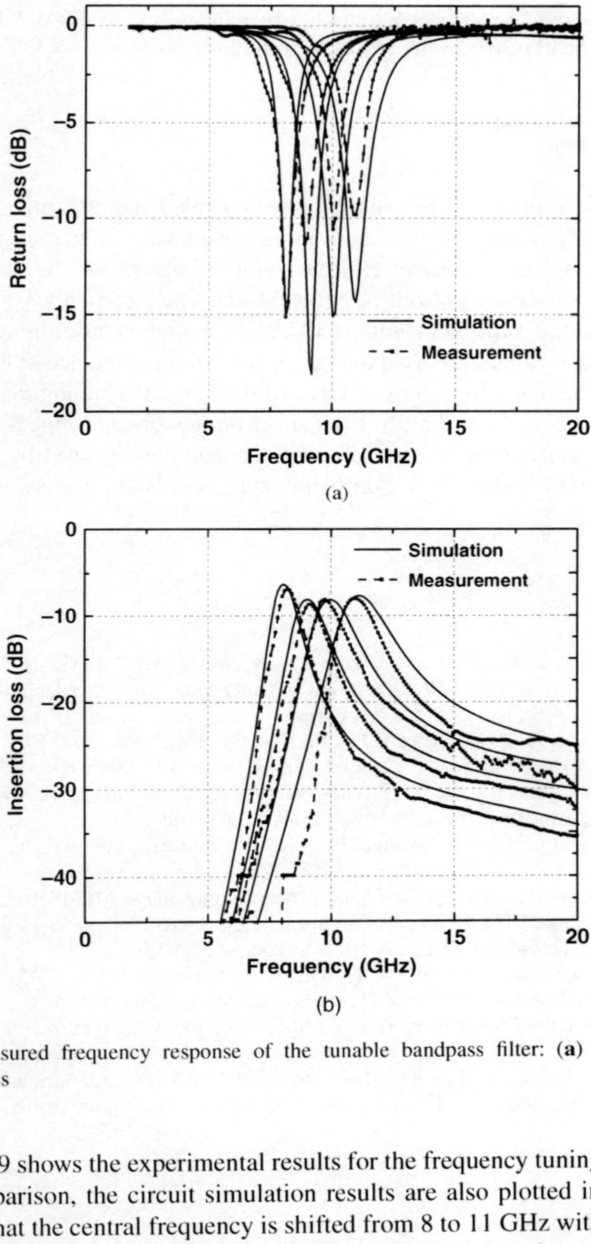

Fig. 8.19 Measured frequency response of the tunable bandpass filter: (**a**) return loss and (**b**) insertion loss

Figure 8.19 shows the experimental results for the frequency tuning. For the purpose of comparison, the circuit simulation results are also plotted in Fig. 8.19. It can be seen that the central frequency is shifted from 8 to 11 GHz with the insertion loss varying from 5.5 to 7.5 dB. The 3-dB bandwidth can be kept at 10% in the whole tuning range. The return loss varies from 15 to 10 dB. The measurement and the simulation results are in good agreement except that the measured return loss is smaller than the simulated results. Such difference can be attributed to the 3% fabrication tolerance and the deviation of the dielectric constant between the nominal and its real values of the SiO_2 used in the MIM capacitor. As a result, the return loss is reduced by nearly 20% at the resonant frequency.

8.3 Summary

This chapter presents two different designs of tunable bandpass filters with different tuning mechanisms. For the capacitance tuning bandpass filter, capacitive switches are used to change the capacitance of the shunt resonators and the coupling capacitors. The filter can be tuned from 10.3 to 4.9 GHz. The insertion loss varies between 5 and 6 dB and the 3-dB bandwidth is 40%. For the inductance tuning bandpass filter, DC-contact switches are used to change the shunt inductance of each resonator. The filter can be tuned from 8 to 11 GHz with nearly 40% tuning range. The insertion loss varies from 5.5 to 7.5 dB in the entire tuning range and the 3-dB bandwidth is 10%. The insertion loss of the tunable filters can be improved by using glass or quartz as substrate and using thicker metal layer as CPW transmission line.

References

1. J. Brank, Z. J. Yao, M. Eberly, A. Malczewski, K. Varian and C. L. Goldsmith, "RF MEMS-based tunable filters," Int. J. RF Microwave Computer Aided Eng. (Special Issue) Vol. 11, 2001, pp. 276–284.
2. A. B. Yu, A. Q. Liu, Q. X. Zhang and H. M. Hosseini, "Micromachined DC contact switch on low-resistivity silicon substrate," Sensor. Actuator. A Vol. 127, 2006, pp. 24–30.
3. G. L. Mattaei, L. Young and E. M. T. Jones, Microwave Filters, Impedance-Matching Networks and Coupling Structures, McGraw-Hill, New York, NY, 1964.
4. A. B. Yu, A. Q. Liu and Q. X. Zhang, "Tunable MEMS LC resonator with large tuning range," IEE Electron. Lett. Vol. 41(15), 2005, pp. 855–857.
5. A. B. Yu, A. Q. Liu and Q. X. Zhang, "Wide tuning range MEMS bandpass filter with inductance change," The 13th International Conference on Solid-State Sensors, Actuators and Microsystems Transducers'05, Seoul, Korea, June 5–9, 2005.
6. I. Zverev, Handbook of Filter Synthesis, Wiley, New York, NY, 1967.
7. Agilent Technology Inc., Advanced Design System Manual, Palo Alto, CA, 2003.
8. G. M. Rebeiz, RF MEMS Theory, Design and Technology, Wiley-Interscience, New York, NY, 2003.
9. S. B. Chen, C. H. Lai, A. Chin, J. C. Hsieh and J. Liu, "RF MIM capacitors using high-K Al_2O_2 and $AlTiO_x$ dielectrics," IEEE MTT-S Int. Microwave Symp. Digest, Seattle, WA, USA, June 2002, pp. 201–204.

Chapter 9
Substrate Transfer Process

In this chapter, a single-mask substrate transfer process for fabrication of high-aspect-ratio (HAR) suspended structures is presented [1–3]. The HAR silicon structures are fabricated using deep reactive ion etching (DRIE) technique and then transferred to a glass wafer though silicon/thin film/glass anodic bonding and silicon thinning techniques. The HAR structures are released using self-aligned wet etching of the glass. Three key processes are presented. First is the silicon/thin film/glass anodic bonding, with special emphasis on the effect of the bonding material on the bonding shear strength. The second is the silicon thinning via aqueous solution of potassium hydroxide (KOH). The last is the shadow mask making and metal deposition methods through the shadow mask [4]. This substrate transfer process has the advantages of low-loss, high-yield, and flexible fabrication technique on the glass substrate.

9.1 Design of the Substrate Transfer Process

The substrate transfer technique uses an 8″ silicon wafer and a glass wafer in the fabrication of high-aspect-ratio suspended silicon structures on a glass. The silicon wafer which is 730-μm thick can be of either low resistivity (ρ = 1–10 Ω cm) or high resistivity (ρ > 4000 Ω cm). The glass wafer is 500-μm- to 1-mm-thick borosilicate Pyrex 7740 glass wafer. The process flow is shown in Fig. 9.1. The details are explained below:

i. A 2-μm-thick SiO_2 is deposited using plasma-enhanced chemical vapor deposition (PECVD) and etched using RIE.
ii. Deep silicon trenches are etched using Surface Technology System's (STS) high-density inductively coupled plasma (ICP) DRIE [5]. The SiO_2 layer is the hard mask. The wide trenches are etched faster than the narrow trenches because of the aspect-ratio-depending-etching (ARDE) effect of the DRIE process. The highest aspect ratio is 30 for a 2-μm-wide, 60-μm-deep trench.

Min Tang and Ai Qun Liu

A.Q. Liu, *RF MEMS Switches and Integrated Switching Circuits*,
MEMS Reference Shelf 5, DOI 10.1007/978-0-387-46262-2_9,

iii. After removing the SiO_2 using RIE, all the exposed structures are covered with a protective thin film – thermal SiO_2 or silicon oxynitride (SiON) – in the furnace to prevent erosion of structures from KOH etching. Therefore, the protective thin film has to be resistant to KOH solution and anodically bondable to the glass. The bonding shear stress has to be high enough to resist the subsequent grinding and polishing process.

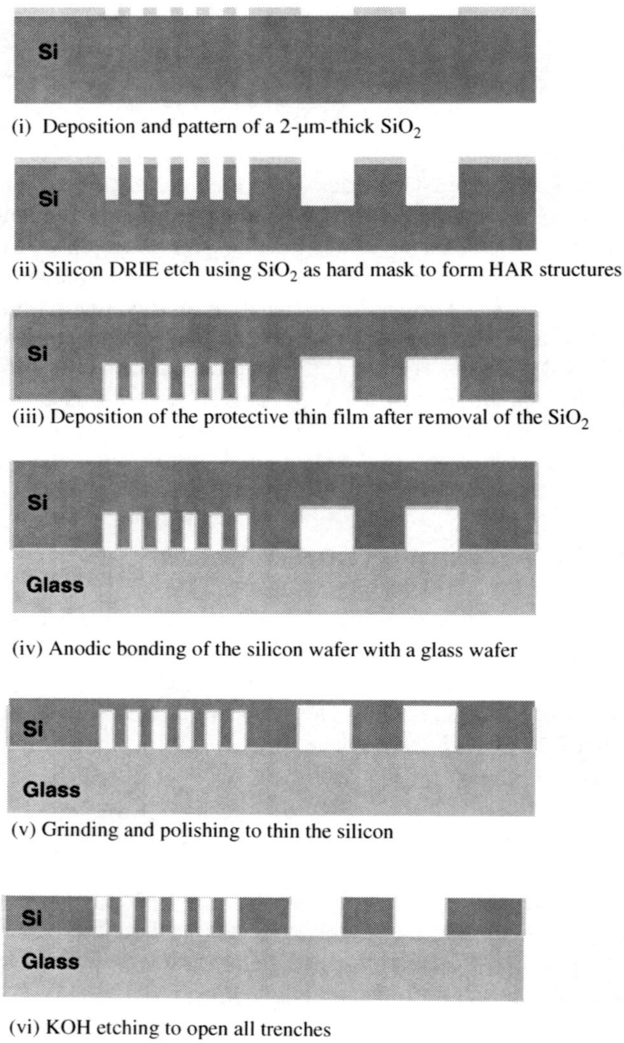

Fig. 9.1 Schematic of the fabrication process flow. Copyright/used with permission of/courtesy of Institute of Physics and IOP Publishing Limited

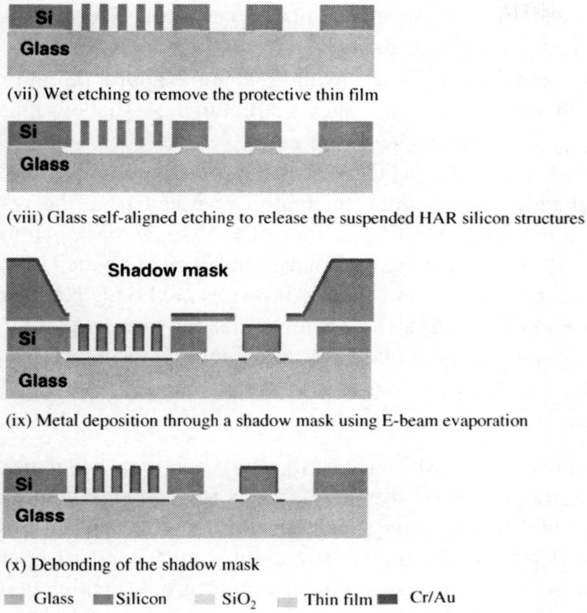

Fig. 9.1 (continued)

iv. The silicon and glass wafers are anodically bonded together using Electronic Vision (EV) bonder EVG520 of EV group. The bonding conditions are temperature of 400ºC, voltage of 1000 V, piston force of 1800 N, chamber pressure of 3 mbar, and bonding time of 1 h and 30 min.

v. The grinding process is employed to thin the silicon wafer rapidly using the GRIND-X GNX200 machine until the deepest trenches are exposed. The grinding rate is 100 μm/min. Then, the polishing process is used to remove grinding marks and smooth the silicon surface. The polishing step sacrifices 5 μm of silicon thickness. Hence, it should be noted that the grinding and polishing process cannot be used to expose all the structures directly since the brush and the slurry used in the grinding and polishing process may damage tiny beams and block small trenches.

vi. The stack is submerged in 35 wt.% KOH aqueous solution at 40ºC to etch silicon until all structures are exposed. The protective thin film can prevent erosion of exposed silicon of sidewall structures from the KOH solution.

vii. The exposed protective thin film is removed using wet etchant. For example, the thermal SiO_2 is etched by buffered oxide etchant (BOE) and the silicon oxynitride is etched by phosphorous acid.

viii. The stack is dipped in the hydrofluoric acid (HF) solution (HF (49%):H_2O = 1:5) for 30 min to etch glass and release movable structures. The silicon structures serve as the hard mask. After rinsing in deionized (DI) water, the wafers were dipped in isopropyl alcohol (IPA) and the stack is dried using critical point drying method.

ix. A layer of metal is deposited using E-beam evaporation and covers the top surface and sidewalls of the silicon structures. A shadow mask can also be used to deposit metal at selective areas.

x. The bonded wafers are heated to 150°C in order to soften the photoresist so that the shadow mask can be manually separated from the wafer stack. The photoresist is wiped off using acetone. The shadow mask is reusable.

xi. The wafer is manually diced into pieces with size of about 2 cm × 3 cm using a diamond cutter. Then, 1-mil-thick gold wires are bonded at the discontinuities of the ground planes of devices using a manual wire-bonding machine. These bond wires can also be replaced by metal interconnections during wafer-level packaging. The device is then ready for testing.

Figure 9.2 shows the SEM image of the cross-sectional view of a silicon trench fabricated. The straight silicon structure is about 50-μm thick. A cavity with sloping sidewalls is formed in the glass substrate after 30-min self-aligned etching. The vertical etching depth and the lateral undercutting are 5.7 and 10.2 μm, respectively. The lateral undercutting is nearly 1.8 times of the vertical etching depth. Ideally, the isotropic chemical wet etching of glass should result in an undercutting similar to the vertical etching depth. The larger undercutting can also be due to the thin layer of native oxide (SiO_2) at the interface, which is formed during the anodic bonding. In the HF solution, the native oxide is etched faster than glass. The HF solution then moves laterally from the bonding interface to attack the native oxide thereby increasing the undercutting; 1000 Å chromium (Cr) and 1 μm gold (Au) are coated

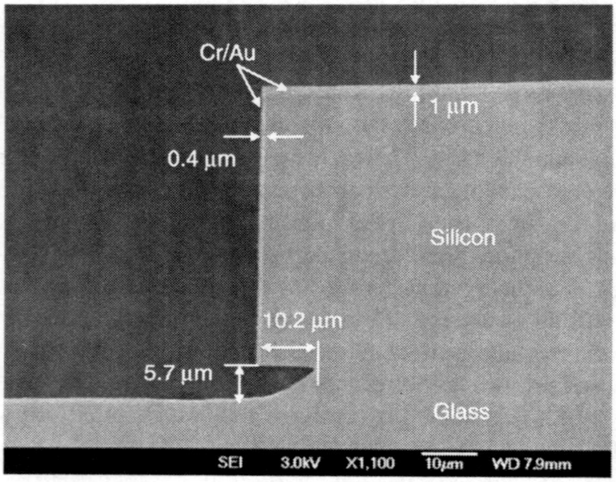

Fig. 9.2 SEM image of the cross-sectional view of a silicon trench after process. Copyright/used with permission of/courtesy of Institute of Physics and IOP Publishing Limited

on the top and sidewalls of the silicon structure uniformly. Due to the step coverage of the metal coating, the metal on the sidewalls is thinner than on the top.

This single-mask substrate transfer process can provide high-aspect-ratio suspended structures on the glass substrate because the DRIE process is fabricated on the bare silicon wafer. The notching effect, which always exits in the conventional SOG-based process, is avoided. Double-side alignment is also avoided since the suspended structures are released by glass self-aligned etching. Hence, only one mask is needed. It is noted that the bonding quality of the silicon/thin film/glass anodic bonding pair is critical for this substrate transfer process as the bonding pair has to resist the subsequent grinding and polishing and the KOH etching processes.

9.2 Silicon and Glass Anodic Bonding

The anodic bonding process is shown in Fig. 9.3. The glass surface is placed against the thin film on silicon surface. This sandwich structure is heated on a hot plate and a negative voltage is applied to the glass. The substrates are heated to 400°C and a bias voltage of 800 V is applied between the glass and the silicon wafer. As a result, an electrostatic force and the migration of ions create an irreversible chemical bond at the interface between the silicon and the glass.

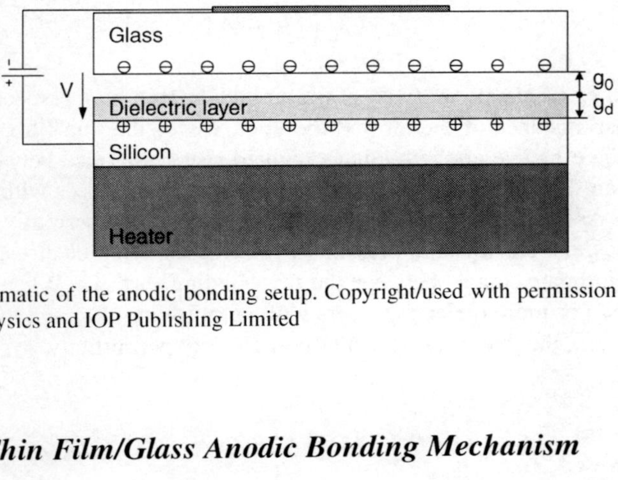

Fig. 9.3 Schematic of the anodic bonding setup. Copyright/used with permission of/courtesy of Institute of Physics and IOP Publishing Limited

9.2.1 *Si/Thin Film/Glass Anodic Bonding Mechanism*

The anodic bonding can be explained by the formation of the Si–O–Si bonds originating from silicon oxidation at the interface [6]. When the temperature increases, the Na^+ ions are more mobile for the Pyrex glass to behave like a conductor. Hence, most of the voltage applied drops across a small gap of a few microns between the two surfaces. The high electric field in this area creates a strong electrostatic pressure, which acts on the two surfaces of the silicon and the glass and effectively pulls them together. Thus, an intimate contact is formed. This contact is necessary to allow the two surfaces to react chemically and form the Si–O–Si bond.

Simultaneously, the Na^+ ions start drifting the cathode, neutralizing the cathode while creating a depletion zone adjacent to the silicon anode. During the charging process of the depletion zone, the electric field is sufficiently high to allow a drift of oxygen anions to the anode. They react with the silicon anode and create a Si–O–Si bond. Subsequently, when the depletion zone becomes too large, the process stops. For the anodic bonding mechanism associated with the reaction between hydroxide groups [7], the anodic bonding consists of the oxidation of silicon and the hydrogen bonding between hydroxy groups. The chemical reaction equations are expressed as follows:

$$\mathrm{Si} + \mathrm{O{-}Si} \rightarrow \mathrm{Si{-}O{-}Si} \tag{9.1}$$

$$\mathrm{Si{-}OH} + \mathrm{HO{-}Si} \rightarrow \mathrm{Si{-}O{-}Si} + \mathrm{H_2O} \tag{9.2}$$

When a thin film of dielectric material with thickness of g_d is coated on the silicon wafer, part of the applied voltage drops in the dielectric film and the other part drops in the air gap, g_0. The voltage that drops in the air gap induces the electrostatic pressure, P, between the two surfaces, which is given by [8]

$$P = \frac{\varepsilon_0 {V_\mathrm{eff}}^2}{2g_0^2\left(1 + \frac{\varepsilon_0 g_\mathrm{d}}{g_0 \varepsilon_\mathrm{d}}\right)^2} \tag{9.3}$$

where ε_0 and ε_d are the permittivity of the air and the thin film, respectively, g_0 and g_d are the gap distance of the air and the thickness of the thin film, respectively, and V_eff is the effective applied voltage applied at the interface between the glass and the silicon. Therefore, the electrostatic pressure, P, increases with the effective voltage, V_eff. As the thickness of the thin film increases or the permittivity of the thin film decreases, the electrostatic pressure, P, decreases. High electrostatic pressure controls the intimate contact between the two bonding surfaces. When the thin film consists of two or more dielectric layers with permittivity of ε_i and thickness of g_i $(i = 1, 2, \ldots, n)$, the thickness, g_d, and the equivalent permittivity, ε_d, are given by

$$g_\mathrm{d} = \sum_{i=1}^{n} g_i \tag{9.4}$$

$$\varepsilon_\mathrm{d} = \frac{g_\mathrm{d}}{\sum_{i=1}^{n} \frac{g_i}{\varepsilon_i}} \tag{9.5}$$

and the ratio of $g_\mathrm{d}/\varepsilon_\mathrm{d}$ is determined by

$$\frac{g_\mathrm{d}}{\varepsilon_\mathrm{d}} = \sum_{i=1}^{n} \frac{g_i}{\varepsilon_i} \tag{9.6}$$

Therefore, for multilayer dielectric deposition, the ratio of g_d/ε_d increases and the electrostatic pressure, P, decreases. The chemical reaction for the bond formation is also affected due to the change of the bonding interface material.

9.2.2 Preparation of Bonding Samples

Three process parameters considered are studied. They are the material of the thin film coated on the silicon wafer, the thickness of the glass wafer, and the resistivity of the silicon wafer. Six types of anodic bonding sample, as listed in Table 9.1, are prepared to determine the effects of different parameters on the bonding quality. Sample A is a bare silicon wafer bonded with a 500-μm-thick glass wafer. The anodic bonding of silicon and glass is very strong. However, when silicon structures are exposed to the KOH solution, the silicon is etched and structures are damaged. Therefore, it is necessary to deposit a protective thin film on silicon structures before anodic bonding, so that it can resist the attack of the KOH solution and protect sidewalls of exposed silicon structures. In samples B, C, and D, the silicon wafer is coated with different thin films. In sample B, the silicon wafer is coated with 1000 Å wet thermal SiO_2. In sample C, the silicon wafer is coated with 1000 Å wet thermal SiO_2, followed by 1000 Å silicon nitride (SiN) deposited using low-pressure chemical vapor deposition (LPCVD). In sample D, the silicon wafer is coated with the same thin film as in sample C and then is put in the wet oxidization furnace at 1000°C for 1 h to oxidize the SiN surface to silicon oxynitride (SiON). In samples E and F, the silicon wafers are coated with the same thin film as in sample B. The glass wafer is 500-μm thick in sample B and 1-mm thick in sample E. The resistivity of the silicon wafer is less than 10 Ω cm in sample B and more than 4000 Ω cm in sample F. Same bonding conditions – the bonding temperature of 400°C, voltage of 800 V, piston force of 1800 N, chamber pressure of 3 mbar, and bonding time of 1 h and 30 min – are used for all samples.

Table 9.1 Tabular sample preparation and average shear strength of anodic bonding

Sample type	Resistivity of silicon (Ω cm)	Thin film on the silicon surface	Thickness of glass (μm)	Average shear strength (MPa)
A	1–10	–	500	16.6
B	1–10	SiO_2	500	13.9
C	1–10	SiO_2 + SiN	500	1.6
D	1–10	SiO_2 + SiN + SiON	500	11.2
E	1–10	SiO_2	1000	13.0
F	>4000	SiO_2	500	13.6

9.2.3 Experimental Results and Discussions

After the anodic bonding, it is observed that samples A, B, D, E, and F are bonded uniformly. All structures are bonded with the glass. There are some small colorful rings with diameters less than 4 mm at the bonding surface, which may be caused by the incomplete cleaning of silicon and glass wafer. However, for sample C, the bonding is relatively poor. There are some bubbles with diameters more than 3 cm in sample C. Many small structures with feature size less than 1 mm are unbonded.

To quantitatively evaluate the bonding quality, the tensile strength measurement is conducted commonly [9, 10]. The shear strength measurement is relatively more important when the bonding wafers need to be ground and polished. Failure usually occurs when the external shear stress exceeds the limitation of the shear bonding strength. Figure 9.4 shows the schematic of the shear test setup using Dage Series 4000 Bond Tester. The glass is faced down and the silicon is faced up. The shear height is set at 200 μm higher than the glass thickness. For instance, the shear height is set at 700 μm when the glass is 500 μm and 1.2 mm when the glass is 1 mm. The bonded wafer pair is diced into dies with 2 mm × 2 mm in size. Seven test dies are prepared for each bonded wafer pair.

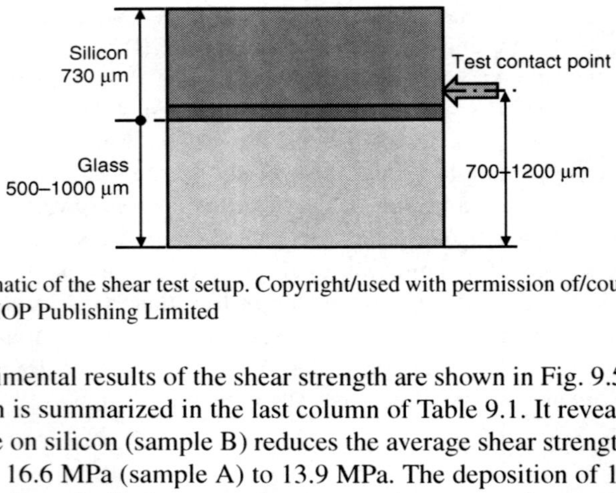

Fig. 9.4 Schematic of the shear test setup. Copyright/used with permission of/courtesy of Institute of Physics and IOP Publishing Limited

The experimental results of the shear strength are shown in Fig. 9.5. The average shear strength is summarized in the last column of Table 9.1. It reveals that 1000 Å thermal oxide on silicon (sample B) reduces the average shear strength of the bonding pair from 16.6 MPa (sample A) to 13.9 MPa. The deposition of 1000 Å SiN on thermal oxide (sample C) decreases the average shear strength to 1.6 MPa. This can be partly explained by the change of the electrostatic pressure, *P*. Multilayer dielectric material coated on the silicon surface increases the ratio of g_d/ε_d and decreases the electrostatic pressure, *P*.

However, wet thermal oxidization of SiN to SiON (sample D) increases the average shear strength to 11.2 MPa. This is lower than silicon–glass (sample A) and silicon–SiO_2–glass (sample B) bonding, but is much higher than silicon–SiO_2–SiN–glass bonding (sample C). Figure 9.6 shows SEM images of the cross-sectional view of samples C and D. Figure 9.6a shows a distinct gap at the interface of sample C. This implies insufficient bond in sample C (silicon + SiO_2 + SiN). However, by oxidizing the surface of SiN into SiON, as in sample D, no gap exists at the interface, as

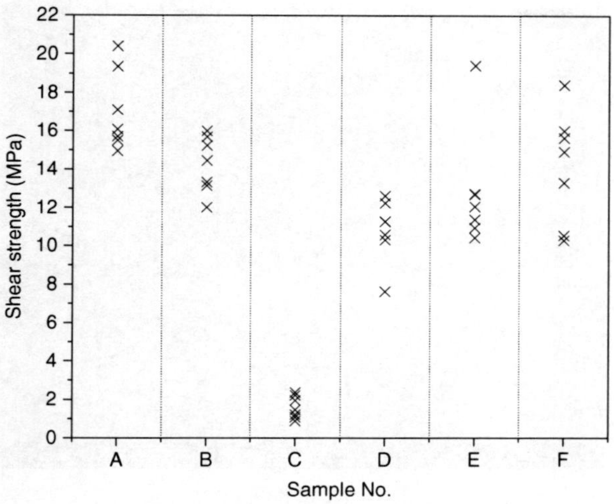

Fig. 9.5 Plot of measured shear strength of different samples. Copyright/used with permission of/courtesy of Institute of Physics and IOP Publishing Limited

shown in Fig. 9.6b. This means that a strong bond is formed in sample D. In sample D, the surface refractive index of silicon with thin film is 1.954 measured using a Rudolph-ellipsometer which is between the refractive index $n = 1.46$ for 1000 Å thermal SiO_2 and $n = 2.02$ for 1000 Å LPCVD SiN. This implies that the SiN surface has changed. Some oxygen (O) atoms have reacted with the SiN and formed a thin layer of silicon oxynitride (SiON). The thickness of SiON layer is about 30 Å. The SiN film is highly dense and the surface terminates with Si–N bond [11]. The Si–N bond is too strong to be broken at 400°C. Therefore, the Si–O–Si bond cannot be formed in sample C. However, nitrogen in the SiN surface layer can be substituted by oxygen in the 1000°C furnace condition and a thin silicon oxide/oxynitride layer is formed. This layer contains a high density of H– and OH– groups at the surface, which helps the interfacial chemical reactions. Therefore, Si–O–Si bond can be formed in sample D effectively with strong bonding.

The average shear strength of sample E (1000-μm-thick glass) is 13 MPa which is slightly lower than sample B (500-μm-thick glass), which implies the bonding strength is slightly weakened when the thickness of the glass increases. This is because thicker glass has higher electrical resistance. Therefore, the effective voltage, V_{eff}, at the interface is reduced, resulting in a lower electrostatic pressure and weaker bonding.

The average shear strength of the sample F ($\rho > 4000\ \Omega$ cm) is 13.6 MPa, which is very close to 13.9 MPa of sample B ($\rho < 10\ \Omega$ cm). That means the resistivity of the silicon wafer has little influence on the bonding shear strength.

The experiment of the grinding process shows that the bonding quality of the sample C is so poor that the pair collapses during the grinding process. Other samples (samples A, B, D, E, and F) go through the grinding process successfully.

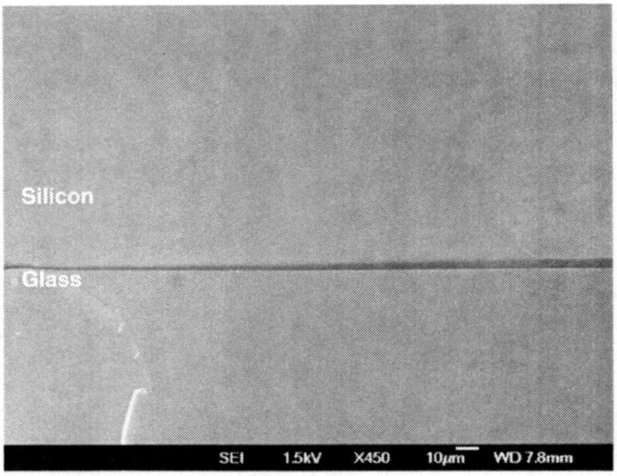

(a) sample C: $SiO_2 + SiN$

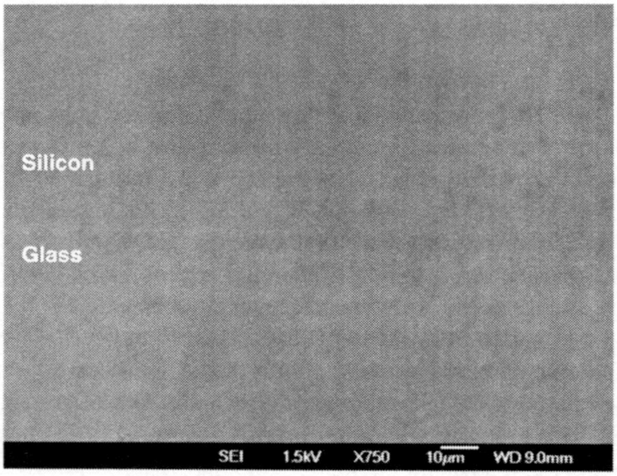

(b) sample D: $SiO_2 + SiN + SiON$

Fig. 9.6 SEM image of the bonding interface region of silicon–thin film–glass bonding pair: (**a**) sample C and (**b**) sample D. Copyright/used with permission of/courtesy of Institute of Physics and IOP Publishing Limited

9.3 Wafer Thinning via KOH Anisotropic Etching

Anisotropic chemical etching using the aqueous solution of KOH has long been used for the fabrication of the MEMS structures due to its excellent repeatability and uniformity and low cost. The etching conditions affect the surface properties of the structures, such as the flatness and the roughness. The aqueous solution of KOH starts to react with the exposed silicon immediately when it reaches the patterned silicon structures. Therefore, silicon structures without any protective thin film may damage immediately. Some protective thin film for the silicon structures, such as thermal oxide and thermal oxide/SiN/SiON, can provide the effective protection to the sidewalls of the silicon structures. The experiment shows that the etching rate of thermal oxide in KOH aqueous solution with 35 wt.% concentration at 40°C is approximately 300 Å/h whereas the etching rate of LPCVD SiN is negligible. Therefore, the use of thermal oxide/SiN/SiON protective thin film is more preferable.

9.3.1 Layout Design Considerations

KOH is an anisotropic etchant of the silicon with different etch rates (E_{R}) on different crystal orientation planes, commonly $E_{\mathrm{R}\,\{110\}} > E_{\mathrm{R}\,\{100\}} > E_{\mathrm{R}\,\{111\}}$ [12, 13]. The protective thin film can only be used to protect the sidewalls of the structures. The top surface is etched continuously by the KOH solution. When the layout of the structures is designed arbitrarily, a V-shaped groove may form on the top of the structures. Therefore, the layout design should consider the KOH crystallography-dependent etching. In KOH etching, the {111} plane is an extremely slow etching plane and accepted to be the non-etching plane. As the {111} planes provide etching stops, it is better not to expose them to the KOH solution during the wet etching so that the etching process can be controlled by adjusting the etching time, with the sidewall remaining vertical. Which orientation plane is introduced initially depends on the geometry and the orientation of the mask feature. In this study, the normal (100) silicon wafers with ⟨110⟩ prime orientation flat is used. When a rectangular wire is accurately aligned with the prime orientation flat, i.e., the ⟨110⟩ direction, only {111} planes are introduced as sidewalls for the KOH etching, which is 54.7° to {100} planes. In this case, the cross section of the etching pit is a V-shaped groove with ⟨110⟩ edges and {111} sidewalls, as shown in Fig. 9.7a. There is one way to etch the silicon uniformly and produce vertical {100} walls on the structures. As shown in Fig. 9.7b, the structures are aligned in such a way that the straight trenches are in 45° angle with the prime orientation flat (⟨110⟩ direction) of the (100) silicon wafer. There are {100} planes perpendicular to the wafer surface and that their intersections with the wafer surface are ⟨100⟩ direction. Consequently, {100} facets are initially introduced as sidewalls. As both the bottom and the sidewall planes belong to the same {100} group, the lateral etching rate equals to the vertical etching rate. Hence, all the exposed silicon can be etched vertically. The layout is turned 45° on the mask to accurately align straight wires with the ⟨100⟩ direction.

Figure 9.8 shows the SEM images of the cross-sectional view of structures with different layout placements. When the straight wires are parallel (0°) or perpendicular (90°) to the ⟨110⟩ flat, a V-shaped groove is formed on the top, as shown in

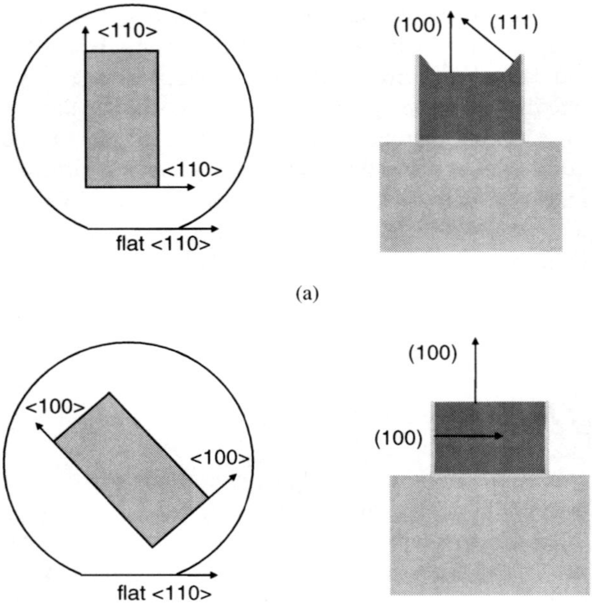

(b)

Fig. 9.7 Schematic top view and cross-sectional view of the silicon etching profile: (**a**) 0° or 90° and (**b**) 45° to ⟨110⟩ prime orientation flat. Copyright/used with permission of/courtesy of Institute of Physics and IOP Publishing Limited

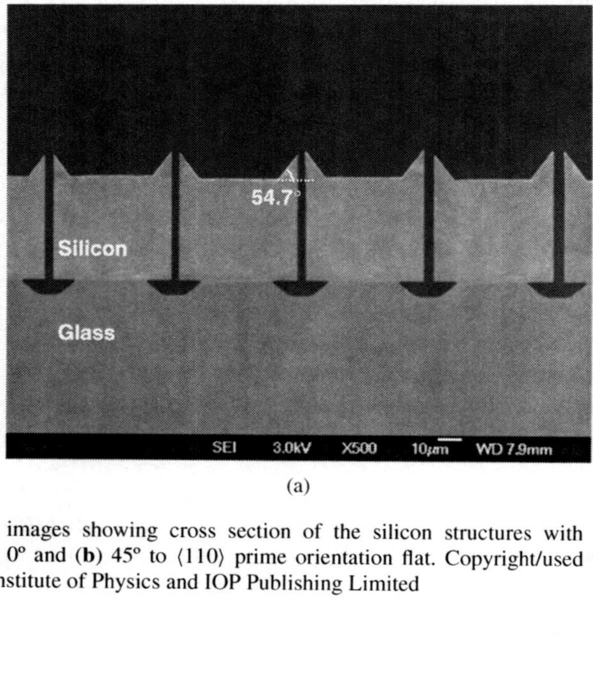

(a)

Fig. 9.8 SEM images showing cross section of the silicon structures with different layout placements: (**a**) 0° and (**b**) 45° to ⟨110⟩ prime orientation flat. Copyright/used with permission of/courtesy of Institute of Physics and IOP Publishing Limited

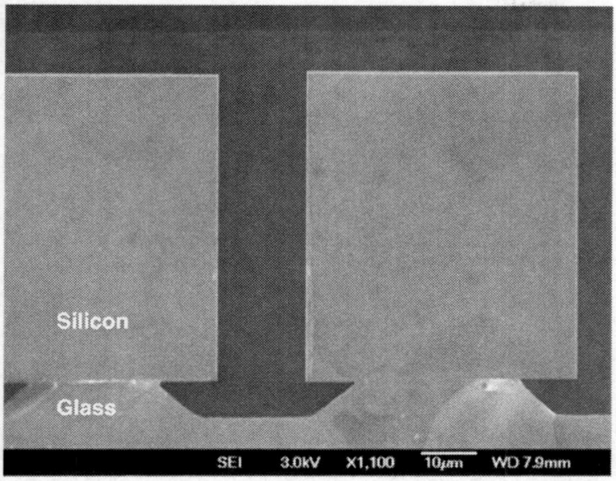

(b)

Fig. 9.8 (continued)

Fig. 9.8a. However, when the straight wires are 45º to the ⟨110⟩ cut plane, the silicon is etched downward and the top etching surface is flat, as shown in Fig. 9.8b.

9.3.2 Etching Temperature Effect

KOH etching is a temperature-dependent process, where the etching rate of silicon increases with the temperature. The etching rate of the silicon (100) in 35 wt.% KOH solution is summarized in Table 9.2. The stabilized etching temperature is ±2°C. The etching rate is approximately 3 μm/h at the room temperature and 62.4 μm/h at 70°C. However, surface roughness changes with the etching temperature too, as shown in Fig. 9.9. It shows that when the etching temperature increases from 40 to 70°C, the surface roughness of the etching surface increases rapidly from 1.317 to 38.003 nm. In consideration of the etching rate and the surface roughness of the silicon etching, the etching temperature of the bonded wafers is 40°C.

Table 9.2 The etching rate and surface roughness of silicon versus etching temperature in 35 wt.% aqueous solution of KOH

Etching temperature (°C)	Etching rate (μm/h)	Surface roughness (nm)
23	2.95	–
40	12	1.317
50	29	6.005
60	40	19.687
70	62.4	38.003

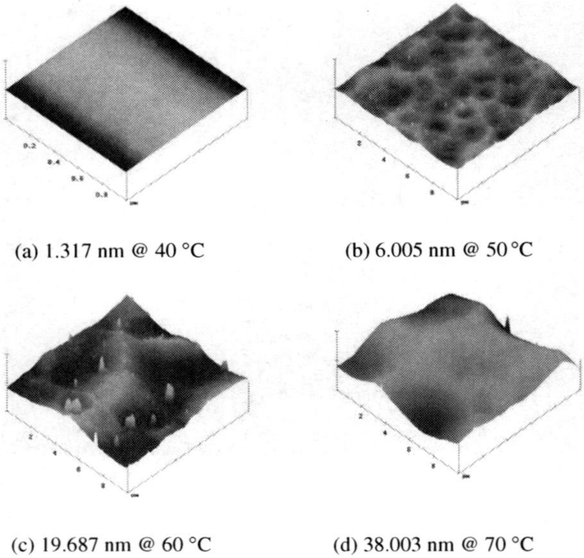

(a) 1.317 nm @ 40 °C (b) 6.005 nm @ 50 °C

(c) 19.687 nm @ 60 °C (d) 38.003 nm @ 70 °C

Fig. 9.9 Experimental results of the surface roughness of silicon etched by 35 wt.% KOH at different temperatures: (**a**) 40ºC, (**b**) 50ºC, (**c**) 60ºC, and (**d**) 70ºC. Copyright/used with permission of/courtesy of Institute of Physics and IOP Publishing Limited

9.4 Shadow Mask Techniques

High-resolution patterning with deep holes is a common problem in the MEMS fabrication process [14] because photoresist spinning cannot be used on such substrate. Shadow mask is useful in cases where normal patterning (lithography) is difficult [15]. In this section, the design, the fabrication, and the usage of the shadow mask are discussed.

9.4.1 Design and Fabrication of the Shadow Mask

The shadow mask is formed using a p-type (100)-oriented silicon wafer. The fabrication process flow is shown in Fig. 9.10a–e. First, 2-μm-thick photoresist is spun on the front side of the wafer and patterned. Then, the DRIE process is used to etch structures to approximately 50-μm deep using photoresist as the mask material. After stripping the photoresist, a dielectric layer consisting of 300-Å thick thermal oxide and 1500-Å thick LPCVD silicon nitride is deposited on both sides of the wafer. Then, the backside of the wafer is patterned using a double-side alignment photolithography technique via Electronic Vision (EV) Aligner. The pattern is transferred to the dielectric layer through RIE and the photoresist is stripped. Next, the wafer is anisotropically etched from the backside using 35 wt.% aqueous solution

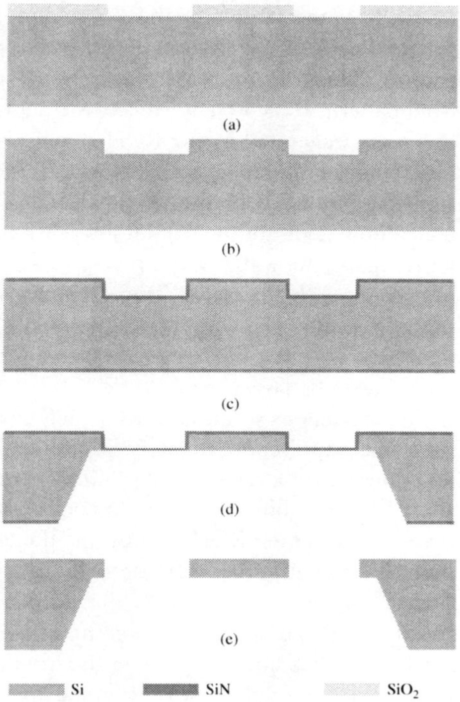

Fig. 9.10 Fabrication process flow of the shadow mask. Copyright/used with permission of/courtesy of World Scientific

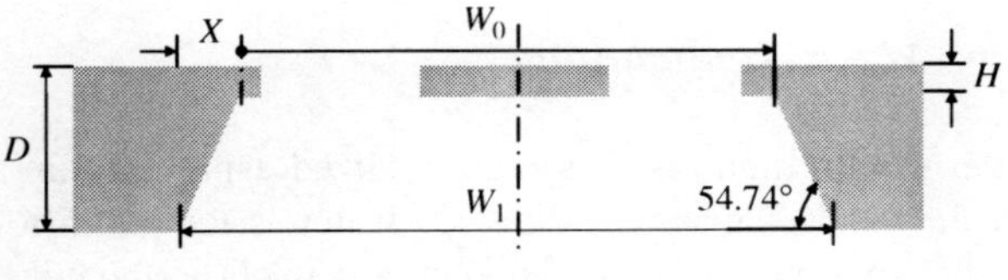

Fig. 9.11 Schematic of the cross-sectional view of the shadow mask. Copyright/used with permission of/courtesy of World Scientific

of KOH at 75°C until the front pattern is visible from the backside clearly. Finally, the remaining silicon oxide and silicon nitride at both sides are etched using RIE.

Figure 9.11 shows the cross-sectional schematic of the shadow mask. High-resolution shadow mask can be fabricated using this fabrication process. The distance between the two holes can be less than 5 μm. The back side of the wafer is etched using the aqueous solution of KOH. The difference between the back side and the front pattern size is $2X$. X can be expressed as

$$X = \frac{W - W_0}{2} = \frac{D - H}{\tan 54.74^\circ} \tag{9.7}$$

where W is the size of the back opening, W_0 is the size of the front pattern, D is the thickness of the shadow mask wafer, and H is the etching depth of the front pattern. Generally, the wafer thickness, D, is 750–800 μm. The etching depth of the front pattern, H, is 50 μm. Hence, X is equal to 495–530 μm. The distance between two holes generated by the KOH etching is larger than $2X$, that is, about 1100 μm. Small holes or holes separated by small distance are combined in one back open and supported by a silicon membrane. The thickness of the membrane is dependent on the etching depth of the front holes. However, if the distances between all holes are larger than 1 mm, the front process can be omitted. Only the KOH etching is used to etch through the wafer. Thus, separate through-holes are formed in the shadow mask. However, the simplified process is obtained at the cost of the resolution. The final hole size, W_0, changes with the wafer thickness and the distance between two holes has to be larger than $2X$.

In the aqueous solution of KOH, large back openings are etched faster than small back openings. Approximately 30-min over-etching time is needed to etch through all patterns. As a result, to shorten the over-etching time, the openings at the back should have a uniform design. In consideration of the membrane tension, the membrane size after the KOH etching should not be too large. Otherwise, the membrane may collapse after the KOH etching. Generally, the membrane size is less than $7 \times 7\ \text{mm}^2$. Figure 9.12 shows the optical photos of the front side of the shadow mask for the single-beam SP2T switching circuit. It can be seen that the backside of the shadow mask is not as smooth as the front side due to the silicon anisotropic etching in the aqueous solution of KOH at the high temperature. The back open is shared by four irregular holes separated by three fixed–fixed beam-mass silicon structures.

9.4.2 Shadow Mask Applications

The metal patterns with high resolution are realized using E-beam evaporation of metal through the shadow mask. The device wafer is temporarily bonded to the shadow mask first. A small amount of photoresist is put at the edge of the wafers as an intermediate material of bonding. Then, the shadow mask is placed on top of the device wafer and aligned using the Electron Vision (EV) aligner. Next, the temporary bonding is implemented by the bonder – EVG520 of EV group. The wafers are bonded under pressure of 1000 N for 10 min at room temperature. After that, the bonded pair is placed in the evaporator and the desired metal (such as Cr/Au, where Cr acts as the adhesive layer of Au) is deposited at selected areas through the shadow mask. After the metal deposition, the shadow mask is de-bonded by heating up the two wafers at 150°C and manually separated, when the photoresist turns soft. Finally, the photoresist residual on the device wafer and the shadow mask is wiped off using acetone. After removing the metal via wet etching, the shadow mask can be recycled.

The coated metal pattern is always larger than the pattern in the shadow mask due to the spread effect, as shown in Fig. 9.13. The enlargement (E) can be calculated as follows:

$$E = \frac{b}{a} = 1 + \frac{d_2}{d_1} \tag{9.8}$$

where a is the pattern size on the shadow mask, b is the pattern size of the metal on the device wafer, d_1 is the distance between the source and the shadow pattern, and d_2 is the distance between the device wafer and the shadow pattern. In the E-beam evaporator, the distance between the source and the wafer holder, d_0, is 0.5 m and the gap between the device wafer and the shadow mask is less than 0.5 mm.

Two different methods can be used to combine the shadow mask and the device wafer, as shown in Fig. 9.14a, b. The first method is to let the backside of the shadow mask facing the device wafer. The second method is to let the front side of the shadow mask facing the device wafer. Comparing the two methods, the distance between the shadow mask pattern and the device wafer, d_2, for the first method is

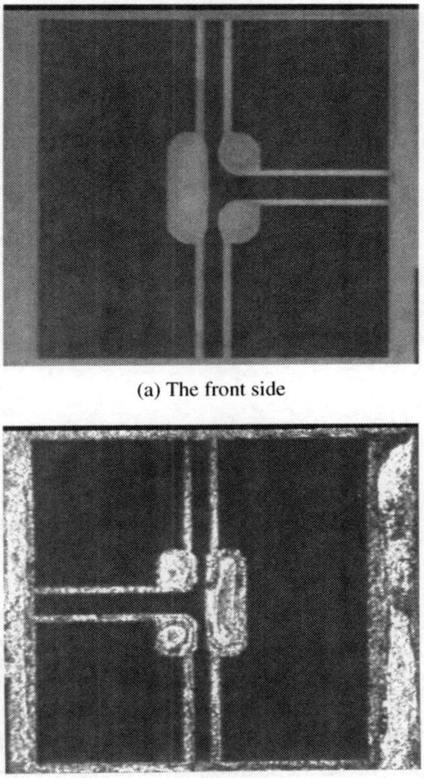

(a) The front side

(b) The back side

Fig. 9.12 Optical photos of the shadow mask of the front side. Copyright/used with permission of/courtesy of World Scientific

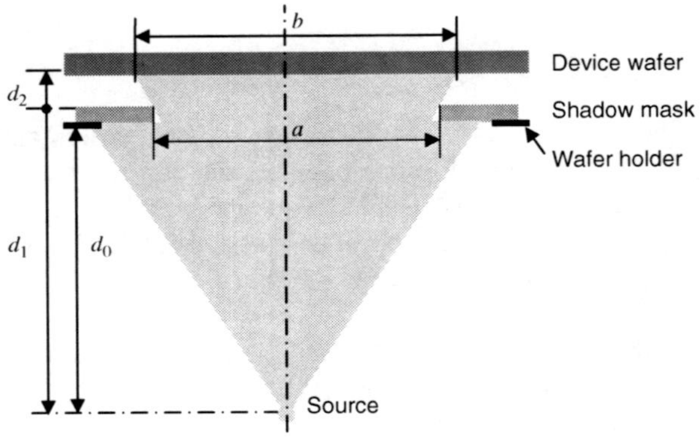

Fig. 9.13 Schematic of the metal deposition through a shadow mask. Copyright/used with permission of/courtesy of World Scientific

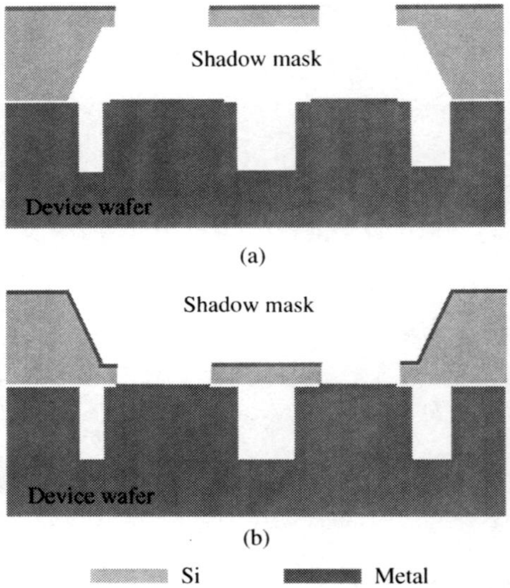

Fig. 9.14 Schematic of the shadow mask: (**a**) the backside of the shadow mask faces the device wafer and (**b**) the front side of the shadow mask faces the device wafer. Copyright/used with permission of/courtesy of World Scientific

smaller than the second method by approximately 700 μm. Therefore, the enlargement, E, of the first method is 1.0024 is significantly larger compared to the second method whose enlargement is 1.001. Hence, the second method is preferred to reduce the metal spreading. In consideration of the spread effect of the metal deposition through the shadow mask, the minimum distance between two holes in the shadow mask is designed to 70 μm.

Figure 9.15 shows the optical photo of the single-beam SP2T switching circuit after the metal deposition through the shadow mask. The bias line is sheltered from the metal deposition with high resistance. Device parts and pads are coated with metal, which guarantees the RF signal propagation and the bias voltage applying. It

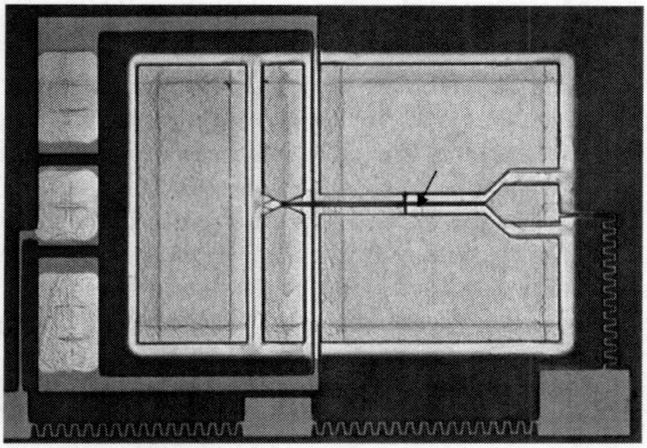

Fig. 9.15 Optical photo of the single-beam SP2T switching circuit after the metal deposition. Copyright/used with permission of/courtesy of Institute of Physics and IOP Publishing Limited

is also noted that the surface color of the device coated with metal is different due to the non-uniformity of metal deposition. This is because of the spread effect at the pattern edge.

Because of the nature of the evaporation process, the metal coated at sidewalls is thinner than that coated on the top surface. Figure 9.16 shows the cross-sectional view illustrating the step coverage of the metal deposition. The thickness of the gold coated on the top surface is 1.2 μm, while the thickness of the gold on the sidewalls is 0.45 μm. It is observed that the metal is tightly coated on the sidewalls and is uniformly covering the height of the entire structures.

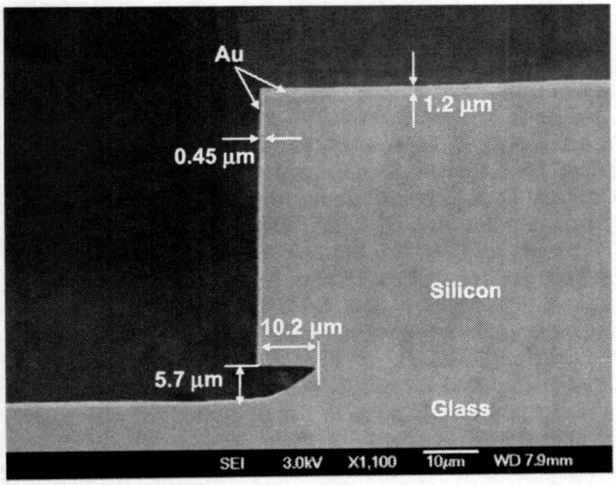

Fig. 9.16 SEM image of cross-sectional view of the metal deposition by the shadow mask

9.5 Summary

This chapter focuses on the fabrication process development of the single-mask substrate transfer technique for the fabrication of a high-aspect-ratio suspended silicon structure on the glass substrate. The processes include silicon DRIE etching process, Si–thin film–glass anodic bonding, silicon wafer thinning, self-aligned glass etching, and metal deposition through a shadow mask. Through the study of the Si–thin film–glass anodic bonding mechanism, thermal oxide, and silicon oxynitride are chosen as the intermediate layer of the Si/thin film/glass anodic bonding because these materials are found to provide sufficient bonding strength with shear strength larger than 11 MPa for the grinding and polishing process. These materials are also strong resistant to KOH etching. The variation of the glass thickness and the silicon resistivity only reduce the bonding strength slightly. The aqueous solution of KOH with concentration of 35 wt.% at 40°C is to thin the silicon wafer, which provides smooth etching surface of the silicon wafer. Based on the analysis of the crystallography-dependent etching of the silicon in the KOH etching, the device layout is adjusted to be 45° to the ⟨110⟩ prime wafer flat in the (100) silicon wafer for the flat etching surface on top of the device. The glass is etched using the silicon structures as the hard mask to release the suspended structures. High-resolution shadow mask is developed by using the silicon DRIE process, followed by KOH etching-through process.

References

1. M. Tang, A. Q. Liu, A. Agarwal and M. H. Habib, "A single-mask substrate transfer technique for the fabrication of high-aspect-ratio micromachined structures," J. Micromech. Microeng. Vol. 17, 2007, pp. 1575–1582.
2. M. Tang, A. Q. Liu and A. Agarwal, "A new approach of high-aspect-ratio MEMS structures fabrication via substrate transfer technology," Asia-Pacific Conference of Transducers and Micro-Nano Technology 2006 (APCOT 2006), Singapore, 25–28 June 2006, Paper No. LS-A0011.
3. A. Q. Liu, A. B. Yu, M. F. Karim and M. Tang, "RF MEMS Switches and integrated switching circuits," J. Semicond. Technol. Sci. Vol. 7(3), 2007, pp. 166–176.
4. M. Tang, A. Agarwal, Q. X. Zhang and A. Q. Liu, "An approach of shadow mask for deep-holes patterning," Int. J. Comput. Eng. Sci. Vol. 4(2), 2003, pp. 315–318.
5. J. Li, A. Q. Liu and Q. X. Zhang, "Tolerance analysis for comb-drive actuator using DRIE fabrication," Sensor. Actuator. A Phys. Vol. 125(2), 2006, pp. 494–503.
6. W. H. Ko, J. T. Sumito and G. H. Yeh, Bonding Techniques for Microsensors, Micromachining and Micropackaging of Transducers, Elsevier, Amsterdam, 1985.
7. C. M. M. Denisse, H. E. Smulders, F. H. P. M. Habraken and W. F. V. D. Weg, "Oxidation of plasma enhance chemical vapour deposited silicon nitride and oxynitride films," Appl. Surf. Sci. Vol. 39, 1989, pp. 25–32.
8. J. A. Plaza, E. Gonzalez and J. Esteve, "FEM of anodic bonding test structures," 6th Int. Conf. on Thermal. Mechanical and Multiphysics Simulation and Experiments in Micro-Electronics and Micro-Systems, EuroSimE, Berlin, 2005, pp. 366–370.
9. J. A. Plaza, J. Esteve and E. Lora-Tamayo, "Effect of silicon oxide, silicon nitride and polysilicon layers on the electrostatic pressure during anodic bonding," Sensor. Actuator. A Vol. 67, 1998, pp. 181–184.

10. T. M. H. Lee, D. H. Y. Lee, C. Y. N. Liaw, A. I. K. Lao and I. Hsing, "Detailed characterization of anodic bonding process between glass and thin-film coated silicon substrates," Sensor. Actuator. Vol. 86, 2000, pp. 103–107.
11. B. Schmidt, P. Nitzsche, K. Lange, S. Grigull, U. Kreissig, B. Thomas and K. Herzog, "In situ investigation of ion drift processes in glass during anodic bonding," Sensor. Actuator. A Vol. 67, 1998, pp. 191–198.
12. I. Zubel and I. Barycka, "Silicon anisotropic etching in alkaline solution I. the geometric description of figure developed under etching Si (100) in various solutions," Sensor. Actuator. A Vol. 70, 1998, pp. 250–259.
13. O. Powell and H. B. Harrison, "Anisotropic etching of {100} and {110} planes in (100) silicon," J. Micromech. Microeng. Vol. 11, 2001, pp. 217–220.
14. P. Kersten, S. Bouwstra and J. W. Petersen, "Photolithography on micromachined 3D surfaces using electrodeposited photoresists," Sensor. Actuator. A Vol. 51, 1995, pp. 51–54.
15. J. Brugger, C. Andreoli, M. Despont, U. Drecsler, H. Rothuizen and P. Vettiger, "Self-aligned 3D shadow mask technique for patterning deeply recessed surfaces of micro-electro-mechanical systems devices," Sensor. Actuator. Vol. 76, 1999, pp. 329–334.

Chapter 10
Surface Planarization Process

Surface planarization process is discussed in this chapter. Compared to bulk micromachining process, surface micromachining process is more suitable for the fabrication of capacitive switches. It is always important to planarize the sacrificial layer under the metal bridges to achieve flat metal bridges. Besides surface planarization, surface roughness is another important consideration to improve the intimate contact between the metal bridges and the dielectric layer when the metal bridges are driven down. Detailed discussion will be given in this chapter.

10.1 Surface Planarization Process

The process flow described in this section is mainly for the fabrication of a single capacitive switch. However, the same process flow can also be used to fabricate DC-contact capacitive switches and tunable band-pass devices.

Since the planarization of the metal bridge plays a significant role in reducing the contact area between the bridge and the dielectric layer, a planarization process is proposed by filling the CPW slot with photoresist which can effectively improve the flatness of the bridge.

Surface micromachining fabrication process with five masks is used to fabricate the switch [1]. High-resistivity (>4000 Ω cm) silicon wafers with thickness of 500 μm are used as substrate. The fabrication process flow is shown in Fig. 10.1 and described as follows: (a) plasma-enhanced chemical vapor deposition (PECVD) 1-μm-thick SiO_2 is deposited as the buffer layer in a Plasma-Therm 790 series RIE system from Material Co. USA; (b) a layer of 2-μm-thick aluminum (Al) thin film is evaporated and patterned as the CPW transmission line with the first mask. The pressure and the deposition rate are 6×10^{-5} Pa and 10 Å/s, respectively. Wet etching is used to pattern the Al metal bridge. The etchant used is a mixture of H_3PO_4:HNO_3:HAc:H_2O = 4:1:4:1, and the etching temperature is 35°C. The etching rate is approximately of 1500 Å/min; (c) a 0.15-μm PECVD SiN is deposited with SiH_4 (diluted in nitrogen at a concentration of 20%) flow rate of 24 sccm and

Aibin Yu and Ai Qun Liu

A.Q. Liu, *RF MEMS Switches and Integrated Switching Circuits*,
MEMS Reference Shelf 5, DOI 10.1007/978-0-387-46262-2_10,

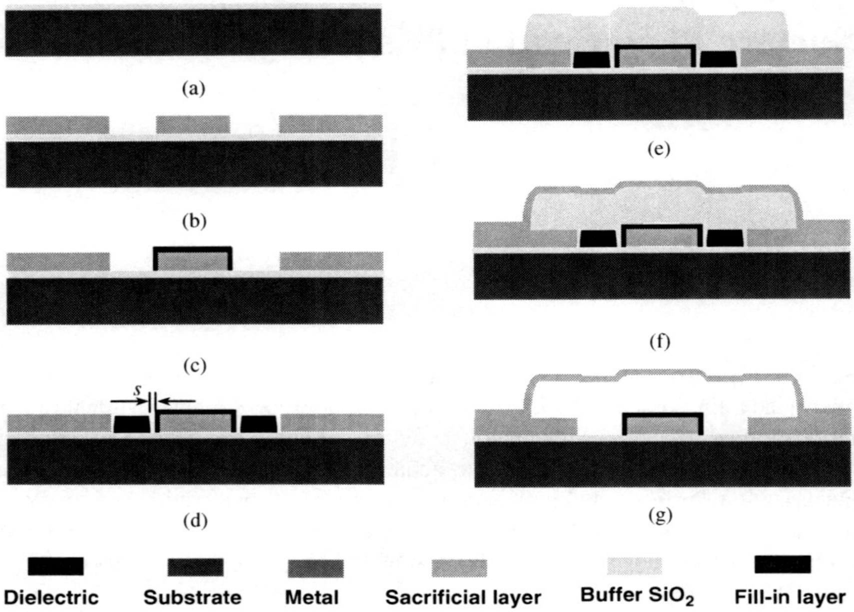

Fig. 10.1 Fabrication process flow [1]. Copyright/used with permission of/courtesy of Elsevier B.V

NH_3 flow rate of 7.1 sccm at a pressure of 1.1 Torr and temperature of 250°C in the same Plasma-Therm 790 series machine. The temperature is 100°C lower than the temperature at normal PECVD SiN deposition. This reduces the surface roughness of the capacitance area. In the process of etching the SiN layer, 2-μm-thick photoresist (Sumitomo PFI26A) is used as soft mask and patterned with the second mask. The SiN layer is etched using reactive ion etching (RIE) with a CHF_3 flow rate of 31 sccm and an O_2 flow rate of 2.6 sccm. The etching power is 235 W and the pressure is 60 mTorr. The self-bias voltage is 350 V and the plasma frequency is 13.56 MHz. The etching rate is 400 Å/min. The photoresist is stripped away after SiN etching with PRS3000 solution; (d) a layer of photoresist with 1.5-μm thick (Sumitomo PFI26A) is spun coated and patterned with the third mask to fill in the CPW slot (refer to this layer of photoresist as fill-in photoresist and this process is known as the planarization step). This is the key step to obtain a flat metal bridge; (e) a 2-μm-thick photoresist (Sumitomo PFI26A) sacrificial layer is spun coated and patterned (refer to this layer of photoresist as sacrificial photoresist); (f) a layer of 0.6-μm Al is evaporated and patterned as the metal bridge with the fourth mask; (g) finally, the metal bridges are released by etching away the photoresist with oxygen plasma also in a Plasma-Therm 790 series machine. After releasing, the metal bridge is free to deform out of plane following the relief of internal stress.

Based on this process flow, after filling the CPW slots with the fill-in photoresist in planarization step (d), the sacrificial photoresist is coated and patterned as in step (e) immediately. Therefore, the hard-bake temperature T_f must be higher than the hard-bake temperature T_s, (e) in order to avoid the peeling-off of the fill-in photoresist. In this process, T_f is 150°C, whereas T_s is 115°C. Another factor that has to be taken into consideration is the mask alignment and tolerance control in the fabrication process. The gap spacing (s) between the patterned fill-in photoresist and the CPW structure, as indicated in Fig. 10.1d, varies between 1 and 3 μm. It is observed that 3 μm is the most optimal as the mask alignment mismatch does not affect the patterns of the fill-in photoresist in this case.

Based on the proposed planarization strategy, two switches, switch A and switch B, are fabricated with different fabrication processes. Switch A is fabricated without the planarization and switch B is fabricated with the planarization. Figure 10.2a, b shows the surface profiles of the two switches. These profiles are obtained after the wet etching and before the release of the bridge. For switch A, the portion of the bridge above the CPW slot is 1.2 μm lower than the portion of the bridge above the CPW central conductor. For switch B, the portion of the bridge above the CPW slot is 0.5 μm higher than the portion of the bridge above the CPW central conductor.

Correspondingly, Fig. 10.3 shows the SEM image of the two switches after releasing. The etching holes on the bridge make the release of the bridge easier and also reduce the damping effect [2]. Figure 10.3a shows that the g_{slot} is smaller

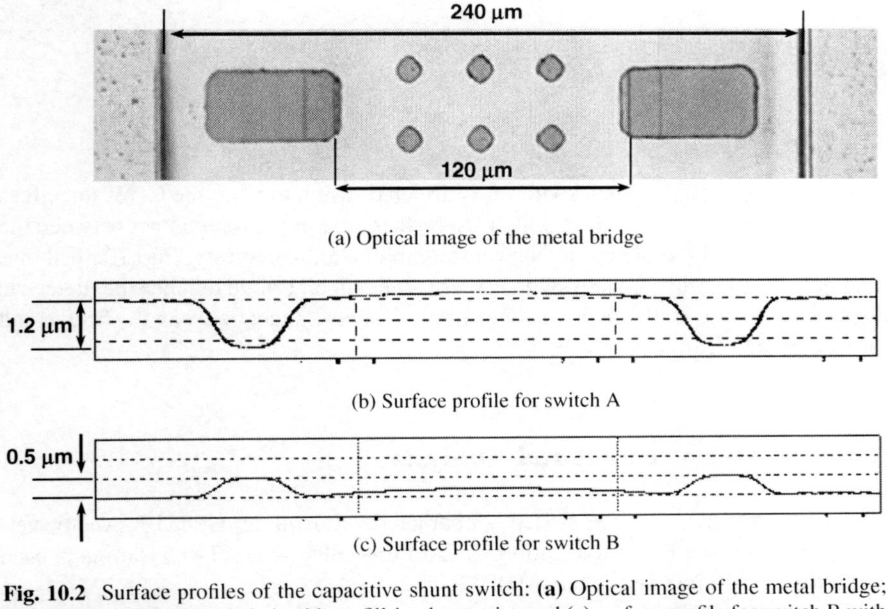

Fig. 10.2 Surface profiles of the capacitive shunt switch: **(a)** Optical image of the metal bridge; **(b)** surface profile for switch A without fill-in photoresist; and **(c)** surface profile for switch B with fill-in photoresist [1]. Copyright/used with permission of/courtesy of Elsevier B.V

Fig. 10.3 SEM images of the capacitive shunt switch **(a)** without fill-in photoresist and **(b)** with fill-in photoresist [1]. Copyright/used with permission of/courtesy of Elsevier B.V

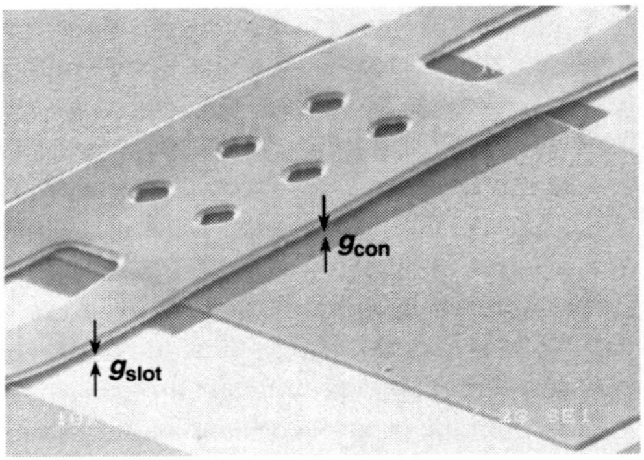

(a)

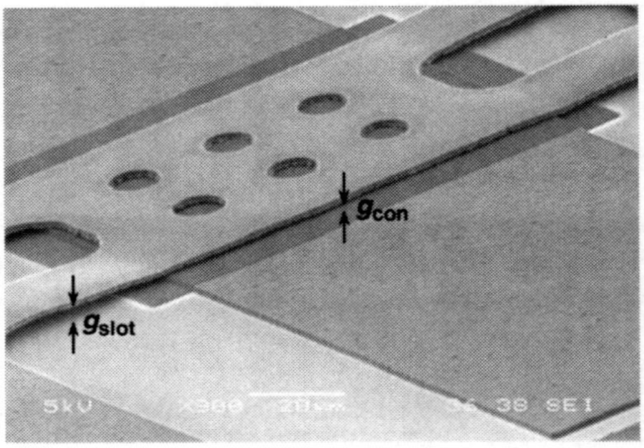

(b)

than g_{con} in switch A. When switch A is driven down, it touches the CPW slot edges first. After that, it is difficult to pull down further, making closer contact between the metal bridge and the dielectric layer impossible. On the contrary, Fig. 10.3b shows that g_{slot} is larger than g_{con} in switch B. Thus, the metal bridge touches the dielectric layer first when it is driven down. This improves the contact between the bridge and the dielectric layer.

10.2 RF Measurement and Analysis

The RF performance of the switch is characterized using an HP8510C vector network analyzer and RF probe station. A full thru–reflect–line (TRL) routine is used to calibrate with NIST software MULTICAL.

Figure 10.4 shows the RF measurement and simulation results of the two switches. All the simulations are conducted using Ansoft's high-frequency simulation software (HFSS) [3]. Figure 10.4a, b shows that the up-state performances of

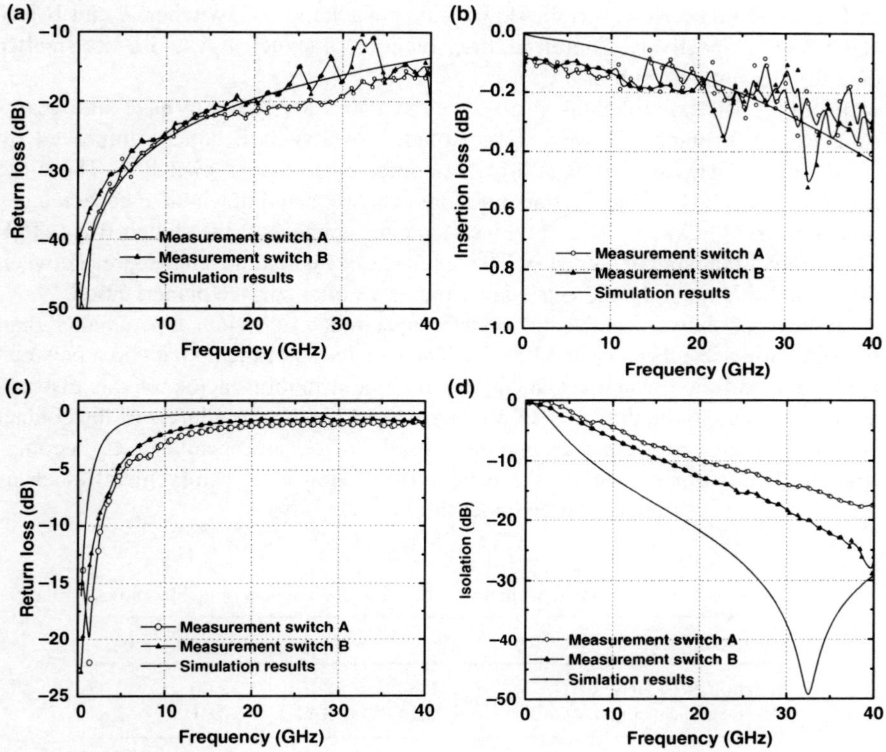

Fig. 10.4 Comparison of simulated and measured *S*-parameters: **(a)** up-state return loss; **(b)** up-state insertion loss; **(c)** down-state return loss; and **(d)** down-state isolation [1]. Copyright/used with permission of/courtesy of Elsevier B.V

switch A are comparable to the simulation results. The extracted up-state capacitance (C_u) is 34.4 fF. However, the down-state isolation of switch A is more inferior compared to the simulation results, as listed in Table 10.1 and shown in Fig. 10.4d. The extracted down-state capacitance of switch A (C_{dA}) is 1 pF, which is 33% ($C_{dA} = 33\%\ C_d$) of the simulation results. In Fig. 10.4d, the simulation results are obtained by assuming that the entire bridge area above the CPW center conductor contacts the dielectric layer when the bridge is pulled down. As a result, the simulated down-state capacitance is 3.01 pF and the resonant frequency of the bridge is

Table 10.1 Comparison of down-state isolation and capacitances

	Switch A	Switch B	Simulation
Isolation at 15 GHz (dB)	7.5	9.7	18.0
Isolation at 40 GHz (dB)	17	27	28
Down-state capacitance (pF)	$C_{dA} = 1.0$	$C_{dB} = 1.3$	$C_d = 3.01$

shifted to 32 GHz. However, the down-state capacitance of switches A and B is 1 and 1.3 pF, respectively. Therefore, the isolations of switches A and B are smaller than the simulation results.

From the RF experimental results, both switches are found to have similar up-state RF performance. However, the isolation of switch B can be improved by 2.2 dB at 15 GHz and 10 dB at 40 GHz compared to that of switch A. These are listed in Table 10.1. Table 10.1 also provides the extracted down-state capacitances of switch B, C_{dB}, at 15 GHz. The down-state capacitance of the switch B is 0.3 pF (30%) larger than that of switch A. The results indicate that the contact area between the metal bridge and the dielectric layer increases after surface planarization.

Table 10.2 shows that the down-state capacitance of switch B is smaller than the simulation results ($C_{dB} = 43\%\ C_d$). The roughness of the contact area between the bridge and the dielectric layer is the main contribution factor for this discrepancy. The average roughness is 24 nm approximately. The roughness of the contact area reduces the capacitance $C_d^{'}$ and the overall down-state capacitance C_d becomes smaller. This surface roughness can be solved using a refractory metal (such as tungsten) as the bottom electrode under the PECVD SiN.

Table 10.2 Down-state capacitances with various diameters of etching holes

Diameter of etching holes (μm)	0	2	4	10
Isolation at 15 GHz (dB)	18.34	18.27	18.00	17.37
Down-state capacitance (pF)	3.14	3.11	3.01	2.8

There are other factors that contribute to the degradation of the down-state capacitance. First, the bridge is warp due to the residual stress in the metal bridge (the residual stress is 20 MPa). This reduces the contact area and increases the gap between the bridge and the dielectric layer. Second, polymer residues due to the incomplete removal of the sacrificial photoresist may also make the intimate contact between the bridge and the dielectric layer difficult. Third, the designed diameter of the etching holes in the metal bridge is 4 μm, but the final diameter of the fabricated etching holes is 10 μm due to over wet etch of the Al metal bridge. The etching holes reduce the ratio of the contact area p. Figure 10.5 shows the simulated down-state isolation in the frequency range of 12–18 GHz for various sizes of the etching holes. The extracted down-state capacitances at 15 GHz from the simulation results are listed in Table 10.2. When the diameter of the holes is 2 μm, the down-state capacitance and isolation are not affected. When the diameter of the holes is 4 μm, the extracted down-state capacitance is 95.9% of the capacitance of the bridge without the holes. This indicates that the effect is still relatively small. However, when the diameter of the holes is 10 μm, the down-state capacitance is only 89.2% of the capacitance of the bridge without the holes. As a result, it can be concluded that the diameter of the etching holes must be carefully considered in the design of the switch.

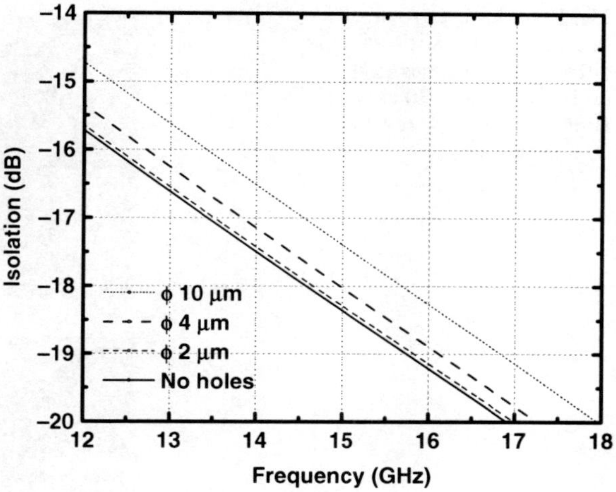

Fig. 10.5 Simulation results for various sizes of the etching holes [1]. Copyright/used with permission of/courtesy of Elsevier B.V

In general, the down-state capacitance of the switch usually varies from 10 to 70% of the expected down-state capacitance [4]. In the current design, this down-state capacitance is 43%. The ratio of the down-state capacitance and the up-state capacitance, C_{dB}/C_u, is approximately 38.

10.3 Effects of Dry Releasing of Metal Bridge

10.3.1 Etch Rate of Photoresist

As mentioned in the process flow in Section 10.1, photoresist is used as the sacrificial layer and oxygen plasma is used to remove the sacrificial photoresist and release the metal bridge. Compared to wet releasing, dry releasing is easier and can prevent the damage of the passivation dielectric and metal thin film by the aggressive etchant.

In this section, the effects of RIE process parameters on the dry releasing process are studied [5]. Figure 10.6a, b shows the optical images of the sacrificial photoresist under the metal bridge before releasing and after 50-min oxygen plasma etch, respectively. The discussion of the fabrication of the lower DC pad and the dielectric layer is not included since they do not have any impact on the sacrificial photoresist releasing. The metal bridges are removed using Al wet etchant. It can be seen clearly that after 50 min of oxygen plasma etch, the edge of the photoresist circle becomes larger and after 70 min of oxygen plasma etch, all the photoresist is etched away and the metal bridge is released. During the oxygen plasma etch, less than 10-nm-thick Al metal bridge is also removed.

For the release process, two main process parameters such as the etching power and the oxygen pressure are investigated. The flow rate of oxygen is fixed at 30 sccm

Fig. 10.6 Surface profile of sacrificial photoresist with different release times: **(a)** before release, **(b)** after 50-min etching, and **(c)** after 70-min etching

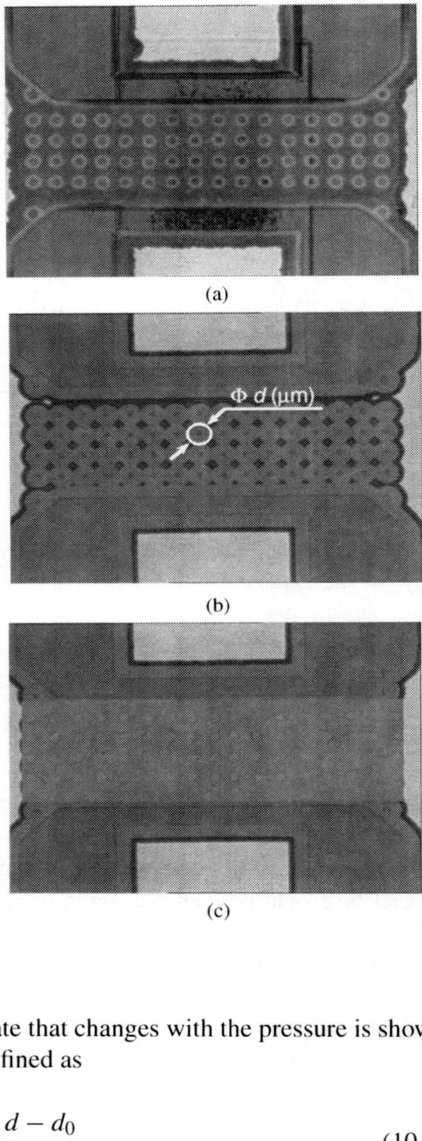

for all experiments. The lateral etching rate that changes with the pressure is shown in Fig. 10.7. The lateral etching rate is defined as

$$R = \frac{1}{2}\frac{d - d_0}{t} \tag{10.1}$$

where d_0 is the diameter of the photoresist circles before oxygen plasma etch and d is the diameter of the photoresist circles after time t (equal to 50 min in this study) of oxygen plasma etch.

It is shown that the lateral etching rate increases slowly with pressure when the pressure is lower than 500 mTorr and becomes constant when the pressure is higher than 500 mTorr. This phenomenon can be explained since the oxygen atom creation rate increases by increasing the oxygen pressure whereas the mean free

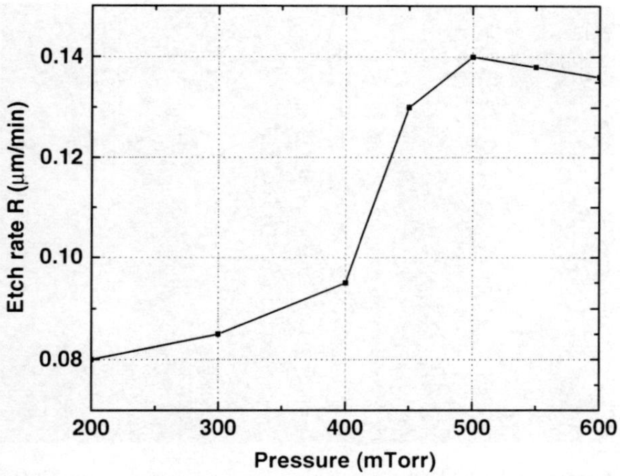

Fig. 10.7 Lateral etching rate versus pressure (power = 430 W; t = 50 min) [5]. Copyright/used with permission of/courtesy of IOP Publishing Limited

path length of the oxygen plasma decreases by increasing the oxygen pressure. As a result, the effect of isotropic etching increases [6] and the lateral etching rate increases first. When the pressure increases further, the oxygen atom recombination increases. At certain pressure, the oxygen atom creation and recombination are balanced. Therefore, there is a trade-off between the pressure and the lateral etching rate.

Another important process parameter that may affect the release process is the power. Figure 10.8 shows that the lateral etching rate increases with the power. Until 600 W, there is no saturation point.

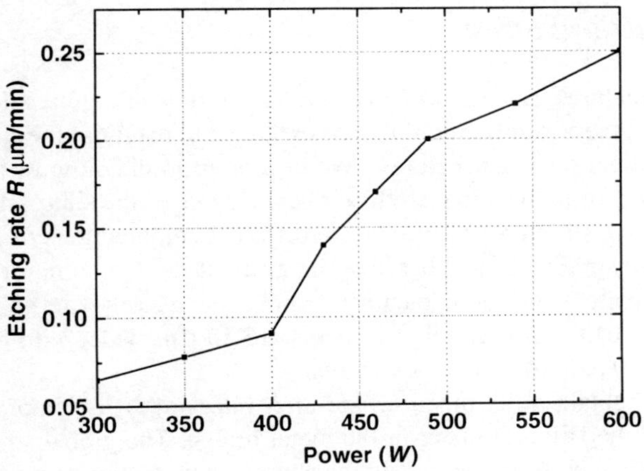

Fig. 10.8 Lateral etching rate versus power (pressure = 500 mTorr; t = 50 min) [5]. Copyright/used with permission of/courtesy of IOP Publishing Limited

Fig. 10.9 Post-buckling of metal bridge after releasing with power of 500 W [5]. Copyright/used with permission of/courtesy of IOP Publishing Limited

However, high power may cause buckling of the metal bridge as shown in Fig. 10.9. Because high plasma power leads to stronger bombardment on the surface of the metal bridge that can change the surface status of the Al thin film, the metal bridge buckles in the process. The compressive stress is due to the bomb of the metal bridge by high energy particles. During the oxygen plasma releasing process, the oxygen absorbed at the surface of the film decreases the surface mobility of the arriving atoms. This results in a less-ordered structure with vacancies and interstitial atoms. Atoms in the grain boundaries and interstitials in the Al thin film produce high compressive stress. Therefore, the etching power in this work is fixed at 430 W.

10.3.2 Structural Stress

Thin film structures can deform undesirably as a result of residual stress; on the other hand, these deformations can be exploited to diagnose the state of stress in the film. Different testing structures have been used to diagnose the state of stress in the film [7]. Normally, microcantilever beam is one of the simplest and most frequently used test structures to investigate different combinations of uniform mean stress and stress gradient [8]. Therefore, an array of micro-cantilevers is fabricated on the same wafer with the capacitive switches to measure the in situ stress, as shown in Fig. 10.10. The cantilever beams are 50-μm wide, with length ranging from 200 to 410 μm in 15-μm increments.

In order to obtain a flat metal bridge after releasing, it is important to control the stress and the stress gradient in the metal bridge. The global stress in the thin film can be extracted from measuring the change in wafer curvature due to the film deposition [9]. However, the stress level in the final structure may be different from

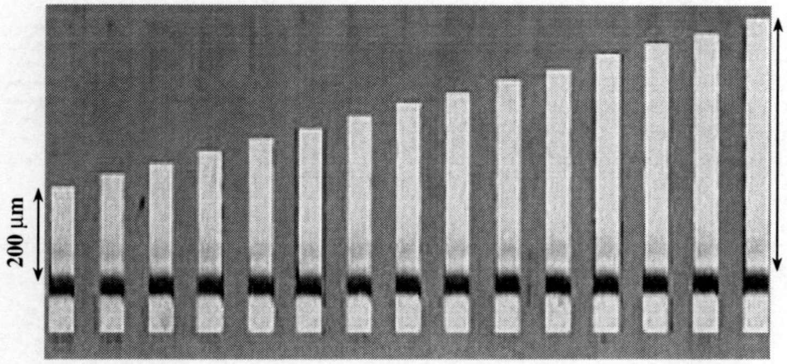

Fig. 10.10 Optical photo of the testing structure of the cantilever beams array [5]. Copyright/used with permission of/courtesy of IOP Publishing Limited

that of the as-deposited state, because the film undergoes various micromachining processes, such as various thermal cycles in different process steps or exposure to plasma bombardment with dry release process. Therefore, instead of measuring the global film stress by measuring the wafer curvature, it is more important to measure the mean stress and the stress gradient in the released structure.

The beam is assumed to be fully clamped at one end and free at the other end. The total stress in the beam can be divided into a constant mean stress and a gradient stress. Assuming a linear stress profile within the beam, the stress can be divided into two components that can be expressed as [7, 9]

$$\sigma(z) = \sigma_0 + \sigma_1 \frac{z}{t/2} \tag{10.2}$$

where $z \in (-t/2,\ t/2)$ is the coordinate normal to the surface of the beam with an origin chosen at the film's mid-plane as shown in Fig. 10.11, t is the thickness of the beam, $\sigma(z)$ is the stress distribution along the thickness of the metal beam, σ_0 is the constant mean stress, and σ_1 is the gradient stress. The uniform stress σ_0 accounts for the in-plane elongation of the beam whereas the stress gradient component σ_1 causes the out-of-plane deflection.

The stress distribution leads to a moment M_0 which can be defined as

$$M_0 = \int_{-t/2}^{t/2} \sigma(z) w z dz \tag{10.3}$$

Substituting Eq. (10.2) into Eq. (10.3), the net moment M_0 acting on the beam can be expressed as

$$M_0 = \frac{wt^2}{6} \sigma_1 \tag{10.4}$$

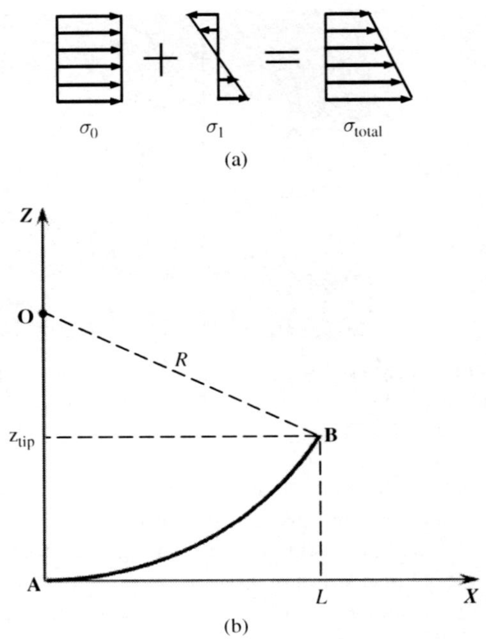

Fig. 10.11 **(a)** Stress in metal bridge and **(b)** schematic drawing of beam deflection [5]. Copyright/used with permission of/courtesy of IOP Publishing Limited

The moment M_0 causes the out-of-plane deflection of the cantilever beams. Assuming the curled cantilever beams are perfectly circular and fixed-support at one end, the radius of curvature can be written as [10]

$$R = \frac{E}{d\sigma/dz} \tag{10.5}$$

Substituting Eq. (10.2) into Eq. (10.5), the radius of the curvature can be given as

$$R = \frac{EI}{M_0} = \frac{Et}{2\sigma_1} \tag{10.6}$$

The relationship between the radius and the displacement throughout the cantilever beam can be calculated from trigonometry as shown in Fig. 10.11.

In Fig. 10.11b, the contour of the released cantilever beam arc AB can be expressed as

$$\sin\left(\mathrm{arctg}\frac{z_{\mathrm{tip}}}{L}\right) = \sqrt{L^2 + z_{\mathrm{tip}}^2}/(2R) \tag{10.7}$$

where L is the length of the deflected cantilever projected on the X-axis and z_{tip} is the deflection of the tip of the cantilever beam.

Substituting Eq. (10.6) into Eq. (10.7), the gradient stress can be expressed as

$$\sigma_1 = \frac{2Et \sin\left(\mathrm{arctg}\left(z_{\mathrm{tip}}/L\right)\right)}{\sqrt{L^2 + z_{\mathrm{tip}}^2}} \tag{10.8}$$

At the same time, the length of the cantilever beam after release s can be calculated from Fig. 10.11b as

$$s = R \times 2 \frac{\pi \arcsin\left(\sqrt{L^2 + z_{\mathrm{tip}}^2}/2R\right)}{180} \tag{10.9}$$

Then, the in-plane residual stress σ_0 can be determined by

$$\sigma_0 = E\varepsilon = E\frac{s - l}{l} \tag{10.10}$$

where ε is the strain of the cantilever beam after release and l is the original length of the cantilever beam.

Based on Eqs. (10.6), (10.7), (10.8), (10.9), and (10.10), the mean stress and the stress gradient can be obtained by curve-fitting method through the following steps: (1) measuring the deflection z of a cantilever beam at different positions and calculating σ_1 according to Eq. (10.8); (2) based on Eq. (10.6) and the calculated σ_1, the radius of the curved cantilever beam R can be calculated; (3) s can be obtained based on Eq. (10.9) and the values of R, L, and z_{tip}; and (4) σ_0 is calculated according to Eq. (10.10).

Figure 10.12 shows the measured 1D profile of the cantilever beam with the length of 400 μm using WYKO NT3300 optical profiler in vertical scanning interferometry (VSI) mode. The WYKO system utilizes white light passed through a

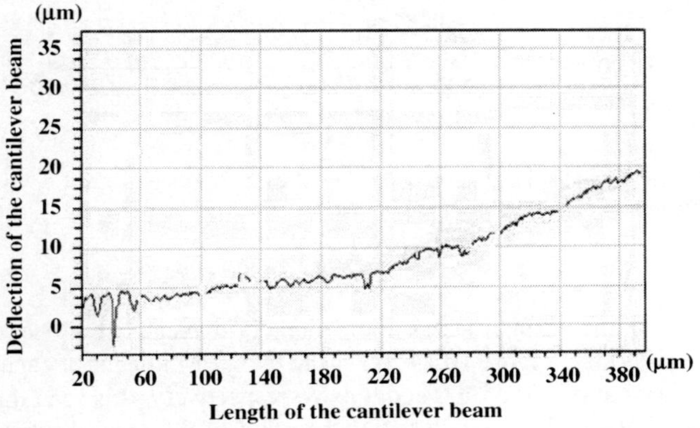

Fig. 10.12 Measured deflection of cantilever beams [5]. Copyright/used with permission of/courtesy of IOP Publishing Limited

beam splitter to form interference fringes as a function of height on the sample; objective height is controlled by a piezoelectric transducer. This is a non-contact method to measure the contour of the structures and it has the advantage of being fast and non-destructive [11, 12]. Other cantilever beams can also be used to analyze the stress status in the thin film. The beam with 400-μm length is selected because its deflection is the largest and its length is similar to the length of the metal bridge. As a result, its measurement results provide useful information for the metal bridge design.

Figure 10.13 shows that the measured and curve-fitted deflection of the cantilever beam varies with the position along the beam. These measured deflections can be obtained from Fig. 10.12. Curve fitting is based on Eqs. (10.6), (10.7), (10.8), (10.9), and (10.10). It is found that the fitted mean stress and the stress gradient are $\sigma_0 =$ 21.8 Mpa and $\sigma_1 = 3.6$ MPa, respectively. Under this stress status, two curves match closely. From Fig. 10.13, it can be seen that for short cantilever beams, deflections of the beam are not sensitive to the stress status since the deflections are small. When the length of the cantilever beam is longer than 150 μm and the deflection is larger than 5 μm, the stress status can be clearly identified by the curve-fitting method.

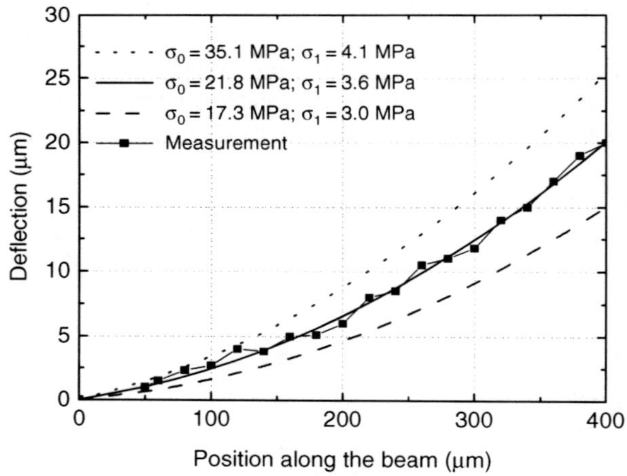

Fig. 10.13 Measured and simulated tip deflection of cantilever beams [5]. Copyright/used with permission of/courtesy of IOP Publishing Limited

As mentioned, the intrinsic stress in the metal bridge can cause buckling or curling of the metal bridge. Figure 10.14 shows the two SEM images of capacitive shunt switches released at power of 500 and 430 W, respectively. It is seen that the metal bridge released at 430 W is much flatter than that of the metal bridge released at 500 W.

The effects of the residual stress on the mechanical characteristics of the RF switch can be expressed as

Fig. 10.14 SEM images of the fabricated capacitive switches: (a) switch with release power of 500 W and (b) switch with release power of 430 W [5]. Copyright/used with permission of/courtesy of IOP Publishing Limited

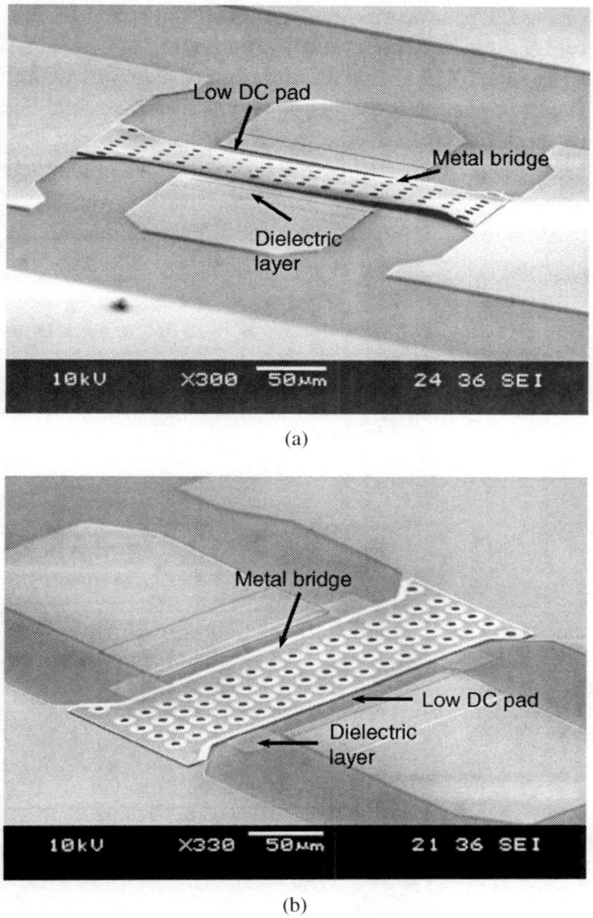

$$V_{PI} = \sqrt{\frac{8k_{eff}g_0^3}{27\varepsilon_0 wW}} \tag{10.11}$$

$$k_{eff} = k' + k'' \tag{10.12a}$$

$$k' = 32Ew\left(\frac{t}{l}\right)^3\left(\frac{27}{49}\right) \tag{10.12b}$$

$$k'' = 8\sigma_0(1-\upsilon)w\left(\frac{t}{l}\right)\left(\frac{3}{5}\right) \tag{10.12c}$$

where V_{PI} is the pull-in voltage of the switch, k_{eff} is the effective spring constant of the metal bridge, k' is the stiffness caused by the restoration force of the metal bridge, and k'' is the stiffness caused by the mean residual stress in the metal bridge, g_0 is the initial gap of the switch, w is the width of the metal bridge, W is the width of the CPW transmission line under the metal bridge, l is the length of the metal bridge, υ is Poisson's ratio of aluminum (Al).

Figure 10.15 shows the variation of k' and k'' with the length of the metal bridge. It can be seen that the spring constant of the metal bridge is mainly determined by the residual stress in the metal bridge. Therefore, the pull-in voltage of the metal bridge is also affected by the residual stress in the metal bridge.

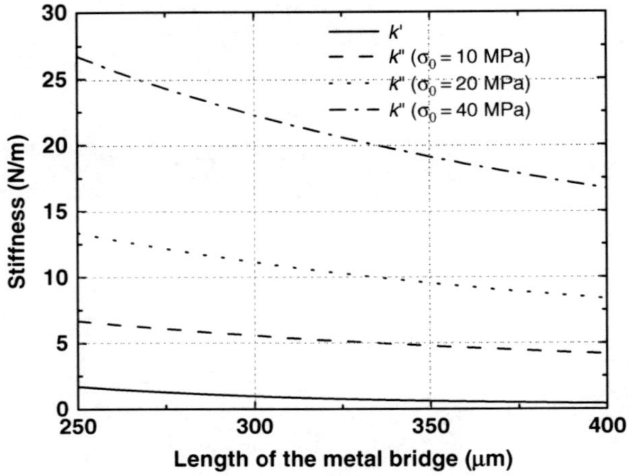

Fig. 10.15 Variation of stiffnesses with length of the metal bridge

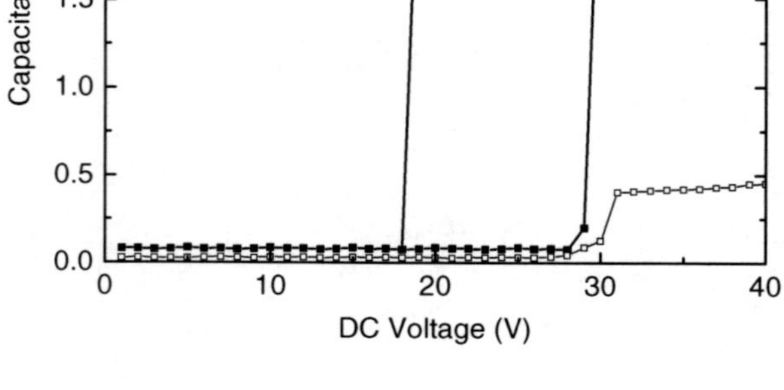

Fig. 10.16 Measured *C–V* curve of the capacitive switch [5]. Copyright/used with permission of/courtesy of IOP Publishing Limited

10.3.3 Characterization of Capacitive Switch

Figure 10.16 presents the measured *C–V* curves of the two switches. The curves are measured with an HP 4275A multi-frequency LCR meter with an internal bias option. For the switch with the release power of 500 W, the up-state capacitance is 35 fF and the down-state capacitance is 420 fF even when the applied voltage is as high as 40 V. The metal bridge cannot bounce back after removing the applied voltage because of the high applied voltage. For the switch with release power of 430 W, the pull-down voltage can be clearly identified as 30 V and the hold-down

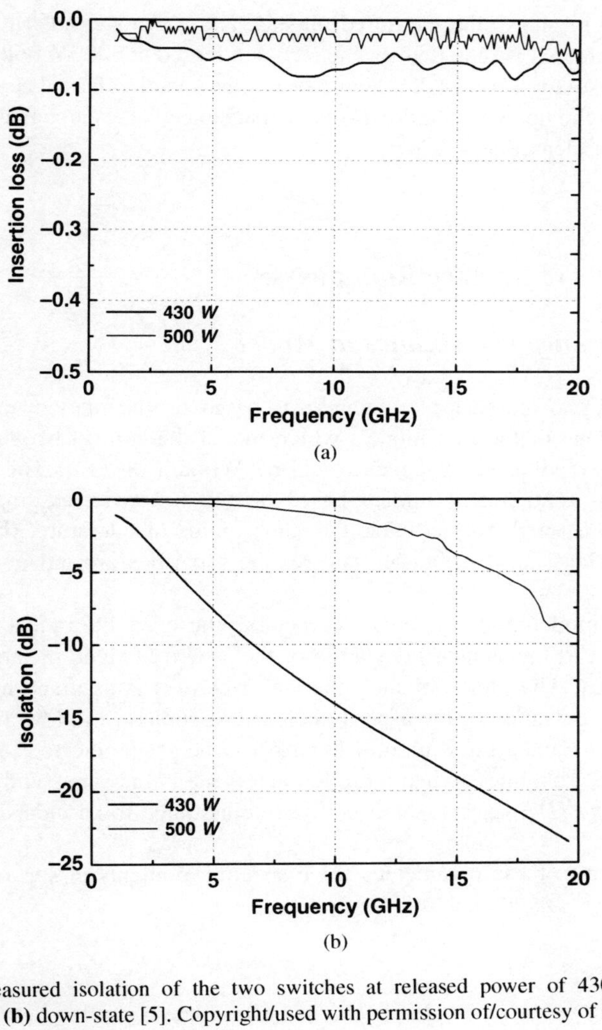

Fig. 10.17 Measured isolation of the two switches at released power of 430 and 500 W: **(a)** up-state and **(b)** down-state [5]. Copyright/used with permission of/courtesy of IOP Publishing Limited

voltage is 20 V. The up-state capacitance is 80 fF and the down-state capacitance is 2.8 pF.

The differences of the capacitance values between the two switches are because that after releasing at 500 W, the metal bridge buckles up due to the compressive stress in it. Therefore, the average gap between the metal bridge and the lower DC pad is larger than the designed value of 2 μm (approximately 4 μm) and the up-state capacitance is only 35 fF. When the metal bridges are driven down, the metal bridge of the switch released at 430 W is flatter than that of the switch released at 500 W; therefore, after the bridge is driven down, the contact between the metal bridge and the lower DC pad is much better for the switch released at 430 W and the down-state capacitance is larger.

Figure 10.17 shows the measured up-state and down-state of RF properties of the two switches. It is seen that for the switch released at 500 W, both the insertion loss and the isolation are smaller than those of the switch released at 430 W. This is because both the up-state and down-state capacitance values are smaller than those of the switch released at 430 W.

10.4 Effects of Surface Roughness

10.4.1 Greenwood–Williamson Model

The modeling of the surface roughness is based on the random process of surface shape. One of the first models which model the contact between two rough surfaces was developed by Greenwood and Williamson [13]. The contact characteristics are statistically studied based on the following assumptions: (a) all asperities on a rough surface have the same radius of curvature, (b) the asperity height is a Gaussian distribution, and (c) the asperities are independent of each other.

The GW model contains three parameters. These are the radius of the spherical asperities, R; the standard deviation of the asperity height, σ_s; and the asperity density, D_{SUM}. The radius of the spherical asperities is assumed to be constant. The standard deviation of the asperity height is assumed to follow a Gaussian distribution. The standard deviation of the asperity height is referred to as root mean square (RMS) roughness when the reference plane is the same as the mean plane [14]. When $\pi R^2 D_{SUM} < 1$, the asperities are assumed to be independent of each other.

In addition to these parameters, three spectral moments of surface topography m_0, m_2, and m_4 are defined as [15, 16]

$$m_k = (2\pi)^n \int_{-\infty}^{\infty} \lambda^n f(\lambda) d\lambda, \quad k = 0,\ 2,\ 4 \tag{10.13}$$

where λ denotes a general spatial frequency in cycles/unit length and $f(\lambda)$ is the spectral density function of the profile. In the space domain, the nth order spectral moments is the mean squared value of the (n/2)th derivative of the surface profile. Normally, m_0 is related to the mean square asperity height, m_2 represents the mean square scope, and m_4 indicates the curvature of the surface profiles.

The relationships between the spectral moments and R, σ_s, and D_{SUM} are expressed as [17]

$$\sigma_s^2 = \left(1 - \frac{0.8968}{\alpha}\right) m_0 \tag{10.14a}$$

$$R = \frac{3\sqrt{\pi}}{8\sqrt{m_4}} \tag{10.14b}$$

$$D_{SUM} = \frac{1}{6\sqrt{3}\pi}\frac{m_4}{m_2} \tag{10.14c}$$

where α is the bandwidth parameter and its value depends on the shape and extent of the spectrum of the roughness profile. It is expressed as [11]

$$\alpha = m_0 m_4 / {m_2}^2 \tag{10.15}$$

The spectral moments m_0, m_2, and m_4 can be first calculated from the measured data of the surface topography, then R, σ_s, and DSUM can be obtained from Eqs. (10.14a), (10.14b), (10.14c), and (10.15) [17].

10.4.2 Analysis of Surface Roughness Effects

Figure 10.18 shows the schematic cross section of the capacitive shunt switch with the metal bridge at up-state and down-state positions. The SEM image of the capacitive switch is shown in Fig. 10.19. The planarization of the metal bridge is not realized by fill in the CPW slot under the metal bridge. However, a thin layer of less than 0.5-μm metal film is used as the lower DC pad under the metal bridge. This lower DC pad is also part of the center conductor of the CPW transmission line, which may cause 0.1- to 0.2-dB insertion loss [18]. The reason to use this planarization method is that different materials can be used for the lower DC pad and provide different surface roughness of the capacitance area. When the metal bridge is driven down, the contact area (capacitance area) between the metal bridge and the dielectric layer is affected by the surface roughnesses of the dielectric layer, as shown in Fig. 10.18b. The characteristic impedance of the CPW transmission line is 50 Ω, with dimensions of $W/S/W = 80/150/80$ μm. The gap between the metal bridge and the asperity mean height plane g_{ma} (which is equal to the thickness of the sacrificial layer) is 2 μm, the thickness of the SiN dielectric layer t_d is 0.15 μm, and the width of the metal bridge w is 100 μm. In the following statistical calculation, the metal bridge is assumed to be perfectly planar and smooth. The holes in the metal bridge do not affect the up-state capacitance because the fringing field covers the holes

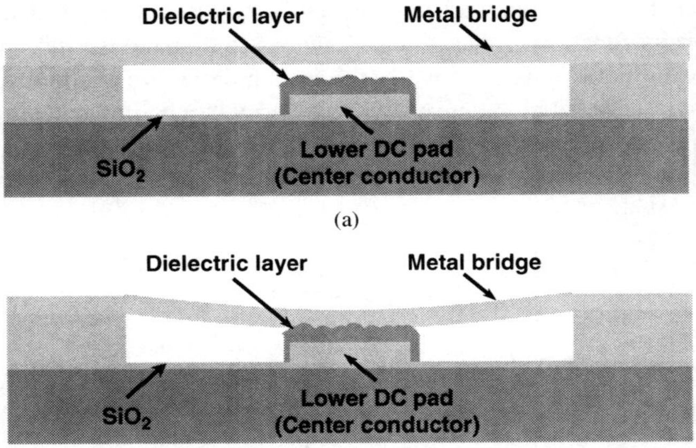

Fig. 10.18 Schematic cross section of capacitive switch: **(a)** up state and **(b)** down state

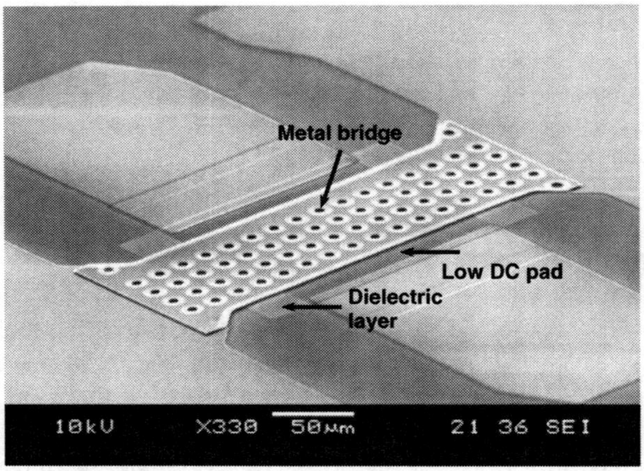

Fig. 10.19 SEM images of capacitive switch [19]. Copyright/used with permission of/courtesy of IOP Publishing Limited

completely; while at the down-state of the switch, the apparent contact area and the designed capacitance have removed the area of the holes [18].

There is no contact between the metal bridge and the dielectric layer when the metal bridge is at its up-state position. As shown in Fig. 10.20a, the real up-state capacitance C_u^r can be defined as [20, 21]

$$C_u^r = D_{SUM} A_0 \int_{-3\sigma_s}^{3\sigma_s} \overline{C}(z)\,\phi(z)\,dz \tag{10.16}$$

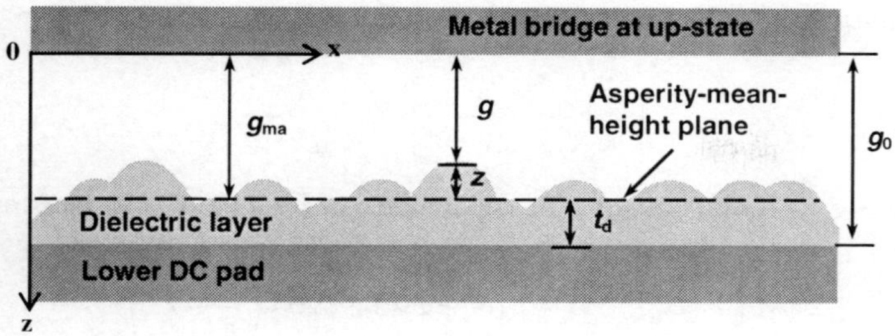

(a)

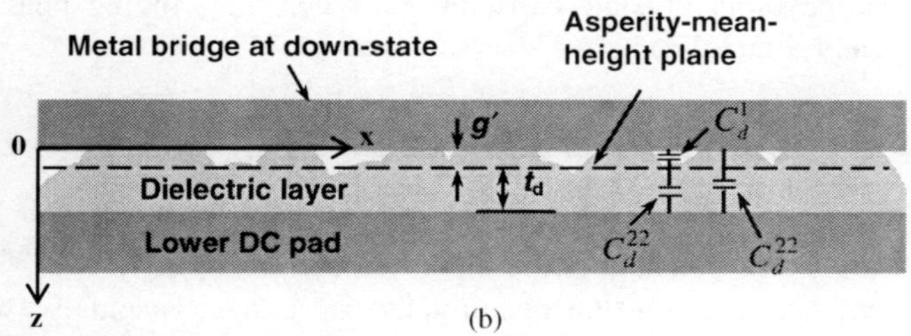

(b)

Fig. 10.20 Schematic view of surface roughness with metal bridge: **(a)** up-state position of the switch (not to scale) and **(b)** down-state position of the switch (not to scale) [19]. Copyright/used with permission of/courtesy of IOP Publishing Limited

where z is the height of the asperity tips which have a Gaussian distribution. Due to the deposition process of the lower DC pad and the SiN dielectric layer, the RMS roughness can be controlled to 100 nm or smaller [19]. A_0 is the apparent overlap area between the metal bridge and the dielectric layer, $\overline{C}(z)$ is the capacitance from one asperity, and $\phi(z)$ is the Gaussian probability density function. $\overline{C}(z)$ and $\phi(z)$ can be expressed as

$$\overline{C}(z) = 2\pi\varepsilon_0 R \ln\left[1 + \frac{R}{g_{\mathrm{ma}} - z}\right] \tag{10.17}$$

$$\phi(z) = \frac{1}{\sqrt{2\pi}\sigma_{\mathrm{s}}} \exp\left[-0.5\left(\frac{z}{\sigma_{\mathrm{s}}}\right)^2\right] \tag{10.18}$$

where ε_0 is the dielectric constant of air, g_{ma} is the initial gap height between the metal bridge and the asperity mean height plane.

According to Eq. (10.16), the limits of integration are selected to be three times of the standard deviation of the height; 99.7% of surface asperities are included in this range for a Gaussian height distribution function. Therefore, the effect of the vast majority of surface asperities is considered [20].

Substituting Eqs. (10.17) and (10.18) into Eq. (10.16) and assuming

$$\ln\left[1+\frac{R}{g_{\mathrm{ma}}-z}\right]\cong\frac{R}{g_{\mathrm{ma}}-z}$$

when $g_{\mathrm{ma}}-z\gg R$, the real up-state capacitance of Eq. (10.16) can be simplified as

$$C_{\mathrm{u}}^{\mathrm{r}}=C_{\mathrm{u}}^{\mathrm{a}}\sqrt{2\pi}D_{\mathrm{SUM}}R^2\left[2.72+2.66\left(\frac{\sigma_{\mathrm{s}}}{g_{\mathrm{ma}}}\right)^2\right] \tag{10.19}$$

where $C_{\mathrm{u}}^{\mathrm{a}}$ is the apparent up-state capacitance. When the fringing field effect is neglected, $C_{\mathrm{u}}^{\mathrm{a}}$ can be calculated as

$$C_{\mathrm{u}}^{\mathrm{a}}=\frac{1}{\frac{g_{\mathrm{ma}}}{\varepsilon_0A_0}+\frac{t_{\mathrm{d}}}{\varepsilon_0\varepsilon_{\mathrm{r}}A_0}}=\frac{\varepsilon_0A_0}{g_{\mathrm{ma}}+\frac{t_{\mathrm{d}}}{\varepsilon_{\mathrm{r}}}}\cong\frac{\varepsilon_0A_0}{g_{\mathrm{ma}}}\qquad\left(g_{\mathrm{ma}}\gg\frac{t_{\mathrm{d}}}{\varepsilon_{\mathrm{r}}}\right) \tag{10.20}$$

where ε_{r} is the dielectric constant of the SiN dielectric and $\varepsilon_{\mathrm{r}}=7$.

When the switch is at the up-state position, the capacitance is actually two capacitors in series. One is due to the SiN dielectric and is equal to $\varepsilon_0\varepsilon_{\mathrm{r}}A_0\,/\,t_{\mathrm{d}}$; another is due to the air gap and is equal to $\varepsilon_0A_0\,/\,g_{\mathrm{ma}}$. However, the capacitance value due to the SiN dielectric is much larger than the capacitance value due to the air gap ($\varepsilon_0\varepsilon_{\mathrm{r}}A_0\,/\,t_{\mathrm{d}}>50\ \varepsilon_0A_0/g_{\mathrm{ma}}$). Therefore, the capacitor due to the SiN dielectric is negligible.

The normalized up-state capacitance is expressed as

$$C_{\mathrm{u}}^{*}=\frac{C_{\mathrm{u}}^{\mathrm{r}}}{C_{\mathrm{u}}^{\mathrm{a}}}=\sqrt{2\pi}D_{\mathrm{SUM}}R^2\left[2.72+2.66\left(\frac{\sigma_{\mathrm{s}}}{g_{\mathrm{ma}}}\right)^2\right] \tag{10.21}$$

Figure 10.21 shows that the normalized up-state capacitance varies with the RMS roughness for different initial gap heights g_{ma}, assuming $R=40$ nm and $D_{\mathrm{SUM}}=100/\mu m_2$. It is shown that C_{u}^{*} increases with the RMS roughness for a specific g_0.

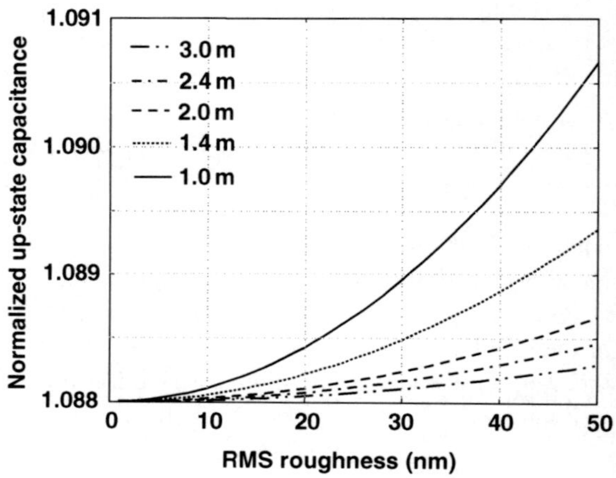

Fig. 10.21 Influence of roughness on the normalized up-state capacitance [19]. Copyright/used with permission of/courtesy of IOP Publishing Limited

This is because when the roughness is larger, the equivalent gap between the metal bridge and the asperity mean height plane g_{ma} becomes smaller, since the tips of the asperities are nearer to the metal bridge.

From Fig. 10.21, it is noted that the smaller the initial gap g_0, the larger the normalized up-state capacitance. Generally, when the metal bridge is at the up-state position, the initial gap varies between 1 and 3 μm and the roughness is less than 10 nm. Therefore, according to Fig. 10.21, the increase in capacitance is approximately 9%.

A higher up-state capacitance results in a larger insertion loss. The relationship between the up-state capacitance and the insertion loss can be expressed as

$$S_{21} = 20\log\left|\frac{1}{1 + j\omega C_u Z_0/2}\right| \tag{10.22}$$

where S_{21} is the insertion loss in decibels, ω is the angular frequency, Z_0 is the characteristic impedance of the CPW line (50 Ω in this example), and C_u is the up-state capacitance.

Figure 10.22a shows the normalized up-state insertion loss for different initial gaps at 10 GHz with the RMS roughness varies from 0 to 20 nm, assuming $A_0 = 100 \times 150\ \mu m^2$. The normalized insertion loss is expressed as S_{21}^{r}/S_{21}^{a}, where S_{21}^{r} is the real insertion loss after considering the effect of the surface roughness and S_{21}^{a} is the insertion loss. It is noted that the normalized insertion loss increases with the RMS roughness. This is because the larger the RMS roughness, the larger the up-state capacitance. Figure 10.22b shows the normalized up-state insertion loss for different initial gaps at 10 GHz with the RMS roughness which is as high as 50 nm. It is observed that the gap lines cross over at 45-nm RMS roughness point. This is because when the RMS roughness is small, the normalized insertion loss is determined mainly by S_{21}^{a}. When the initial gap is smaller, the insertion loss S_{21}^{a} is larger. This results in smaller normalized insertion loss for smaller RMS roughness. As the RMS roughness increases, the normalized insertion loss increases with the real insertion loss S_{21}^{r} and the increase is more significant for a smaller initial gap. At a specific RMS roughness point, the gap lines cross over at a point (45 nm in this chapter) for different initial gaps.

Figure 10.20b shows a schematic cross section when the metal bridge is driven down. The metal bridge contacts high points on the surface of the dielectric layer first, producing real contact at a finite number of asperities.

Based on the GW model, the real contact area with respect to the apparent contact area A_0 can be expressed as [15]

$$A^* = \frac{A_c}{A_0} = 0.064(\alpha - 0.8968)^{1/2} \times \int_{g'}^{\infty} \left(\frac{z - g'}{\sigma_s}\right) \phi(z)dz \tag{10.23}$$

where A^* is the normalized contact area, A_c is the real contact area, and g' is the separation between the metal bridge and the asperity mean height plane. As the metal bridge approaches the dielectric layer, g' decreases.

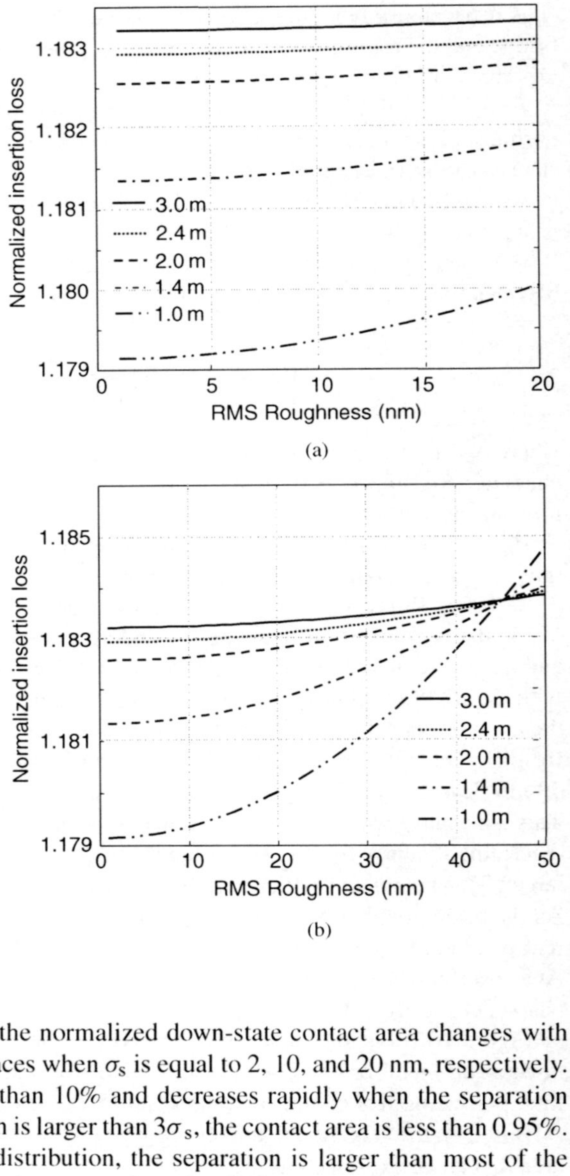

Fig. 10.22 Influence of *RMS* roughness on the normalized up-state insertion loss: RMS roughness varies **(a)** from 0 to 20 nm and **(b)** from 0 to 50 nm [19]. Copyright/used with permission of/courtesy of IOP Publishing Limited (Copyright IOP Publishing Limited)

Figure 10.23a shows how the normalized down-state contact area changes with the separation of the two surfaces when σ_s is equal to 2, 10, and 20 nm, respectively. It is observed that A^* is less than 10% and decreases rapidly when the separation increases. When the separation is larger than $3\sigma_s$, the contact area is less than 0.95%. According to the roughness distribution, the separation is larger than most of the RMS roughness and therefore the two surfaces hardly have any contact.

The separation g' is related to the applied load. The total load P with respect to the apparent contact area A_0 can be calculated as [15]

$$P^* = \frac{P}{A_0} = 0.0333E^* m_2^{1/2}(\alpha - 0.8968)^{3/4} \times \int_{g'}^{\infty} \left(\frac{z - g'}{\sigma_s}\right)^{3/2} \phi(z)dz \quad (10.24a)$$

Fig. 10.23 **(a)** Variation of normalized contact area with separation and **(b)** variation of load with separation [19]. Copyright/used with permission of/courtesy of IOP Publishing Limited

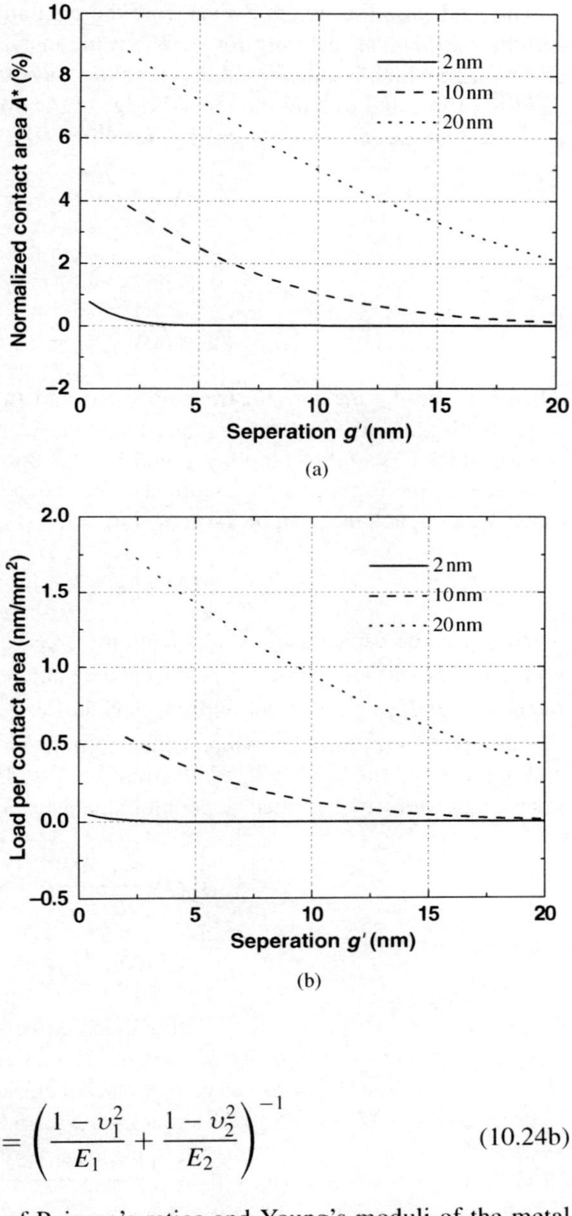

$$E^* = \left(\frac{1 - \upsilon_1^2}{E_1} + \frac{1 - \upsilon_2^2}{E_2} \right)^{-1} \tag{10.24b}$$

where E^* is the combination of Poisson's ratios and Young's moduli of the metal bridge and the SiN layer. In the study, $E_1 = 70$ GPa, $\upsilon_1 = 0.35$ for the metal bridge (Al) [22] and $E_2 = 295$ GPa, $\upsilon_2 = 0.25$ for the SiN dielectric layer [23].

The effect of the applied load on the separation g' is shown in Fig. 10.23b. The separation is smaller when the applied load is larger. This is because a larger applied load forces the asperities to deform elastically and pushes the two surfaces nearer.

The load P on the switch comes from the combination of the electrostatic force and the mechanical restoring force. When the metal bridge is driven down, a DC bias voltage which is called hold-down voltage is needed to balance the mechanical restoring force and to hold the metal bridge at the down-state position. As a result, the load P on the metal bridge can be simplified as

$$P = F_{\mathrm{e}} - F_{\mathrm{k}} \tag{10.25a}$$

$$F_{\mathrm{e}} = \frac{1}{2} C_{\mathrm{d}}^{\mathrm{r}} \frac{V^2}{t_{\mathrm{d}}} \tag{10.25b}$$

$$F_{\mathrm{k}} = k(g_0 - t_{\mathrm{d}} - g') \tag{10.25c}$$

where F_{e} and F_{k} are the electrostatic force and the mechanical restoring force, respectively. $C_{\mathrm{d}}^{\mathrm{r}}$ is the real down-state capacitance, k is the effective spring constant, t_{d} is the thickness of dielectric layer, and V is the applied hold-down voltage.

When all the asperities are assumed to be independent of one another, the real down-state capacitance can be expressed as

$$C_{\mathrm{d}}^{\mathrm{r}} = C_{\mathrm{d}}^{1} + C_{\mathrm{d}}^{2} \tag{10.26}$$

where C_{d}^{1} is the capacitance results from the area where the metal bridge contacts with the SiN surface, while C_{d}^{2} is the capacitance results from the area where the metal bridge does not contact with the SiN surface. The C_{d}^{2} can be further divided into two parts, C_{d}^{21} and C_{d}^{22}, which come from the air gap between the metal bridge and the SiN and the SiN itself, as illustrated in Fig. 10.20b. For simplicity, all these micro-capacitors are assumed as parallel plate capacitors that can be expressed as

$$C_{\mathrm{d}}^{1} = \frac{\varepsilon_0 \varepsilon_{\mathrm{r}} A_0 A^*}{t_{\mathrm{d}} + g'} \tag{10.27}$$

$$\frac{1}{C_{\mathrm{d}}^{2}} = \frac{1}{C_{\mathrm{d}}^{21}} + \frac{1}{C_{\mathrm{d}}^{22}} \tag{10.28a}$$

where

$$C_{\mathrm{d}}^{21} = \frac{\varepsilon_0 A_0 (1 - A^*)}{g'} \tag{10.28b}$$

$$C_{\mathrm{d}}^{22} = \frac{\varepsilon_0 \varepsilon_{\mathrm{r}} A_0 (1 - A^*)}{t_{\mathrm{d}}} \tag{10.28c}$$

where ε_{r} is the dielectric constant of SiN.

Therefore, the normalized down-state capacitance can be expressed as

$$C_{\mathrm{d}}^{*} = \frac{C_{\mathrm{d}}^{\mathrm{r}}}{C_{\mathrm{d}}^{\mathrm{a}}} \tag{10.29}$$

where C_d^a is the apparent down-state capacitance which can be expressed as

$$C_d^a = \frac{\varepsilon_0 \varepsilon_r A_0}{t_d} \tag{10.30}$$

Based on Eqs. (10.23), (10.24a), (10.24b), (10.25a), (10.25b), (10.25c), (10.26), (10.27), (10.28a), (10.28b), (10.28c), (10.29), and (10.30), Fig. 10.24 shows how the normalized contact area changes with the applied hold-down voltage. The effective stiffness k is assumed to be 40 N/m. It is seen that the normalized contact area increases with the applied hold-down voltage for the same roughness; when the applied hold-down voltage is kept constant, the larger the roughness, the smaller the normalized contact area.

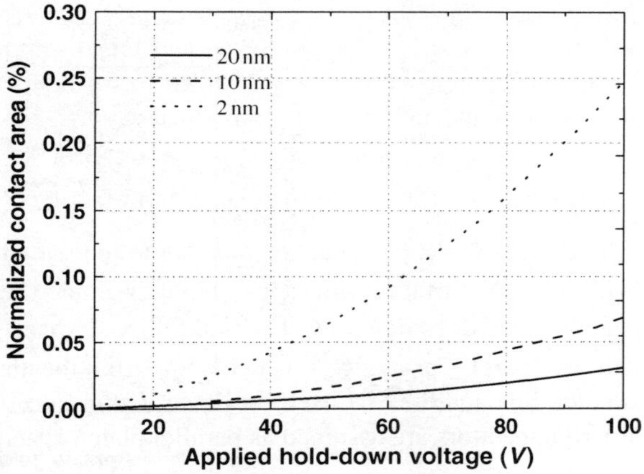

Fig. 10.24 Variation of normalized contact area with applied hold-down voltage [19]. Copyright/used with permission of/courtesy of IOP Publishing Limited

Figure 10.25a shows the variation of C_d^1 with the applied hold-down voltage when the RMS roughness σ_s is equal to 2 nm, Fig. 10.25bb presents C_d^{21}, C_d^{22}, and C_d^2 change with the applied hold-down voltage, and Fig. 10.25c shows the variation of C_d^r and C_d^* with the applied hold-down voltage. As a result, it is interesting to conclude that (a) the down-state capacitance increases with the applied hold-down voltage and (b) C_d^2 is much larger than C_d^1; therefore, C_d^r is mainly determined by C_d^2, which means that the down-state capacitance of the capacitive switch is actually determined by the capacitance area where the metal bridge does not contact with the SiN surface.

Figures 10.24 and 10.25 can be explained as the average load P added between the metal bridge and the rough surface increases with the applied hold-down voltage. Therefore, when the gap g' becomes smaller the down-state capacitance becomes larger.

The variations of C_d^1, C_d^2, C_d^r, and C_d^* with the applied hold-down voltage for different RMS roughnesses σ_s are shown in Fig. 10.26. It can be concluded that

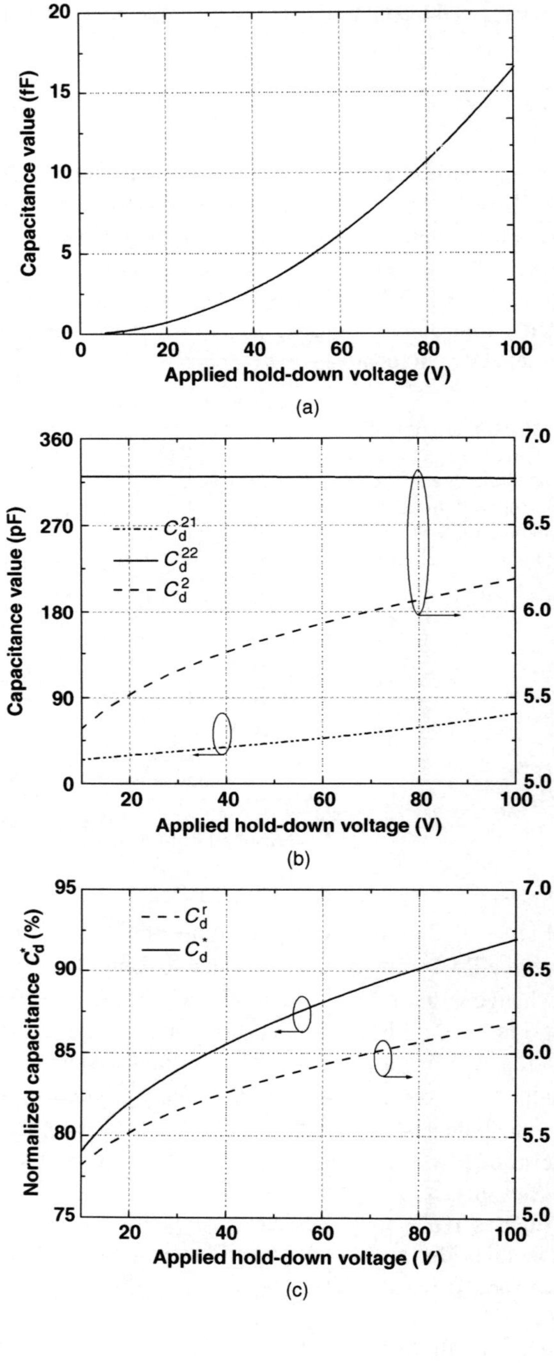

Fig. 10.25 Variation of down-state capacitance with applied hold-down voltage for 2 nm RMS roughness: **(a)** C_d; **(b)** C_d^{21}, C_d^{22}, and C_d^{2}; and (**c**) C_d^{r} and C_d^{*} [19]. Copyright/used with permission of/courtesy of IOP Publishing Limited

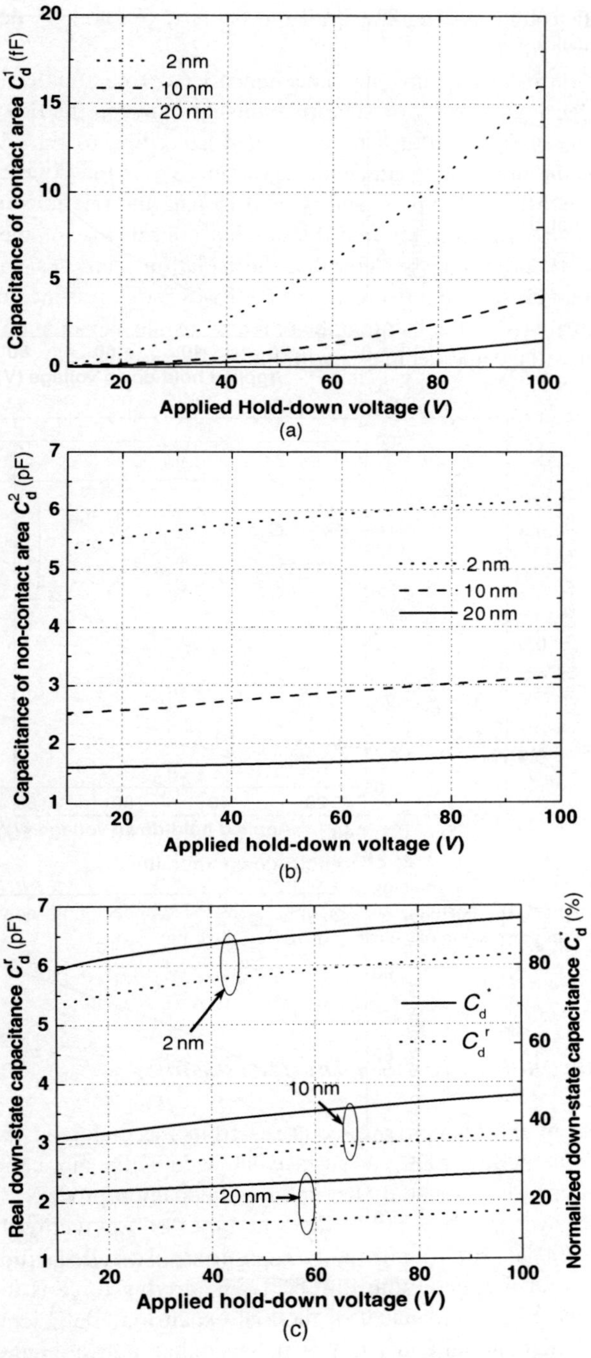

Fig. 10.26 Variation of down-state capacitance with the applied hold-down voltage: **(a)** C_d^1, **(b)** C_d^2, and **(c)** C_d^r and C_d^* [19]. Copyright/used with permission of/courtesy of IOP Publishing Limited

when the RMS roughness increases, all the C_d^1, C_d^2, C_d^r, and C_d^* decrease for the same hold-down voltage.

The reduction in the down-state capacitance leads to a smaller isolation when the metal bridge is driven down. The relationship between the down-state capacitance and the isolation can also be calculated according to Eq. (3.6), where C_u is replaced by the down-state capacitance C_d and S_{21} is the down-state isolation. Assuming $A_0 = 100 \times 150\ \mu m^2$ and $t_d = 0.15\ \mu m$, the relationship between the normalized down-state isolation and the applied hold-down voltage at 10 GHz is shown in Fig. 10.27. It can be seen that the isolation increases with the applied hold-down voltage slowly but decreases as the RMS roughness increases. This variation can be attributed to the fact that the real down-state capacitance increases with the hold-down voltage and decreases with the RMS roughness.

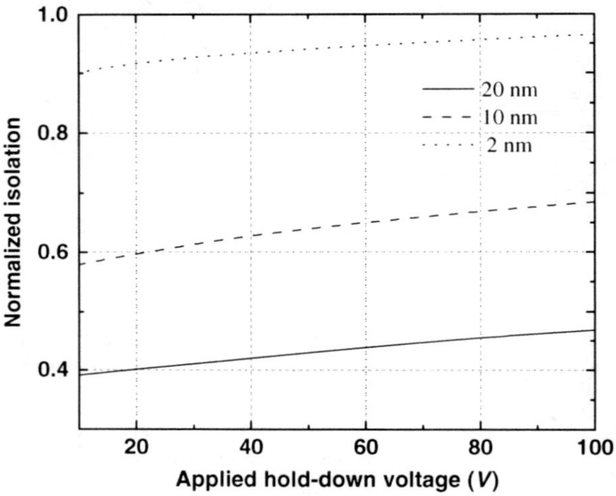

Fig. 10.27 Variation of normalized isolation with the applied hold-down voltage [19]. Copyright/used with permission of/courtesy of IOP Publishing Limited

10.4.3 Experimental Results and Discussions

The roughness of the dielectric layer arises from the dielectric itself and, more importantly, from the lower DC pad surface damaged by the fabrication steps at elevated temperature, such as the dielectric layer deposition process. The commonly used dielectric layer is SiN which is deposited by PECVD within the temperature range of 250–300°C. High temperature leads to the so-called hillock [24–26] or larger grain size of the metal thin film [27]. The driving force is the compressive stress caused by a large mismatch of thermal expansion coefficients between the metal film and the material under it. It is believed that high strength material, such as Ti and refractory metals, deposited with free of pinholes can reduce this effect.

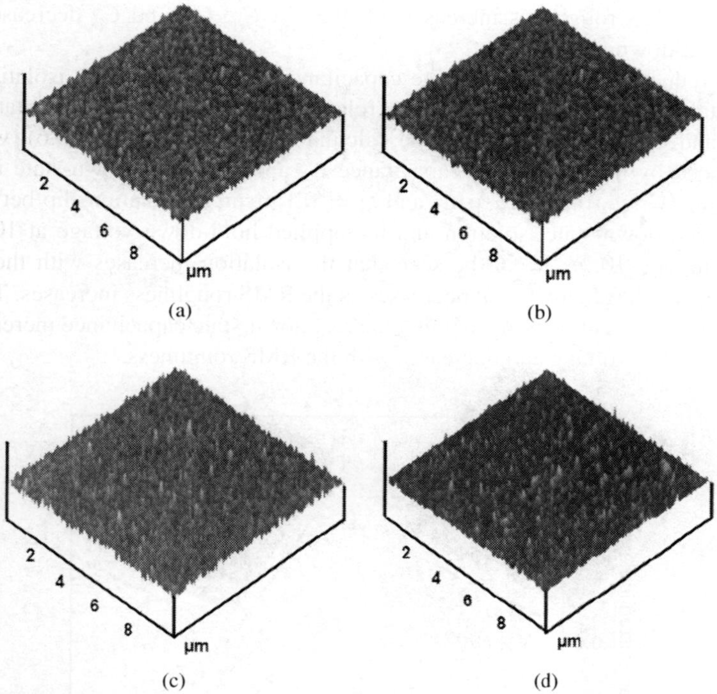

Fig. 10.28 Surface roughness of different metal thin films after sputtering: **(a)** Ti, **(b)** Cr, **(c)** Al, and **(d)** Cu (the thickness of all metal films is 0.5 μm) [19]. Copyright/used with permission of/courtesy of IOP Publishing Limited

However, the refractory metal usually has lower resistivity and hence leads to a higher insertion loss in the CPW transmission line.

Figure 10.28 shows the surface roughness for different metal thin films after sputtering. The surface roughness is examined with an atomic force microscopy (AFM). The substrates used are Si substrate with 2-μm PECVD SiO_2 deposited on them. The metal thickness is 0.5 μm. It can be seen that all these metal surfaces have similar roughness under the state of sputtering. However, their surface roughness changes differently after a 0.15-μm SiN is deposited on these metal films that used as lower DC pad as shown in Fig. 10.29. For Cr and Ti thin films, the roughness is almost the same, while for Al and Cu, the roughness becomes larger. The increase in the roughness of Al thin film is mainly caused by the "hillock" produced in the film whereas for the Cu thin film it is mainly because of the agglomeration and coalescence of grains. All the parameters of the metal layer are calculated using the GW model based on the measured roughness data listed in Table 10.3.

Figure 10.30a shows the measurement results of the insertion loss of the switches with different types of metallic thin film under the SiN layer. The insertion losses have been de-embedded by subtracting the insertion loss of the CPW transmission

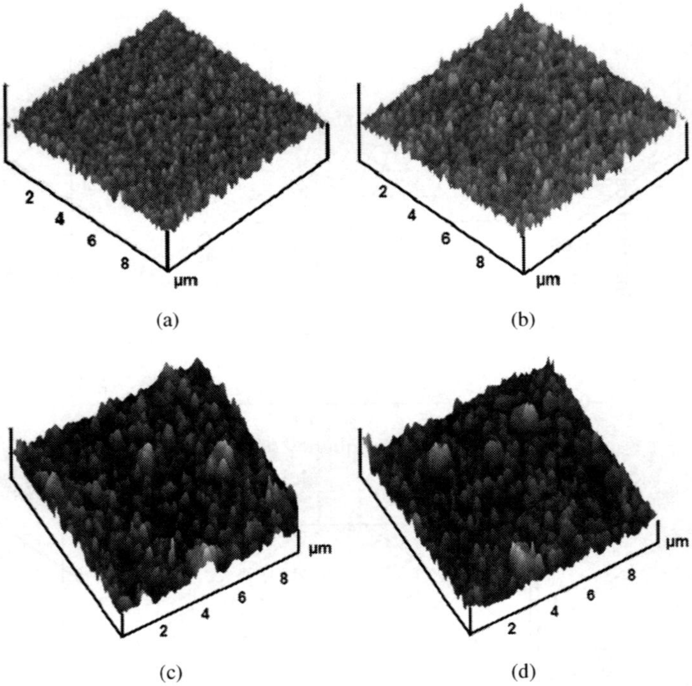

Fig. 10.29 Surface roughness of different metal thin films after SiN deposition: **(a)** Ti, **(b)** Cr, **(c)** Al, and **(d)** Cu (the thickness of SiN is 0.15 μm) [19]. Copyright/used with permission of/courtesy of IOP Publishing Limited

Table 10.3 Parameters of surface roughness characterization for different metal thin films before and after SiN deposition

		σ_s (nm)	R (nm)	D_{SUM} (/μm^2)	m_0 (nm^2)	m_2	m_4 (/μm^2)
Before	T_i	2.162	30.05	160	169	0.0935	489
	C_r	3.067	35.50	101	262	0.1060	350
	Al	4.196	39.50	99	260	0.0875	283
	C_u	4.314	48.15	95	270	0.0618	192
After	T_i	2.192	31.11	158	159	0.0831	456
	C_r	4.471	41.05	96	266	0.0835	262
	Al	12.29	59.18	50	398	0.0771	126
	C_u	14.81	63.22	41	450	0.0828	111

line without the MEMS switches; therefore, these losses are mainly due to the up-state capacitance value. It can be seen that the larger the RMS roughness of the film, the higher the insertion loss. This is consistent with the theoretical analysis results. The ripples in the measured insertion loss are due to the measurement noise. Table 10.4 lists the calculated and measured normalized up-state capacitance and insertion loss. It can be seen that the measured up-state capacitance and isolation

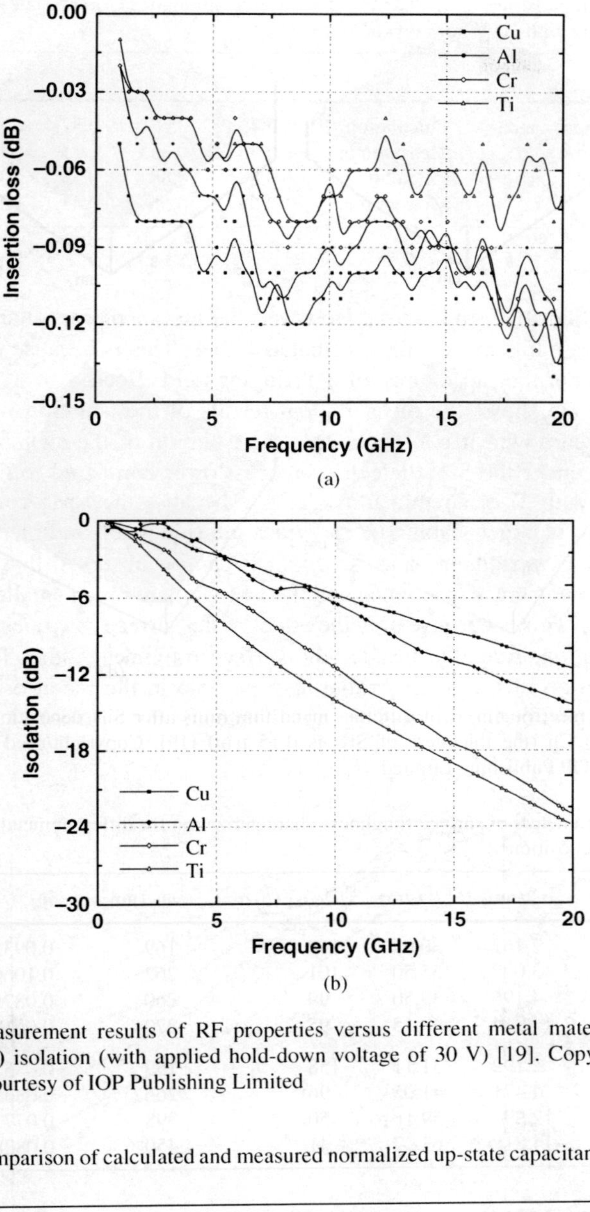

Fig. 10.30 Measurement results of RF properties versus different metal materials: **(a)** insertion loss and **(b)** isolation (with applied hold-down voltage of 30 V) [19]. Copyright/used with permission of/courtesy of IOP Publishing Limited

Table 10.4 Comparison of calculated and measured normalized up-state capacitance and insertion loss (10 GHz)

		Ti	Cr	Al	Cu
Capacitance	Calculation	1.08	1.21	1.41	1.39
	Measurement	1.21	1.24	1.47	1.51
Insertion loss	Calculation	1.04	1.10	1.19	1.18
	Measurement	1.47	1.51	2.16	3.25

Table 10.5 Comparison of calculated and measured normalized down-state capacitance and isolation (10 GHz with 30-V hold-down voltage)

		Ti	Cr	Al	Cu
Capacitance	Calculation	0.75	0.68	0.32	0.29
	Measurement	0.53	0.35	0.25	0.22
Isolation	Calculation	0.88	0.83	0.53	0.50
	Measurement	0.69	0.56	0.41	0.37

increase with the RMS roughness. However, the measured capacitance values and insertion loss are higher than the calculated values. This is because the calculation does not consider the capacitance of the fringing field effect.

Figure 10.30b shows the measurement results of the isolation when the metal bridge is driven down. It is obvious that the isolation of the switches with Cu or Al thin films under the SiN dielectric layer is lower compared to the isolation of the switches with Ti or Cr thin films. This is because the RMS roughness of Cu or Al thin film is larger. Table 10.5 provides the calculated and measured normalized down-state capacitance and isolation. It can be observed that the measured normalized down-state capacitance and the isolation are much smaller than the calculated values. This is because only the effect of the surface roughness is considered in the calculation. Actually, besides the surface roughness, other factors such as non-planarization of the metal bridge and via hole in the metal bridge also have significant impact on the down-state capacitance and the isolation.

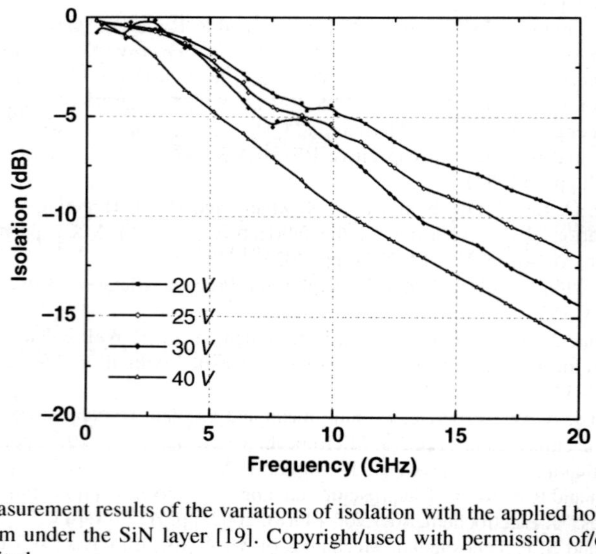

Fig. 10.31 Measurement results of the variations of isolation with the applied hold-down voltage with Al thin film under the SiN layer [19]. Copyright/used with permission of/courtesy of IOP Publishing Limited

Figure 10.31 indicates the relationship between the measured isolation and the hold-down voltage when Al metal film is deposited under the SiN layer. It is shown that the higher the applied hold-down voltage, the higher the isolation. This further re-enforces the prediction in Fig. 10.27.

10.5 Summary

The fabrication process of the capacitive switches, especially the dry etching of the sacrificial photoresist with the oxygen plasma, is characterized and optimized. For the dry etching process, the etching rate of the sacrificial photoresist increases with etching power and oxygen pressure. However, when the etching power is higher than 500 W, the metal bridges are easily damaged. The intrinsic stress in the metal film can be divided into the mean stress and the stress gradient which are determined by the curve fitting of the measured deflection of the cantilever beams and the stress model. For the dry releasing process, when the release power is higher than 500 W, the negative stress gradient in the metal bridge leads to the warp up of the metal bridge and hence increases the pull-down voltage of the metal bridge. The optimized etching power and oxygen pressure used in the releasing process are 430 W and 500 mTorr, respectively.

References

1. A. B. Yu, A. Q. Liu, Q. X. Zhang, A. Alphones, L. Zhu and S. A. Peter, "Improvement of isolation MEMS capacitive switch via membrane planarization," Sensor. Actuator. A: Phys. Vol. 119, 2005, pp. 206–213.
2. T. Veijola, T. Tinttunen, H. Nieminen, V. Ermolov and R. Ryhanen, "Gas damping model for a RF MEMS switch and its dynamic characteristics," *Microwave Symposium Digest. IEEE MTT-S International Conference*, Seattle, WA, USA, June 2002, pp. 1213–1216.
3. Ansoft Corporation, Ansoft Maxwell HFSS, Release 8.0.25, www.ansoft.com
4. W. M. van Spengen, R. Puers, R. Mertens and I. De wolf, "A comprehensive model to predict the charging and reliability of capacitive RF MEMS switches," J. Micromech. Microeng. Vol. 14, 2004, pp. 514–521.
5. A. B. Yu, A. Q. Liu, J. Oberhammer, Q. X. Zhang and H. M. Hosseini, "Characterization and optimization of dry releasing for the fabrication of RF MEMS capacitive switches," J. Micromech. Microeng. Vol. 17, 2007, pp. 2024–2030.
6. D. L. Smith, Thin Film Deposition: Principles and Practice, McGraw-Hill, Inc., New York, 1995, pp. 453–555.
7. B. P. van Drieenhuizen, J. F. L. Goosen, P. J. French and R. F. Wolffenbuttel, "Comparison of techniques for measuring both compressive and tensile in thin films," Sensor. Actuator. A: Phys. Vol. 37–38, 1993, pp. 756–765.
8. W. Fang and J. A. Wickert, "Determining mean and gradient residual stresses in thin films using micromachined cantilevers," J. Micromech. Microeng. Vol. 6, 1996, pp. 301–309.
9. A. K. Chinthakindi, D. Bhusari, B. P. Dusch, J. Musolf, B. A. Willemsen, E. Prophet, M. Roberson and P. A. Kohl, "Electrostatic actuators with intrinsic stress gradient I. Materials and structures," J. Electrochem. Soc. Vol. 149(8), 2002, pp. H139–H145.
10. J. M. Gere and S. P. Timoshenko, Mechanics of Materials, 3rd edn., PWS-KENT, Boston, MA, 1990, pp. 301–308.

11. J. W. Weigold, W. H. Juan, S. W. Pang and J. T. Borenstein, "Characterization of bending in single crystal Si beams and resonators," J. Vac. Sci. Technol. B Vol. 17(4), 1999, pp. 1336–1340.
12. P. J. Caber, "Interferometric profiler for rough surfaces," Appl. Opt. Vol. 32, 1993, pp. 3438–3441.
13. J. A. Greenwood and J. B. P. Williamson "Contact of nominally flat surfaces," Proc. R. Soc. Lond., Ser. A Vol. 295, 1966, pp. 300–319.
14. C. Y. Poon and B. Bhushan, "Comparison of surface roughness measurements by stylus profile, AFM and non-contact optical profiler," Wear Vol. 190, 1995, pp. 76–88.
15. J. I. McCool, "Comparison of models for the contact of rough surfaces," Wear Vol. 107, 1986, pp. 37–60.
16. J. I. McCool, "Finite difference spectral moment estimation for profiles: the effect of sample spacing and quantization error," J. Precis. Eng. Vol. 4, 1982, pp. 181–184.
17. R. W. Carpick, E. E. Flater, J. R. Van Langendon and M. P. de Boer "Friction in MEMS: from single to multiple asperity contact," Int. Congr. Exposition Exp. Appl. Mech. 2002 pp. 282–287.
18. G. M. Rebeiz, RF MEMS Theory, Design and Technology, Wiley-Interscience, New York, NY, 2003.
19. A. B. Yu, A. Q. Liu, Q. X. Zhang and H. M. Hosseini, "Effects of surface roughness on electromagnetic characteristics of capacitive switches," J. Micromech. Microeng. Vol. 16, 2006, pp. 2157–2166.
20. L. Kogut, "The influence of surface topography on the electromechanical characteristics of parallel-plate MEMS capacitors," J. Micromech. Microeng. Vol. 15, 2005, pp. 1068–1075.
21. S. Hudlet, J. M. Saint, C. Guthmann and J. Berger J, "Evaluation of the capacitive force between an atomic force microscopy tip and a metallic surface," Eur. Phys. J. B Vol. 2, 1998, pp. 5–10.
22. http://www.webelements.com. Accessed 1993–2010.
23. O. P. Hugh, Handbook of Thin-film Deposition Processes and Techniques: Principles, Methods, Equipment and Applications, 2nd edn., Noyes Publications, Park Ridge, NJ, 2002.
24. D. K. Kim, B. Heiland, D. N. William et al., "Microstructure of thermal hillocks on blanket Al thin films," Thin Solid Films Vol. 371, 2000, pp. 278–282.
25. M. Zaborowski and P. Dumania, "Kinetics of hillock growth in Al and Al-alloys," Microelectron. Eng. Vol. 50, 2000, pp. 301–309.
26. S. Aceto, C. Y. Chang and R. W. Vook, "Hillock growth on aluminum and aluminum alloy films," Thin Solid Films Vol. 219, 1992, pp. 80–86.
27. J. C. Lin and C. Lee "Nucleation and growth of Cu thin films on silicon wafers deposited by radio frequency sputtering," Thin Solid Films Vol. 307, 1997, pp. 96–99.

Index

A.-Q. Liu, *RF MEMS Switches and Integrated Switching Circuits*, MEMS Reference Shelf 5, DOI 10.1007/978-0-387-46262-2,

About the Author

Ai-Qun Liu (A.-Q. Liu) received his PhD degree from National University of Singapore in 1994. He obtained his MSc degree from Beijing University of Posts & Telecommunications in 1988 and BE degree from Xi'an Jiaotong University in 1982. Currently, he is an associate professor at the Division of Microelectronics, School of Electrical & Electronic Engineering, Nanyang Technological University. He was the recipient of the *IES Prestigious Engineering Achievement Award* in 2006 and won the *University Scholar Award* in 2007. He has also written "*Photonic MEMS – Design, Fabrication and Control*".

About the Author

Lightning Source UK Ltd.
Milton Keynes UK
01 December 2010

163754UK00009B/7/P

9 780387 462615